Discovery EDUCATION | MATH TECHBOOK

GEOMETRY WITH HONORS

FLORIDA INTERACTIVE STUDENT EDITION

Log in to Math Techbook at
www.DiscoveryEducation.com

ISBN 13: 978-1-68220-451-1

Library of Congress Cataloging-in-Publication Data available upon request from the Library of Congress.

Printed in the United States of America.

1 2 3 4 5 6 7 8 9 10

Discovery EDUCATION™

800-323-9084

One Discovery Place, Silver Spring, Maryland 20910

Table of Contents

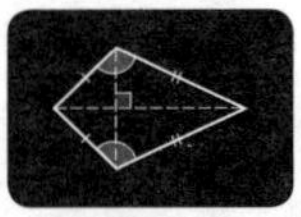

UNIT 9: Quadrilaterals and Other Polygons

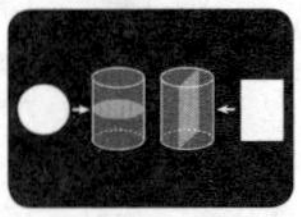

UNIT 10: 3-D Figures

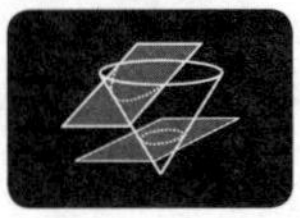

UNIT 11: Conic Sections

Letter to the Student

Dear Student,

You are about to experience mathematics like you never have before! In this class, you'll be using Math Techbook™—a comprehensive, digital math instructional program developed by the educators and designers at Discovery Education. Math Techbook is full of interactives, videos, digital tools, and game-like activities to help you learn mathematical concepts and apply them to the real world. You won't just study the concepts; you'll develop real-world skills and tackle relevant problems that are worth solving. There are multiple pathways for you to work at your own pace and ability. You'll even be able to monitor your progress in real time using the Student Learning Dashboard.

As a print resource to use in conjunction with Math Techbook, this *Interactive Student Edition (ISE)* allows you to practice the skills you've learned in class and online. You can use this resource to record questions, find connections to the digital content, and develop your own mathematical understanding. Math Techbook is divided into units, and each unit is divided into concepts. Each concept is then divided into three tabs: Discover, Practice, and Apply. This print resource is organized by concept and includes the following:

- OVERVIEW: Lesson Objectives, Essential Questions, key vocabulary, and short descriptions of the online investigations will help you make connections to the mathematical content.
- DISCOVER: This page includes QR codes that connect to the online Discover investigations and provides a place for you to take notes and record important information.
- CHECKS FOR UNDERSTANDING: There is a Check for Understanding page for each investigation within Discover. The exercises in each Check for Understanding are your chance to show what you've learned. The final question in each set asks you to justify your reasoning and communicate mathematically.
- SUMMARY: The Summary is a written description of what you've learned in each investigation and encourages you to make connections.
- PRACTICE EXERCISES: These exercises provide additional review material to sharpen your skills. You can find more problems in the online Practice section, with Coach exercises that provide feedback and support and Play exercises that allow you to earn badges while practicing what you've learned.
- APPLY: The Apply problem uses real-world examples to put your knowledge to the test. A list of criteria and a rubric will give you direction on what's required.

Within each section of this resource, you'll find QR codes that take you to the corresponding online section of Math Techbook for that concept. For instance, the QR codes on the Discover page provide a direct link to each online investigation, and the QR codes on the Apply page connect you to related videos and additional problems. Once inside Math Techbook, you'll have access to activities that develop conceptual understanding in Discover, practice exercises that develop procedural fluency in Practice, and real-world problems worth solving in Apply.

We hope this program offers you the chance to dive deeper into math and have fun, too! Best of luck for a fantastic year!

Sincerely,

The Discovery Education Math Team

Letter to the Parent/Guardian

Dear Parent/Guardian,

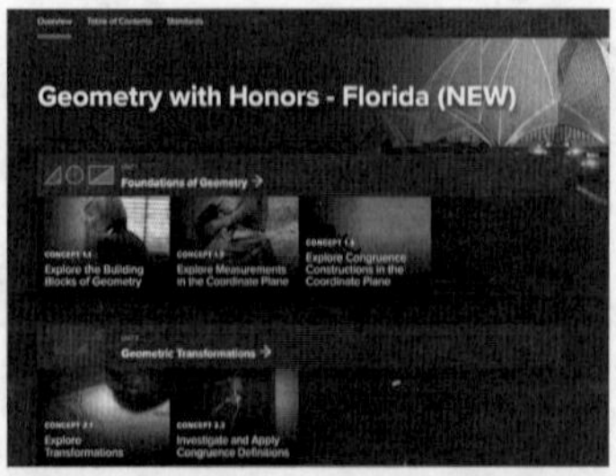

This year, your student will be using Math Techbook™, a comprehensive, digital program developed by the educators and designers at Discovery Education. Math Techbook is an innovative program that offers engaging, real-world problems to help your student master key mathematical concepts and procedures. In class, students experience dynamic content, interactives, videos, digital tools, and game-like activities that increase their motivation to learn math.

As a print resource to use in conjunction with Math Techbook, this *Interactive Student Edition (ISE)* offers a way for students to practice the skills they've learned in class and online. Students are encouraged to use this resource to capture important ideas, seek connections, and develop their own mathematical understanding. Math Techbook is divided into units, and each unit is divided into concepts. Each concept is then divided into three tabs: Discover, Practice, and Apply. This print resource is organized by concept and includes the following:

- OVERVIEW: Students learn mathematics through inquiry and active participation. In the overview for each concept, the Essential Questions and short descriptions of the in-class investigations provide a roadmap for what students are expected to learn and help them make connections to the mathematical content.
- DISCOVER: The Discover page is a place for students to record important information that they learn while completing the activities within the online Math Techbook. QR codes link directly to the online investigations.
- CHECKS FOR UNDERSTANDING: There is a Check for Understanding page for each investigation within Discover. The exercises in each Check for Understanding give students an opportunity to demonstrate what they've learned. The final question in each set allows them to justify their reasoning and communicate mathematically.
- SUMMARY: Math Techbook is aligned with math standards that require deep exploration of critical topics and encourage students to make connections. The Summary provides a reference for what students have learned in each concept.
- PRACTICE EXERCISES: Students need additional review material to practice what they've learned. In addition to the Coach and Play items found in the online Practice section, the Practice Exercises in this print resource allow students to sharpen their skills and show what they've learned.
- APPLY: Students experience learning when they apply their mathematical thinking to problems worth solving, rooted in real-world contexts that are relevant to their lives. Online, students have access to several Apply problems. In this print resource, one of these in-depth tasks is provided with a rubric and evaluation criteria.

Within each section, QR codes will take you and your student to the corresponding section of Math Techbook. Once inside, your student will have access to the investigations that develop conceptual understanding in Discover, practice exercises that develop procedural fluency in Practice, and real-world problems worth solving in Apply.

While this print resource can be used at home to complete assignments when the Internet is not available, it is only through Math Techbook that students develop the type of deep conceptual understanding that fosters mathematical proficiency. QR codes are provided throughout the *ISE* to redirect students to Math Techbook's digital resources.

To use the QR codes, you will need a QR reader. Readers are available for phones, tablets, laptops, desktops, and virtually any device in between. Most use the device's camera, but there are some that scan documents that are on your screen. Download a free QR reader in the App Store or Google Play. To access Math Techbook resources, follow these steps:

1. Open the QR code reader on your device.
2. Hold your device so the QR code is visible within your device's screen. One of two things will happen:
   - The device may automatically scan the code; or,
   - The device will scan the code when you press a button, similar to taking a picture.
3. Once scanned, the QR code will direct you to a page or resource on the web.
4. For resources in Math Techbook, you may need to log in with your username and password the first time you access a QR code. After that, you won't need to log in again, unless you sign out or remain inactive for too long.

Scan this QR code to access a video that provides a deeper introduction to Math Techbook:

We encourage you to support your student in using the interactive materials provided in Math Techbook, as well as the exercises and material in this print resource at home. Together, may you and your student enjoy a fantastic year of mathematics!

Sincerely,

The Discovery Education Math Team

Math Tools

Within Math Techbook, you'll have access to the following tools:

Calculator: Solve addition, subtraction, multiplication, and division problems with large numbers or decimals.

Scientific Calculator: Solve complex problems involving multiple operations, exponents, and functions.

Graphing Calculator: Display equations and functions on a coordinate plane.

Whiteboard: Draw lines and shapes, insert images and grids, display text and graphics together.

Dynamic Geometry Tool: Create more advanced shapes, polygons, and conics, transform them, and discover the equations behind them.

Unit Converter: Easily change units of measurement for length, mass, time, temperature, electric current, data storage, and angles from one to another.

Construction Tool: Practice your freehand drawing skills with a ruler and compass for a digital pencil.

Matrix Solver: Solve complex problems involving one or multiple matrices by changing the size and inputting values.

Glossary: View the definitions, key context, animations, and related videos for key mathematical terms.

Interactives: Investigate mathematical problems with game-like activities.

Board Builder: Use images and text to create presentations with this handy tool.

UNIT 1: Foundations of Geometry

1.1 Explore the Building Blocks of Geometry

photo: Getty Images

Lesson Objectives

- Interpret the axioms and undefined terms of geometry.
- Use undefined terms and axioms to formally define ray, angle, circle, and line segment.
- Differentiate between conjectures, theorems, and axioms.
- Verify that a statement is false using a counterexample.

Essential Question

- What are the fundamental parts of geometry, and how can they be used to create valid arguments?

Investigations

The Case for Geometry

What makes a strong argument? Geometric proofs are the mathematician's version of making a case.

Research and Analyze Evidence

Take on the role of a geometry lawyer. Collect pertinent evidence and present your findings.

An Educated Guess

Look for patterns and use them to make conjectures. Do you have enough evidence to build a theorem?

Explore Axioms

What past rulings can you use to build your case? Create Exhibit C to show the evidence.

Theorems on Trial

Prove your first case! Analyze and defend court briefs. Present evidence to prove a theorem.

photo: Getty Images

Key Vocabulary

angle, axiom, circle, collinear, complementary angles, congruent, conjecture, counterexample, intersection, line, line segment, linear pair, midpoint (of segment), parallel, perpendicular, plane, point, postulate, radius, ray, skew lines, supplementary angles, theorem, undefined term

Discover

As you complete Engage and the investigations, record the most important ideas you've learned.

Engage	
Investigation 1	**Investigation 2**
Investigation 3	**Investigation 4**

Name ______________________ Date __________

Check for Understanding

Explore the Building Blocks of Geometry: Investigation 1

1. Use the clues to create a geometric diagram in the coordinate plane.

- $\overline{PQ}$ intersects line n at R.
- M, R, and T are collinear.
- T lies on $\odot R$.
- Q is coplanar, but not collinear, with R and T.
- $R\left(1, \frac{1}{2}\right)$

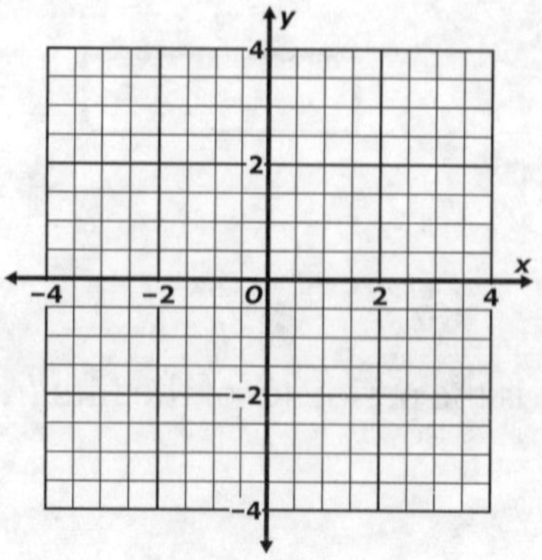

2. Which of the following is another name for $\angle VWX$? *Select all that apply.*

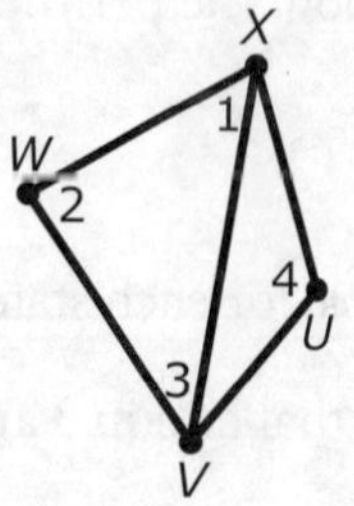

A. $\angle W$

B. $\angle WVX$

C. $\angle XVW$

D. $\angle 2$

E. $\angle XWV$

F. $\angle UVWX$

3. Identify each statement as either true or false.

A. Through any three collinear points, there is exactly one plane. ______

B. Through any two points, there is exactly one line. ______

C. A point is an exact location that has no size or dimension. ______

D. A circle is the set of all points in a plane that are the same distance from a given point called the center of the circle. ______

4. Why is precision in notation, descriptions, and drawings important in the preparation of a mathematical argument?

Name ______________________________ Date ______________

Check for Understanding

Explore the Building Blocks of Geometry: Investigation 2

The figures below show the first three figures in a pattern.

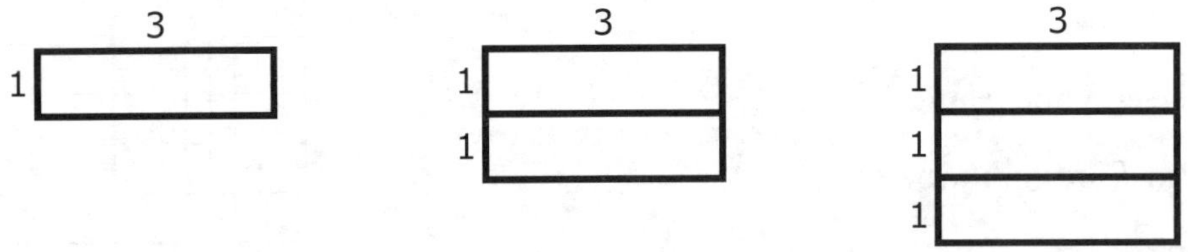

1. What is the perimeter of the fifth figure in this pattern?

2. Make a conjecture about the perimeter of the nth figure in the pattern.

3. Give a counterexample for each statement.

A. All numbers that end in 3 are prime numbers.

B. If three points are in the same plane, they are collinear.

4. How would you explain the difference between a postulate, a conjecture, and a theorem?

Name ______________________________ Date __________

Check for Understanding

Explore the Building Blocks of Geometry: Investigation 3

You are given the following clues:

- *M*, *N*, and *P* are three collinear points.
- $MP > MN$ and $MP > NP$.
- $MP = 18 - 3.8x$, $MN = 7.74$, and $NP = 1.9x$.

1. Draw a diagram to represent this situation and label each segment with the given expressions.

2. Determine the values of x and MP.

3. Miguela is creating a geometric design by combining different polygons. How does the angle addition postulate relate to Miguela's design? Include at least one specific example in your explanation.

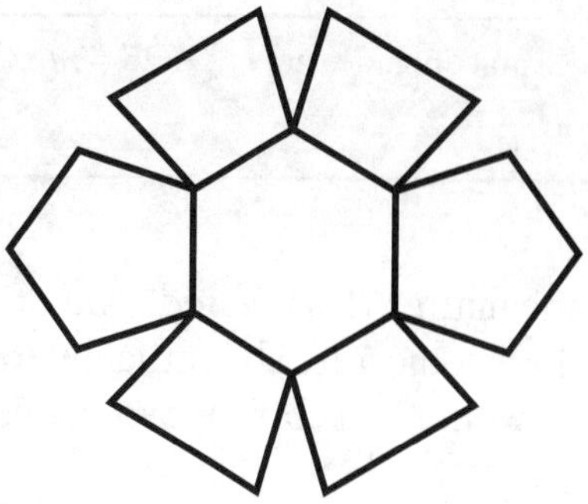

Name ______________________________ Date ______________

Check for Understanding

Explore the Building Blocks of Geometry: Investigation 4

1. In the figure, $m\angle AOC = m\angle BOD$. Lisa claimed that based on the given information, $m\angle AOB = m\angle COD$. Remember that to prove a conjecture is true beyond a doubt, you can only use evidence, or reasons, in the form of givens, definitions, postulates, properties, and previously proved theorems. Examine the argument Lisa presented. *Circle the reason Lisa used for each statement of her defense.*

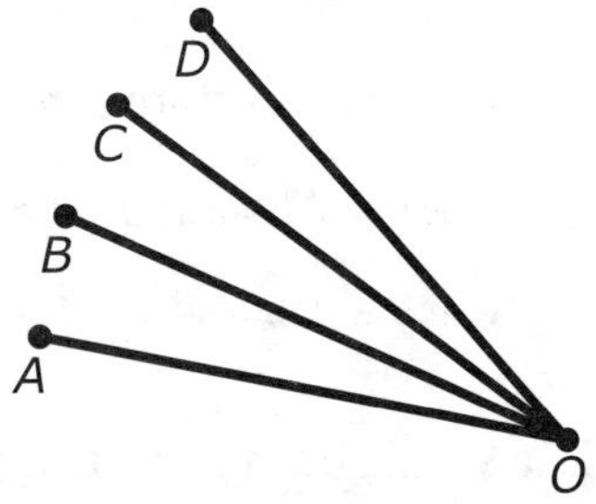

Statement	Reason
1. I know that $m\angle AOC = m\angle BOD$.	**A.** angle addition postulate **B.** given **C.** definition of $\angle AOD$
2. I can rewrite the angle measures as $m\angle AOC = m\angle AOB + m\angle BOC$ and $m\angle BOD = m\angle BOC + m\angle COD$.	**A.** angle addition postulate **B.** substitution **C.** definition of $\angle AOD$
3. In other words, my original statement becomes $m\angle AOB + m\angle BOC = m\angle BOC + m\angle COD$.	**A.** definition of $\angle AOD$ **B.** symmetry property **C.** substitution
4. Notice that both sides of the equation include $m\angle BOC$. I know that $m\angle BOC = m\angle BOC$.	**A.** substitution **B.** reflexive property **C.** symmetric property
5. I can conclude that $m\angle AOB = m\angle COD$.	**A.** given **B.** angle addition postulate **C.** subtraction property

2. In the figure, M is the midpoint of $\overline{AC}$ and $\overline{BD}$, and $AC = BD$. Make a conjecture about the different segments in the figure. How would you convince someone your conjecture is correct?

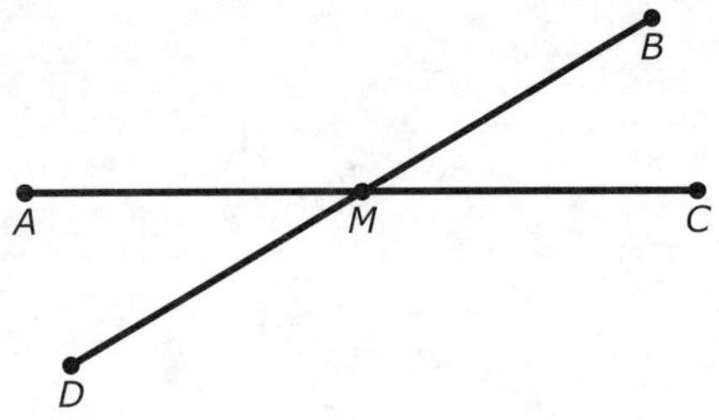

Summary

Before you attempt the Practice Exercises, review what you've learned.

Undefined Terms

Undefined terms are the most basic building blocks of geometry.

- A **point** is an exact location that has no size or dimension.
- A **line** is a one-dimensional path that extends infinitely in both directions.
- A **plane** is a flat two-dimensional surface that extends forever.

Defined Terms

A **definition** in geometry gives the meaning of a term based only on commonly understood words, undefined terms, or other previously defined terms.

- A **line segment** is a part of a line between two points on the line called endpoints. (Notice that this definition makes use of the undefined terms **point** and **line**.)
- A **distance** along a line is measured between two points on the line.
- A **ray** is a part of a line that extends from one endpoint infinitely in one direction.
- An **angle** is formed by two rays with the same endpoint. The two rays are called the sides of the angle, and the shared endpoint is called the vertex.
- A **circle** is the set of all points in a plane that are the same distance from a given point called the center of the circle.

Conjecture and Counterexample

A **conjecture** is statement based on known information or observation that is believed to be true but not yet proven.

For example, triangular numbers form the pattern: 1, 3, 6, 10, 15, . . . After careful observation and generalizations, a conjecture might be that the nth term in this pattern is equal to $\frac{n(n+1)}{2}$. The statement remains a conjecture until it is proved true using definitions, properties, axioms, or previously proved theorems.

A **counterexample** is an example that is used to disprove a statement; it contradicts a statement or argument.

For example, someone might examine the following partial list of primes:

3, 5, 7, 11, 13

After examining this partial list, the person might conjecture that all prime numbers are odd. A counterexample to this conjecture is the prime number 2. A single counterexample is enough to prove that a conjecture is not true.

My Notes

Summary *(continued)*

Postulates

A **postulate** or **axiom** is a statement that is accepted in geometry without proof. In the same way that you can use undefined terms to help define other terms, you can use postulates to help prove other statements.

Unique Line Postulate

Through any two points, there is exactly one line.

Unique Plane Postulate

Through any three points not on the same line, there is exactly one plane.

Line Intersection Postulate

If two lines intersect, then their intersection is exactly one point.

Parallel Line Postulate

If there is a line and a point not on the line, then there exists exactly one line through the point that is parallel to the given line.

My Notes

Plane Intersection Postulate

If two planes intersect, then their intersection is a line.

Segment Addition Postulate

If A, B, and C are collinear and B is between A and C, then $AB + BC = AC$.

Angle Addition Postulate

If R is in the interior of $\angle PQS$, then $m\angle PQR + m\angle RQS = m\angle PQS$.

Theorems

In geometry, a **theorem** is a statement that has been proved using definitions, postulates, properties, previously proven theorems, and logical reasoning.

Congruent Complements Theorem

If two angles are complementary to the same angle, then the angles are congruent.

Midpoint Theorem

If M is the midpoint of $\overline{AB}$, then $\overline{AM} \cong \overline{MB}$.

Congruent Supplements Theorem

If two angles are supplementary to the same angle, then the angles are congruent.

Linear Pair Theorem

If two angles form a linear pair, then the angles are supplementary.

Practice Exercises

Review what you've learned using these practice problems. For practice problems with feedback, try the Coach and Play items in the Practice section online.

1. Which statements accurately describe Euclidean geometry? *Select all that apply.*

A. Parallel lines stay the same distance apart and never intersect.

B. A point has length and width.

C. A line segment is determined by its end points.

D. An angle consists of two rays that share an end point.

2. In the diagram, C is a point on $\overline{AB}$.

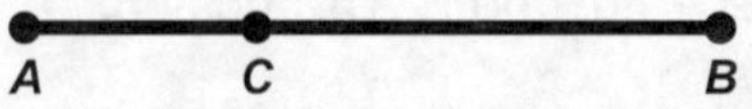

If $AC = 2x - 13$ centimeters, $BC = 3x + 4$ centimeters, and $AB = 36$ centimeters, what is the length of $\overline{BC}$?

$BC =$ ____________ cm

3. If $\angle GFH$ measures 35° and $\angle GFJ$ measures 85°, what is the measure of $\angle HFJ$?

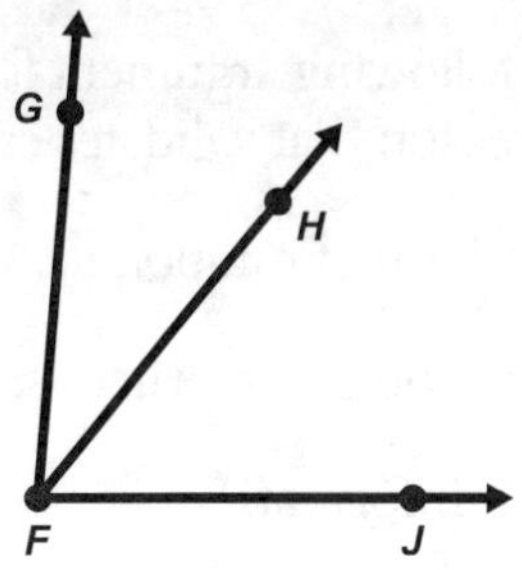

A. 35°

B. 40°

C. 45°

D. 50°

E. 120°

4. Examine the angles formed by intersecting rays in the diagram. Which statement is not true about the relationships among the angles in the diagram?

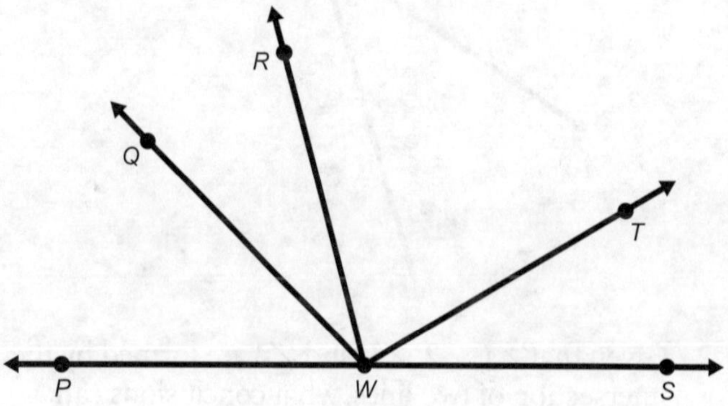

A. $m\angle RWT + m\angle TWS = m\angle RWS$

B. $m\angle TWQ - m\angle TWR = m\angle RWQ$

C. $m\angle PWR + m\angle RWS = m\angle PWS$

D. $m\angle PWT - m\angle QTW = m\angle PWQ$

5. If D is between B and C, which statement is true?

A. $BC + DB = DC$

B. $CD - DB = CB$

C. $CB - CD = DB$

D. $CD + CB = DB$

6. Three points are shown.

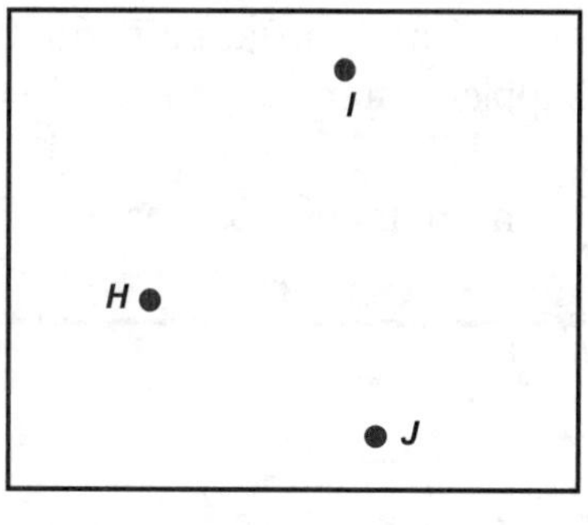

In total, how many lines can be drawn through any two or more of the given points?

There are ____________ lines that can be drawn through two or more of the points H, I, and J.

Practice Exercises *(continued)*

7. The diagram shows two intersecting lines.

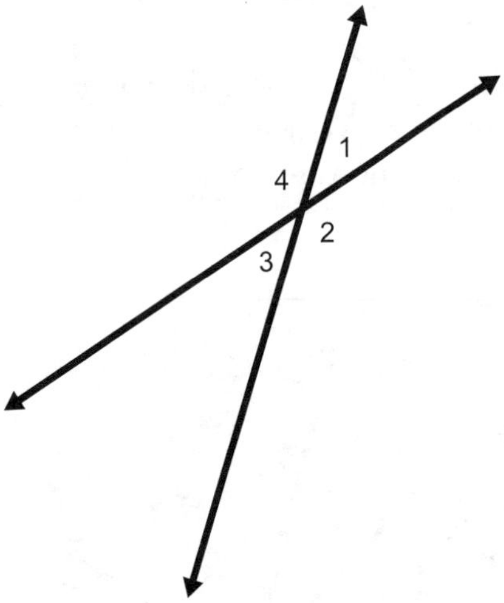

Given that ∠1, ∠2, ∠3, and ∠4 are formed by the intersection of two lines, what conclusions can you draw and why? *Select all that apply.*

A. $m\angle 1 + m\angle 2 + m\angle 3 + m\angle 4 = 360°$, because ∠1 and ∠2 are a linear pair and ∠3 and ∠4 are a linear pair.

B. $m\angle 1 = m\angle 2$, because the sum of $m\angle 1 + m\angle 2$ equals the sum of $m\angle 3 + m\angle 4$. It then follows by the subtraction property of equality that $m\angle 1 = m\angle 2$.

C. $m\angle 1 = m\angle 3$, because the sum of $m\angle 1 + m\angle 2$ equals the sum of $m\angle 3 + m\angle 2$. It then follows by the subtraction property of equality that $m\angle 1 = m\angle 3$.

D. $180 - m\angle 4 = m\angle 2 + m\angle 3$, because ∠2 and ∠3 are a linear pair and, therefore, supplementary.

8. If M is the midpoint of $\overline{AB}$, what is AB?

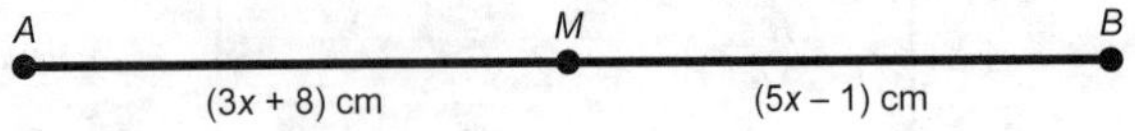

$AB =$ ____________ cm

9. How many lines can be drawn through point K, perpendicular to line HJ?

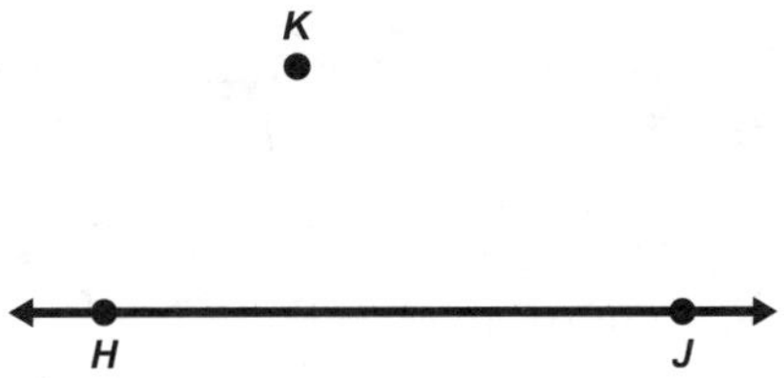

A. 0

B. 1

C. 2

D. any number of lines

10. If M is the midpoint of $\overline{AB}$, $AB = 6x + 40$, $MB = 2x + 30$, find x, AB, and MB.

$x =$ ____________

$AB =$ ____________

$MB =$ ____________

11. In the diagram, M is the midpoint of $\overline{AB}$.

Support the following argument for $\overline{AM} \cong \overline{MB}$ by writing the reason that validates each statement.

A. Definition of midpoint

B. Definition of congruent segments

C. Given information

We know that M is the midpoint of $\overline{AB}$. ______

This means that $AM = MB$. ______

Therefore, $\overline{AM} \cong \overline{MB}$. ______

Apply

Where's Geometry?

Geometry is all around us—in human-made and natural structures. By studying an interesting scene or structure, you can uncover lines, angles, and other geometric building blocks and learn more about the relationships among them.

Study various photographs (your own or from the Internet). Choose one from which you can identify at least seven of the following geometric objects and/or postulates.

1. Parallel lines
2. Perpendicular lines
3. Line intersection postulate
4. Plane intersection postulate
5. Perpendicular planes
6. Congruent segments
7. Congruent angles
8. Complementary angles
9. Supplementary angles
10. Midpoint of a segment
11. Circle

Once you have chosen your photo:

- Label the objects/postulates that you identified by either drawing directly on the photo or uploading the photo and using the Whiteboard.
- Write a report that provides evidence proving why the identified objects in the photo are in fact the geometric objects you claim and/or why they illustrate postulates.
- Explain why this two-dimensional photo does or does not accurately represent the three-dimensional world.
- Use the definitions of geometric objects and the postulates to support or deny the existence of these objects/postulates in the world.

Show what you've learned by completing the other performance tasks in the online Apply section.

Apply *(continued)*

Your answer to Apply will be assessed on the following criteria:

1. Submitting a labeled photo that supports claims made in the report
2. Reporting seven or more geometric objects or postulates from the list with evidence from the uploaded photo
3. Describing how viewing a 3-D structure in a 2-D photo affects how geometric objects appear in the photo
4. Commenting on the existence of geometric objects in the physical world

Criteria \ Scale	4	3	2	1	0
	Exceeds Criteria	**Meets Criteria**	**Progressing to Criteria**	**Below Expectations**	**No Expectation**
Photo	Submits a photo with detailed labels that support seven or more geometric objects/ postulates.	Submits a photo with labels that only support four to six geometric objects/ postulates.	Submits a photo with labels that only support one to three geometric objects/ postulates.	Submits a photo without labels.	Does not submit a photo.
Geometric Objects	Presents evidence for all submitted geometric objects and uses geometric terms correctly (e.g., line *AB*, angle *CDE*, etc.).	Presents evidence for fewer than 75% of the submitted geometric objects or uses a geometric term incorrectly.	Presents evidence for fewer than 50% of the submitted geometric objects or uses several geometric terms incorrectly.	Presents evidence for fewer than 25% of the submitted geometric objects or uses most of the geometric terms incorrectly.	Does not provide evidence of geometric objects.
Effect of a 2-D Photo on a 3-D Place	Clearly describes the effect of taking a 2-D photo of geometric objects found in a 3-D location.	Describes the effect of taking a 2-D photo of geometric objects found in a 3-D location with a minor error.	Describes the effect of taking a 2-D photo of geometric objects found in a 3-D location with a few errors.	Incorrectly describes the effect of taking a 2-D photo of a geometric objects found in a 3-D location.	Does not describe the effect.
Existence of Geometric Objects	Provides a clear argument that supports or denies that the geometric objects in the list are found in the physical world.	Provides an argument that contains a minor error.	Provides an argument that contains a few errors.	Provides an argument that is illogical and unsupported.	Does not address the existence of geometric objects.

UNIT 1: Foundations of Geometry

1.2 Explore Measurements in the Coordinate Plane

photo: Getty Images

Lesson Objectives

- Apply the distance formula to calculate perimeter and area in the coordinate plane.
- Derive formulas to partition segments equally.
- Determine the position of a point relative to another point or set of points.

Essential Questions

- How can you use the structure of the coordinate plane to calculate measurements in geometric figures and justify properties of figures algebraically?
- What is the distance formula and how is it related to the Pythagorean theorem?

photo: Discovery Education

Investigations

Exploring an Archaeological Dig

Help a team of archaeologists find a long-lost statue. Put coordinates to work!

Is It the Long-Lost Temple?

Lean on Pythagoras to find the area and perimeter of a structure found at the dig site.

Dividing Segments

Break a board in a 1 : 1 ratio. Use coordinates to specify the exact location of the break.

Dividing Segments: The Sequel

Use precise karate chops to break boards at given ratios. Can a formula help you?

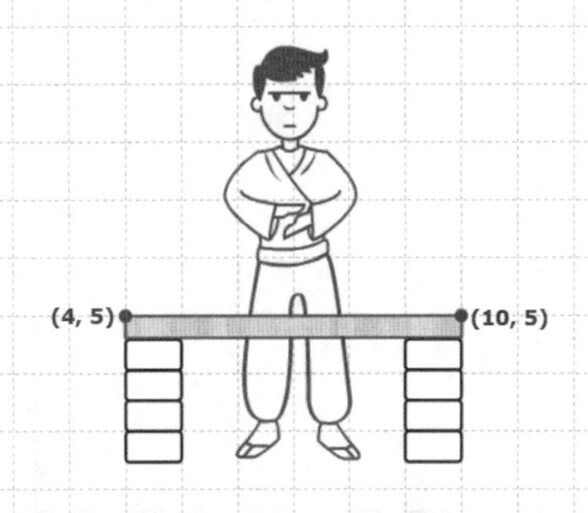

Key Vocabulary

dimension, distance formula, endpoint, fractal, hexagon, midpoint (of segment)

Discover

As you complete Engage and the investigations, record the most important ideas you've learned.

Engage	Investigation 1
Investigation 2	Investigation 3

Name ______________________________ Date ____________

Check for Understanding

Explore Measurements in the Coordinate Plane: Investigation 1

1. Calculate the perimeter and area of the triangle formed by the coordinates $K(-4, -1)$, $L(-2, 2)$, and $M(3, -1)$.

Perimeter = ____________ Area = ____________

2. The town of Lakehorn is built on a grid system. Town hall is located downtown at point (0, 0). A new school is located 3 miles north and 2 miles east of town hall. Only students who live outside a 5-mile radius from the school are eligible to ride the school bus. Which of the following students are eligible to ride the bus? *Select all that apply.*

A. Charity lives 1 mile due south of town hall.

B. Thomas lives 3 miles south and 4 miles east of the school.

C. Kaleb lives 2 miles south and 4 miles east of town hall.

D. Marinna lives 2 miles north and 3 miles west of town hall.

3. Two students calculated the area of a polygon in the coordinate plane using different strategies. Explain the two methods. Which method do you prefer, and why?

Student 1

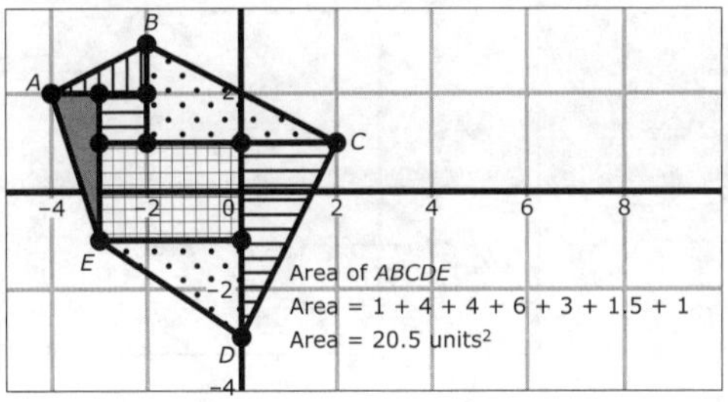

Student 2

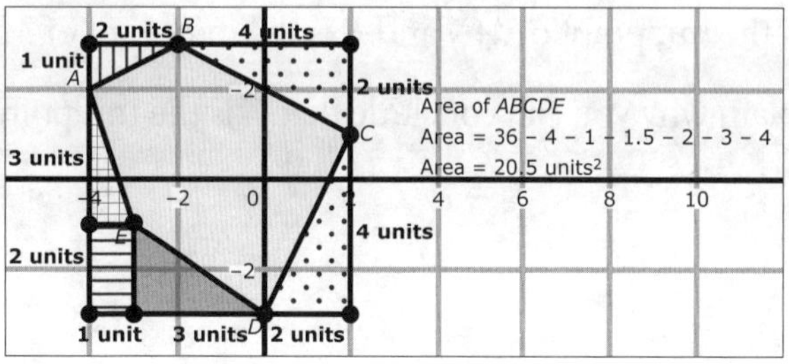

Name ______________________________ Date ____________

Check for Understanding

Explore Measurements in the Coordinate Plane: Investigation 2

1. In the diagram, M is the midpoint of $\overline{AB}$. Which of the following statements must be true? *Select all that apply.*

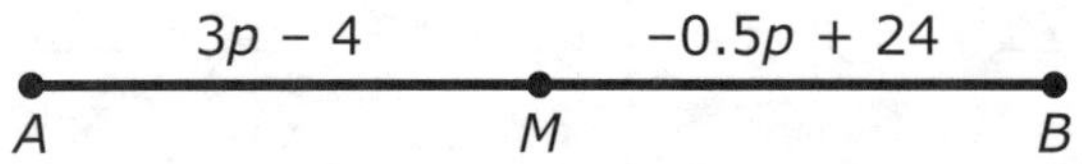

A. $AB = 2(3p - 4)$

B. $3p - 4 = -0.5p + 24$

C. $p = 11.2$

D. $3.5p = 20$

E. $MB = 20$

2. $A(4.25, -2.5)$ is the midpoint of a segment with endpoints $C(5.3, -1)$ and $T(x, y)$. What are the coordinates of T?

____

3. Examine rectangles $MNPR$ and $MACB$.

A is the midpoint of $\overline{MN}$ and B is the midpoint of $\overline{MR}$.

Explain why you can conclude that C is the midpoint of $\overline{MP}$.

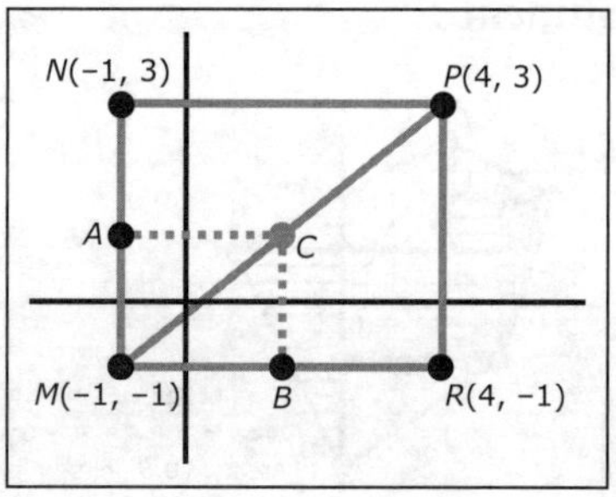

Name ______________________________ Date ____________

Check for Understanding

Explore Measurements in the Coordinate Plane: Investigation 3

1. F divides $\overline{HJ}$ in the ratio 3 : 1. If $H = (-11, 7)$ and $J = (5, 3)$, what are the coordinates of F?

____

2. $T(4.6, 5.7)$ is a point on $\overline{AB}$ given by $A(3, 6.5)$ and $B(7, 4.5)$. In which ratio does T divide $\overline{AB}$?

   A. 3 : 2

   B. 2 : 3

   C. 1 : 4

   D. 4 : 1

3. Josh calculated the ratio in which the line $y = -3x - 8$ divides the directed line segment $\overline{AB}$ with endpoints $A(-5, -13)$ and $B(6, 9)$. However, his answer does not match those of his friends. Examine Josh's work. Which statement is incorrect?

   A. A and B lie on $y = 2x - 3$, because $m = \frac{-13 - 9}{-5 - 6} = 2$ and $9 = 2(6) - b$ means $b = -3$.

   B. The line and the segment intersect when $2x - 3 = -3x - 8$.

   C. When $2x - 3 = -3x - 8$, then $5x = -5$, so $x = -1$, and $y = 2(-1) - 3 = -5$.

   D. The distance from $(-1, -5)$ to A is $\sqrt{80} = 4\sqrt{5}$. The distance from $(-1, -5)$ to B is $\sqrt{245} = 7\sqrt{5}$.

   E. The ratio into which the line $y = -3x - 8$ divides $\overline{AB}$ is 4 : 11.

4. Explain the difference between partitioning a directed line segment $\overline{MN}$ in a 2 : 7 ratio versus a 7 : 2 ratio. Include a specific example as part of your explanation.

Summary

Before you attempt the Practice Exercises, review what you've learned.

Coordinate geometry is based on the location of points, lines, and other figures on the coordinate plane.

Distance Formula

The distance d between any two points with coordinates (x_1, y_1) and (x_2, y_2) is given by $d = \sqrt{(x_2 - x_1)^2 + (y_2 - y_1)^2}$.

To find the coordinates of the midpoint of a segment, average the coordinates of its endpoints.

Midpoint

For a segment with endpoints (x_1, y_1) and (x_2, y_2), the coordinates of its midpoint M are $\left(\frac{x_1 + x_2}{2}, \frac{y_1 + y_2}{2}\right)$.

EXAMPLE: Finding the Midpoint of a Segment

Find the coordinates of the midpoint of $\overline{AB}$.

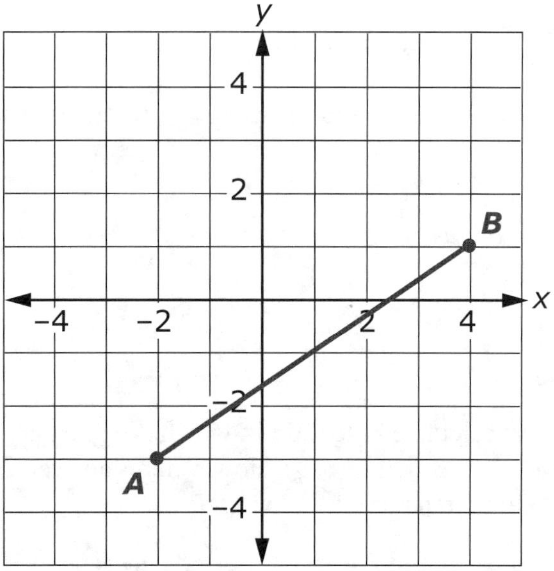

SOLUTION:

Use $A(-2, -3)$ as (x_1, y_1) and $B(4, 1)$ as (x_2, y_2). Substitute into the midpoint formula, and then simplify.

$$M\left(\frac{x_1 + x_2}{2}, \frac{y_1 + y_2}{2}\right) = M\left(\frac{-2 + 4}{2}, \frac{-3 + 1}{2}\right) = M(1, -1)$$

The coordinates of the midpoint of $\overline{AB}$ are $(1, -1)$.

My Notes

Summary *(continued)*

When you **partition** a segment on the coordinate plane, you divide it into parts. You can partition segments so that the lengths of the parts have a given ratio. For example, if point P partitions $\overline{AB}$ in the ratio 4 : 1, this means that the ratio of AP to PB is 4 to 1.

Partition Formula

If point P partitions a segment with endpoints (x_1, y_1) and (x_2, y_2) in the ratio, $a : b$ then the coordinates of point P are $\left(x_1 + \frac{a}{a+b}(x_2 - x_1), y_1 + \frac{a}{a+b}(y_2 - y_1)\right)$.

EXAMPLE: Partitioning a Segment

Find the coordinates of point P so that it partitions $\overline{AB}$ in the ratio 4 : 1.

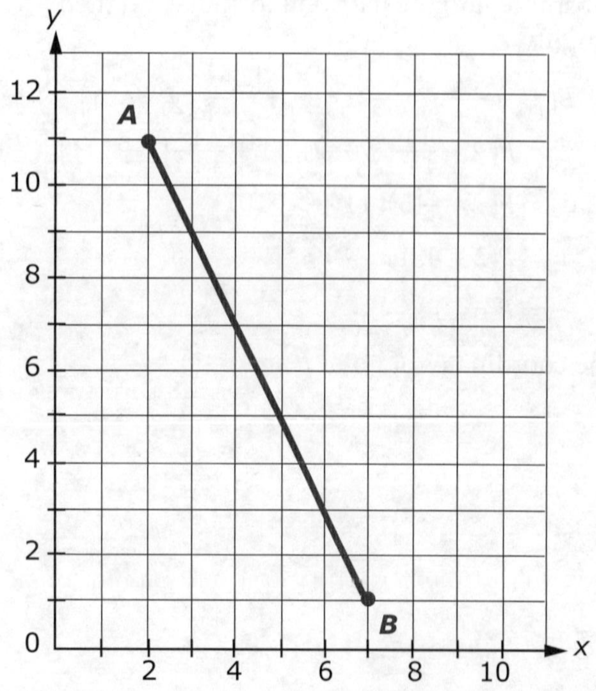

My Notes

Summary *(continued)*

SOLUTION:

Use $A(2, 11)$ as (x_1, y_1) and $B(7, 1)$ as (x_2, y_2). The partition ratio is 4 : 1, so let $a = 4$ and $b = 1$.

Substitute into the partition formula and then simplify.

$$P\left(x_1 + \frac{a}{a+b}(x_2 - x_1), y_1 + \frac{a}{a+b}(y_2 - y_1)\right)$$
$$= P\left(2 + \frac{4}{4+1}(7 - 2), 11 + \frac{4}{4+1}(1 - 11)\right)$$
$$= P\left(2 + \frac{4}{5}(5), 11 + \frac{4}{5}(-10)\right)$$
$$= P(2 + 4, 11 - 8)$$
$$= P(6, 3)$$

The coordinates of point P are (6, 3).

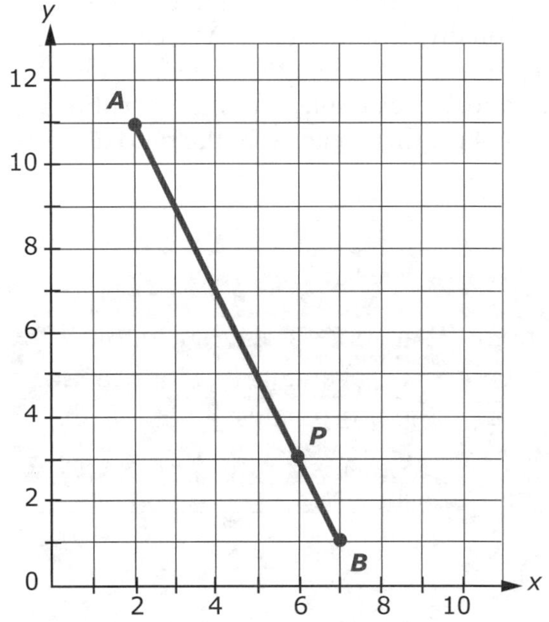

My Notes

Practice Exercises

Review what you've learned using these practice problems. For practice problems with feedback, try the Coach and Play items in the Practice section online.

1. Consider how to partition a line segment into two segments having a length ratio of 2 : 9.

   What might be a first step in this process?

   A. Divide the line segment into 11 equal parts.

   B. Divide the line segment into 9 equal parts.

   C. Divide the line segment into 7 equal parts.

   D. Divide the line segment into 2 equal parts.

2. Marlee has to find the point that divides the line segment between the points $P(-6, 10)$ and $R(2, -4)$ exactly in half. Marlee states: "the answer is $M(-2, 3)$. Since a midpoint divides a line segment into 2 equal parts, I used the midpoint formula." Is this a logical argument?

   A. Yes; using the midpoint formula will determine the middle point of a line segment.

   B. No; she should have used the distance formula.

   C. No; she should have used the quadratic formula.

   D. No; she should have used slope-intercept formula $y = mx + b$.

3. Consider the given end points of a segment.

   (8, −2) and (5, 3)

   Complete the statement.

   To find the midpoint of the segment, calculate

   $M = \left(\frac{8 + \square}{2}, \frac{-2 + \square}{\square}\right).$

4. Which of the following statements describe how to locate the midpoint of a segment with endpoints (1, 3) and (5, 7)? *Select all that apply.*

   A. Subtract the x-coordinates and the y-coordinates and divide both differences by 2.

   B. Add the x-coordinates and the y-coordinates and divide both sums by 2.

   C. Find the average value of the x- and y-coordinates of the end points of the segment.

   D. Use the Pythagorean theorem and the end points of the segment to find the midpoint.

   E. Use the distance formula and the end points of the segment to find the midpoint.

5. *ABCD* is a square.

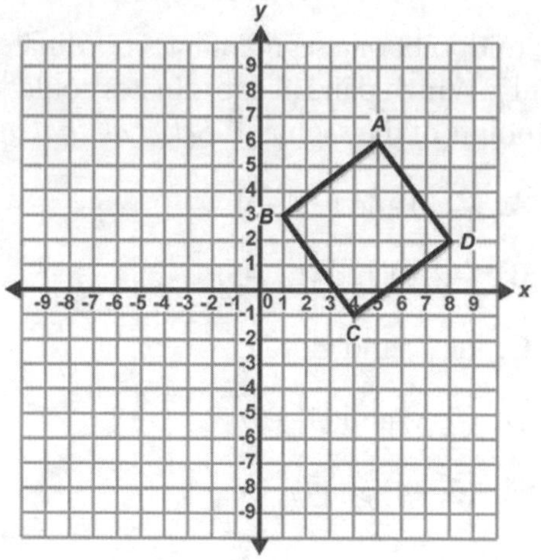

   What is the perimeter of *ABCD*?

   A. 12 units

   B. 16 units

   C. 20 units

   D. 25 units

Practice Exercises *(continued)*

6. Consider $\overline{FJ}$ shown in the figure.

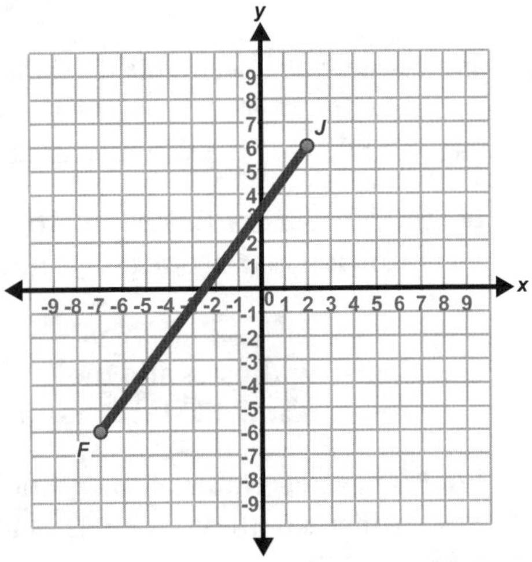

Which of the following points partition the segment into a ratio of either 1 : 2 or 2 : 1? *Select all that apply.*

A. (−4, −2)

B. (−1, 2)

C. (−2.5, 0)

D. (0, 3.25)

7. The midpoint of a segment has coordinates (5, −1). Which pairs of coordinates could be the endpoints of the segment? *Select all that apply.*

A. (7, 3) and (−2, −5)

B. (−2, 3) and (12, −5)

C. (6, 7) and (4, −7)

D. (9, 5) and (1, −6)

E. (3, −4) and (7, 2)

8. Consider the rectangle shown on the coordinate plane.

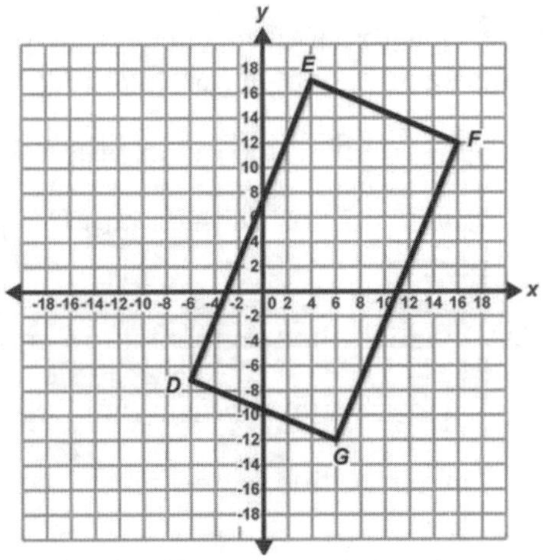

The area of rectangle *DEFG* is ____________ square units.

9. A line segment on a coordinate plane has a slope of 1 and end points at (1, *a*) and (9, *b*). Two points on this line segment partition it in a ratio of 1 : 3 and 3 : 1.

Which of the following are the *x*-coordinates of these two points?

A. 2

B. 3

C. 4

D. 5

E. 6

F. 7

G. 8

10. If Memphis is *M*(2, 7) and Chicago is *C*(4, 15) on a map, which city is located at the midpoint?

A. Poplar Bluff (2, 8)

B. Marion (2, 4)

C. Mount Vernon (3, 11)

D. Peoria (4, 3)

Apply

Why Is This Shape Special?

Since ancient times, people have been fascinated with the algebraic and geometric properties of the rectangle below. See if you can find out why.

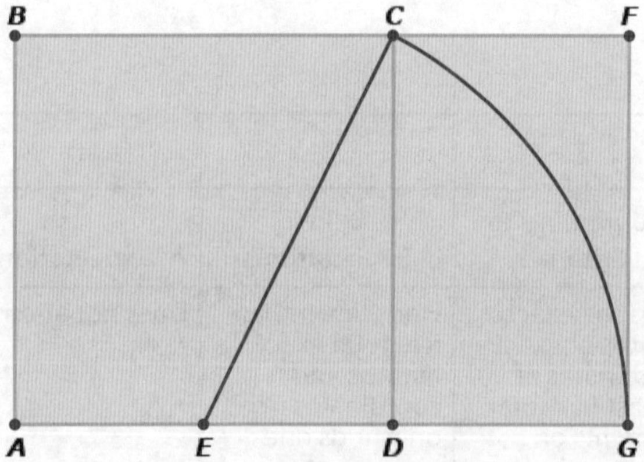

Given: *ABCD* is a square and *E* is the midpoint of *AD*.

- Superimpose a grid on the figure.
- Find the length of each segment and the exact coordinates of each point.
- Calculate the following ratios of lengths, and research the values you find:
  - $BF : BC$
  - $BC : CF$
  - $AG : GF$
  - $CD : DG$
- Explain why this rectangle is so special.

Show what you've learned by completing the other performance tasks in the online Apply section.

Apply *(continued)*

Your answer to Apply will be assessed on the following criteria:

1. Finding the exact lengths of the segments and the exact coordinates of the points
2. Calculating the length ratios
3. Researching the values found and including citations
4. Explaining why this rectangle is special

Criteria \ Scale	4 Exceeds Criteria	3 Meets Criteria	2 Progressing to Criteria	1 Below Expectations	0 No Expectation
Lengths and Coordinates	Finds the exact lengths of all segments and the exact coordinates of all points.	Finds the exact lengths of at least six of nine segments and the exact coordinates of at least five of seven points.	Finds the exact lengths or exact coordinates of at least five total segments or points, including one of each type (segment and point).	Finds fewer than five total exact lengths and coordinates or only finds one type: lengths or coordinates.	Does not attempt task.
Length Ratios	Correctly calculates all four length ratios.	Correctly calculates three of the four length ratios.	Correctly calculates two of the four length ratios.	Correctly calculates one of the four length ratios.	Does not attempt task.
Research and Citations	Correctly identifies the major historical contexts of the ratios and provides a valid citation.	Correctly identifies the major historical contexts of the ratios, but fails to provide a valid citation.	Partially identifies some of the correct historical context of the ratios and provides a valid citation.	Partially identifies some of the correct historical context of the ratios and fails to provide a valid citation.	Does not attempt task.
Explanation	Clearly identifies and explains a significant, unique characteristic of the rectangle.	Identifies a significant and unique characteristic of the rectangle, but does not fully or clearly explain it.	Identifies a characteristic that is not significant or unique, but fully explains this characteristic.	Identifies a characteristic that is not significant or unique without a valid explanation.	Does not attempt task.

UNIT 1: Foundations of Geometry

1.3 Explore Congruence Constructions in the Coordinate Plane

photo: Getty Images

Lesson Objectives

- Construct congruent angles and line segments.
- Verify congruence in the coordinate plane and prove simple theorems.

Essential Questions

- How can you construct congruent figures precisely without measuring tools?
- How can you use algebra to prove properties of geometric figures in the coordinate plane?

Investigations

Roller Coasters

Help maintain a roller coaster. How can you get an exact match to replace a cross tie?

Creating Congruent Figures

Given limited tools, how can you create exact replicas?

The Geometry of Miniature Golf

Miniature golf is a game of angles. Apply what you learn about angles to defend a billiard's claim.

Analyzing a Roller Coaster Design

Help a roller coaster engineer ensure that the coaster's support structure is square.

photo: Getty Images

Key Vocabulary

arc, congruent, consecutive, midpoint (of segment), parallelogram, perpendicular, protractor, quadrilateral, rectangle, rhombus, segment, square, trapezoid

Discover

As you complete Engage and the investigations, record the most important ideas you've learned.

Engage	**Investigation 1**
Investigation 2	**Investigation 3**

Name ______________________________ Date __________

Check for Understanding

Explore Congruence Constructions in the Coordinate Plane: Investigation 1

1. Examine the construction. What can you conclude about *MN*?

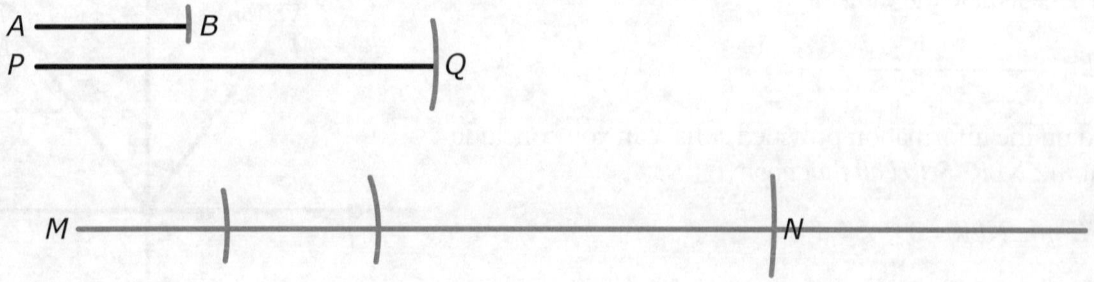

A. $MN = AB + 2PQ$

B. $MN = 3AB$

C. $MN = 2PQ - AB$

D. $MN = PQ + 2AB$

2. How is a construction similar to and different from a drawing and a sketch? *Select all that apply.*

A. Both sketches and constructions are approximations of geometric figures.

B. You can use a protractor to make a drawing, but not to make a construction.

C. You can use tracing paper to help make a sketch, but not to make a construction.

D. Both drawings and constructions depend upon the precision of measurement tools.

E. Unlike a drawing, a construction is an accurate representation of a geometric figure.

F. Both sketches and constructions can be used to show relationships within geometric figures.

3. Explain how you might use constructions to copy a triangle.

Name ______________________________ Date ____________

Check for Understanding

Explore Congruence Constructions in the Coordinate Plane: Investigation 2

1. A ray of light creates the angles of incidence and reflection shown. Calculate the value of n.

   $n =$ ____________

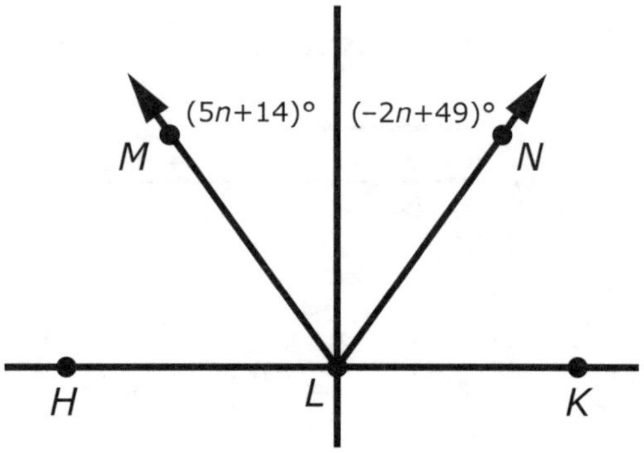

2. Based on the information provided, what can you conclude about $m\angle NLK$? *Select all that apply.*

   A. $m\angle NLK = 39°$

   B. $m\angle NLK = 51°$

   C. $m\angle NLK = m\angle MLH$

   D. $m\angle NLK = (5n + 14)°$

   E. $m\angle NLK + (5n + 14)° = 90°$

3. Michelle and William each used a different solution strategy to calculate the value of w, based on the following information:

   $\angle GFJ$ is the angle of incidence created by the path of a ball as it follows the path beginning at G, to F, and then toward H. $m\angle GFH = 95°$. Compare their solution strategies, and explain how they use the angle addition postulate in their solutions.

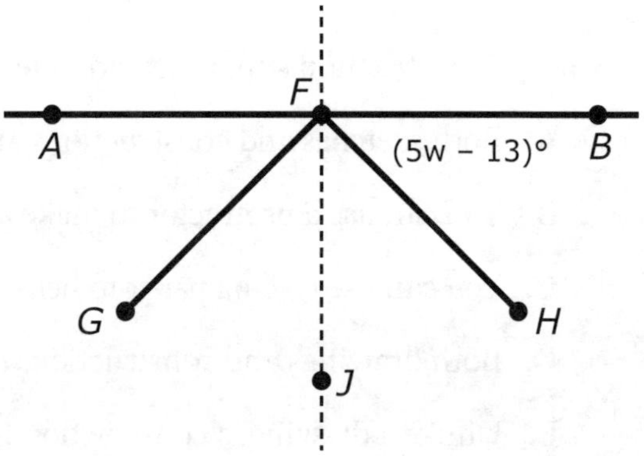

Michelle:	**William:**
$5w - 13 = 90 - \frac{1}{2}(95)$	$2(5w - 13) = 180 - 95$
$5w - 13 = 42.5$	$10w - 26 = 85$
$5w = 55.5$, so $w = 11.1$	$10w = 111$, so $w = 11.1$

Name ______________________________ Date __________

Check for Understanding

Explore Congruence Constructions in the Coordinate Plane: Investigation 3

1. A circle in the coordinate plane has a diameter with endpoints $M(-2, -3)$ and $R(7, 9)$. Examine each statement, and write whether the statement is *True* or *False*.

- The center of the circle is at $Q(2.5, 3)$, because Q is the midpoint of $\overline{MR}$. __________
- $B(4, 5)$ partitions the diameter $\overline{MR}$ at a ratio of 2 : 1, because $\frac{MB}{BR} = 2$. __________
- The line $y = \frac{4}{3}x + 1$ divides the circle in half, because $\overline{MR}$ is part of this line. __________

2. $M(2, 0.5)$, $T(3.5, -0.5)$ and $H(4.5, 1)$ are plotted in the coordinate plane. Based on the information provided, what can you conclude about $\triangle MTH$? Use the coordinate plane provided as needed. *Select all that apply.*

A. $m\angle T = 90°$

B. $MH = 6.5$ units

C. $\left(\frac{13}{4}, 1\right)$ is the midpoint of $\overline{MH}$.

D. $\triangle MTH$ is isosceles.

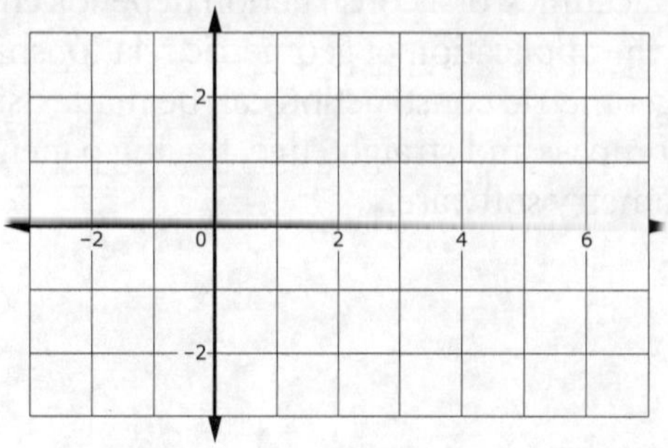

3. Circle A is drawn in the coordinate plane so that the center is at $(1, 2)$. The length of the radius of the circle is $\sqrt{13}$ centimeters. $P(4, 5)$ is then added to the diagram. Explain how you would use the distance formula or the Pythagorean theorem to prove that P must lie outside of the circle.

Summary

Before you attempt the Practice Exercises, review what you've learned.

You can create geometric figures in different ways.

- A **sketch** is a quick picture of a figure. It is not necessarily accurate in measurements, but it can be used to show relationships within the figure.
- A **drawing** can make use of measurement tools, such as rulers and protractors. Any drawing is still approximate because it depends on the accuracy and precision of the measurement tools used to make it.
- A **construction** is an accurate representation of a geometric figure made without using rulers, protractors, or other measurement tools. Instead, the accuracy of a construction depends entirely on the application of geometric relationships. A geometric construction can be made using a compass and straightedge, tracing paper, or geometry software.

EXAMPLE: Constructing Congruent Segments

Construct $\overline{RS}$ so that it is congruent to $\overline{PQ}$.

Construction Tool: Segment Tutorial: View this video segment to see how to use the Construction Tool to create segments.

EXAMPLE: Creating Congruent Angles

Construct $\angle JKL$ so that it is congruent to $\angle FGH$.

Construction Tool: Angle Tutorial: View this video segment to see how to use the Construction Tool to create angles.

My Notes

Summary *(continued)*

EXAMPLE: Verifying That Segments Are Congruent

Verify that $\triangle ABC$ has exactly two congruent sides.

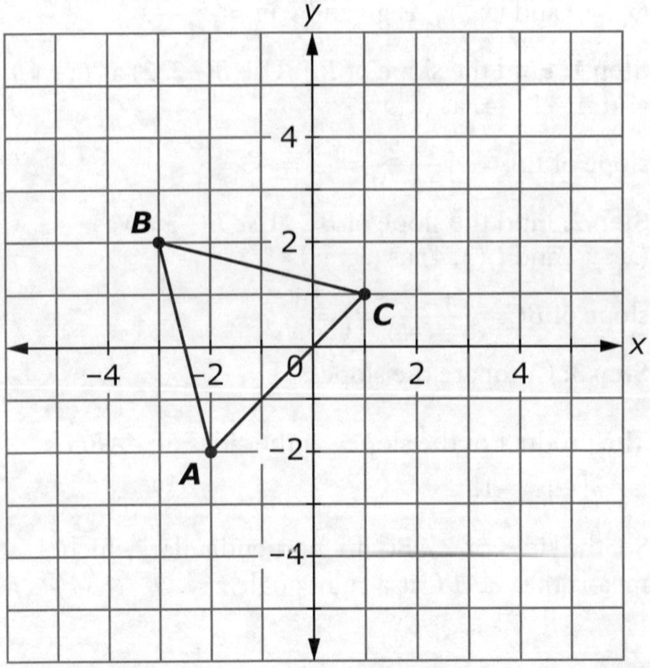

SOLUTION:

Use the distance formula to find the length of each side of the triangle. The distance between any two points (x_1, y_1) and (x_2, y_2) is given by $\sqrt{(x_2 - x_1)^2 + (y_2 - y_1)^2}$.

Find *AB*. Use $A(-2, -2)$ as (x_1, y_1) and $B(-3, 2)$ as (x_2, y_2).

$$AB = \sqrt{(-3 - (-2))^2 + (2 - (-2))^2}$$
$$= \sqrt{(-1)^2 + 4^2}$$
$$= \sqrt{17}$$

Find *BC*. Use $B(-3, 2)$ as (x_1, y_1) and $C(1, 1)$ as (x_2, y_2).

$$BC = \sqrt{(-1 - (-3))^2 + (1 - 2)^2}$$
$$= \sqrt{4^2 + (-1)^2}$$
$$= \sqrt{17}$$

Find *AC*. Use $A(-2, -2)$ as (x_1, y_1) and $C(1, 1)$as (x_2, y_2).

$$AC = \sqrt{(-1 - (-2))^2 + (1 - (-2))^2}$$
$$= \sqrt{3^2 + 3^2}$$
$$= \sqrt{18}$$

$\overline{AB}$ and $\overline{BC}$ have the same length, so $\overline{AB} \cong \overline{BC}$. $\overline{AC}$ has different length, so it is not congruent to the other sides. Therefore, the triangle has exactly two congruent sides.

You can show that an angle on the coordinate plane is a right angle by showing that the sides of the angle are perpendicular. For an angle with non-vertical sides, the sides are perpendicular if their slopes are opposite reciprocals. Two numbers are opposite reciprocals if their product is −1.

My Notes

Summary *(continued)*

EXAMPLE: Verifying That an Angle Is a Right Angle

Verify that $\angle ABC$ is a right angle.

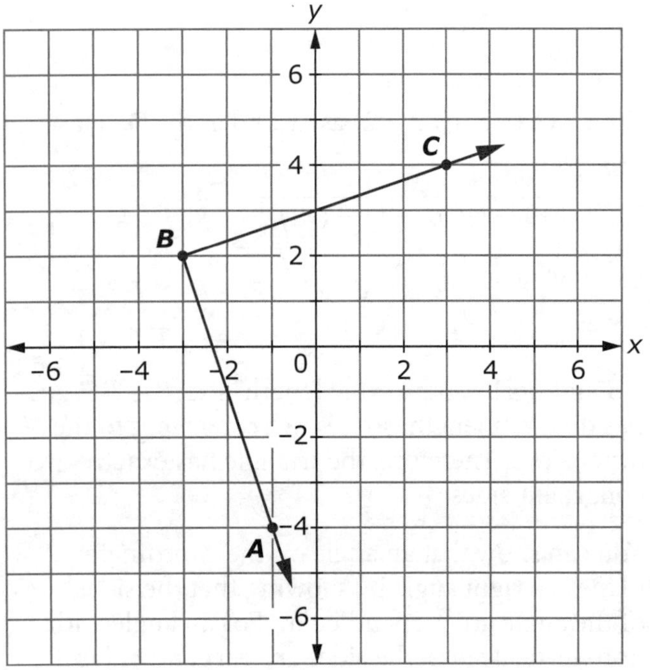

SOLUTION:

Find the slope of each side of the angle. The slope m of a line or part of a line that passes through points (x_1, y_1) and (x_2, y_2) is given by $m = \frac{y_2 - y_1}{x_2 - x_1}$.

Step 1: Find the slope of $\overrightarrow{BA}$. Use $B(-3, 2)$ as (x_1, y_1) and $A(-1, -4)$ as (x_2, y_2).

slope of $\overrightarrow{BA} = \frac{-4 - 2}{-1 - (-3)} = \frac{-6}{2} = -3$

Step 2: Find the slope of $\overrightarrow{BC}$. Use $B(-3, 2)$ as (x_1, y_1) and $C(3, 4)$ as (x_2, y_2).

slope of $\overrightarrow{BC} = \frac{4 - 2}{3 - (-3)} = \frac{2}{6} = \frac{1}{3}$

Step 3: Compare the slopes.

The product of the slopes of the sides of $\angle ABC$ is $-3\left(\frac{1}{3}\right) = -1$.

So, the sides of $\angle ABC$ are perpendicular, which means that $\angle ABC$ is a right angle.

My Notes

Practice Exercises

Review what you've learned using these practice problems. For practice problems with feedback, try the Coach and Play items in the Practice section online.

1. An isosceles triangle has vertices at $A(8, -1)$, $B(4, 2)$, $C(7, 6)$. Sides $\overline{AB}$ and $\overline{BC}$ are congruent. You want to determine if the triangle is an isosceles right triangle.

   Write the numerical answers for the questions below in fractional form.

   What is the slope of $\overline{AB}$? ______

   What is the slope of $\overline{BC}$? ______

   Are sides $\overline{AB}$ and $\overline{BC}$ perpendicular?

   ______

   Is the triangle an isosceles right triangle?

   ______

2. A circle is graphed on a coordinate system. The center is at the origin, and the point (0, 3) lies on the circle.

   You want to prove or disprove that the point $(1, \sqrt{8})$ lies on the circle.

   What is the radius of the circle? ______

   What is the distance from the center to $(1, \sqrt{8})$?

   ______

   Does the point $(1, \sqrt{8})$ lie on the circle?

   ______

3. In the diagram, $\overleftrightarrow{TR}$ and $\overleftrightarrow{PS}$ intersect at Q to form four angles. *The diagram is not to scale.*

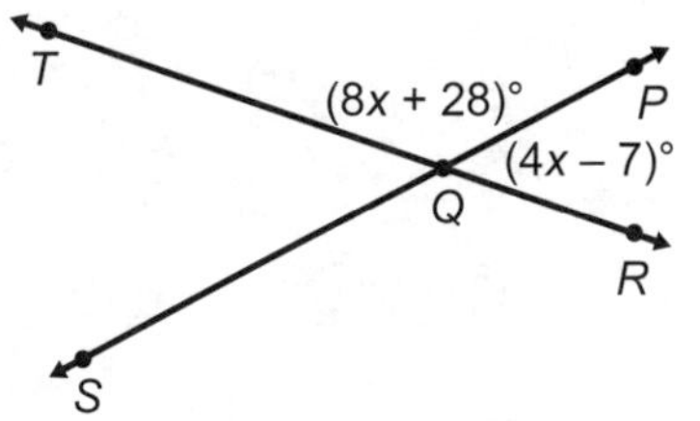

   Use the information in the diagram to calculate each value:

   $x =$ ______

   $m\angle TQS =$ ______

4. You can test whether a quadrilateral is a square without measuring any of its side lengths. If a quadrilateral has four right angles and its diagonals are perpendicular, then it must be a square.

   Quadrilateral $MNPQ$ contains four right angles. The vertices of quadrilateral $MNPQ$ lie at $M(-2, 4)$, $N(4, 7)$, $P(7, 1)$, and $Q(1, -2)$.

   Prove or disprove that the quadrilateral is also a square by determining if the diagonals are perpendicular.

   Simplify fractional answers as necessary.

   What is the slope of $\overline{MP}$? ______

   What is the slope of $\overline{NQ}$? ______

   Are diagonals $\overline{MP}$ and $\overline{NQ}$ perpendicular?

   ______

   Is the quadrilateral a square? ______

5. Which of the following sets of four coordinates could not be the four vertices of a square?

   A. (–1, 0), (–1, 2), (–3, 0), (–3, 2)

   B. (1, –3), (3, –3), (1, –5), (3, –5)

   C. (1, 0), (4, 3), (4, –3), (7, 0)

   D. (0, –2), (0, –4), (–3, –2), (–3, –4)

Practice Exercises *(continued)*

6. A perpendicular bisector is a line that is both

- perpendicular to the given segment and
- contains the midpoint of the given segment.

Apply the given definition to graph the perpendicular bisector of $\overline{AB}$ on the coordinate plane.

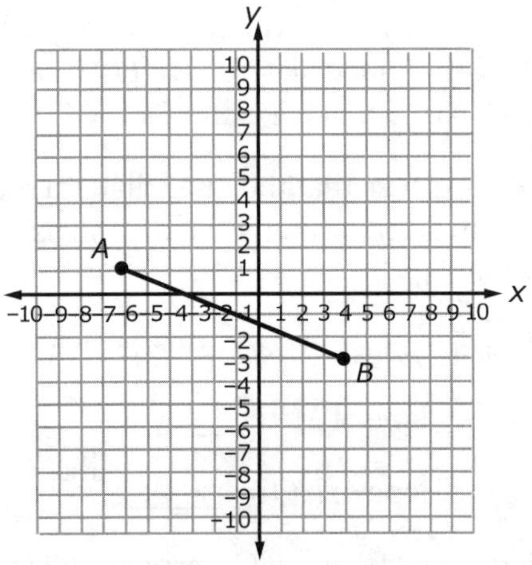

7. David copied $\overline{CD}$ and $\overline{GH}$ to construct $\overline{PQ}$. Examine his construction in the diagram.

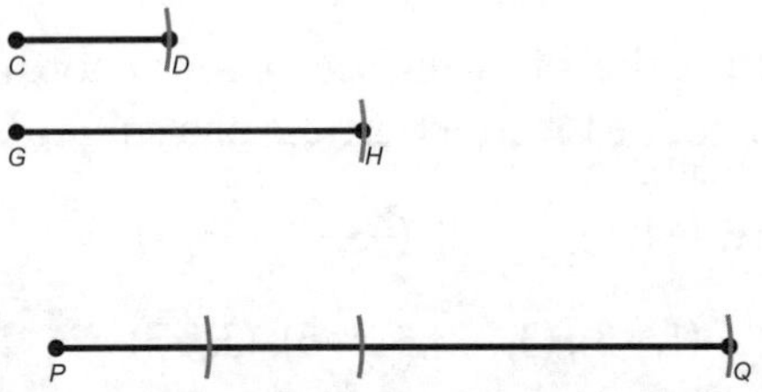

Which statements are true based on David's construction? *Select all that apply.*

A. $PQ = 2GH + CD$

B. $PQ = 2CD + GH$

C. $GH + CD = PQ - CD$

D. $CD + GH = PQ - GH$

E. $PQ = 2GH$

F. $PQ = 2CD$

8. The diagram shows the construction of ΔABC.

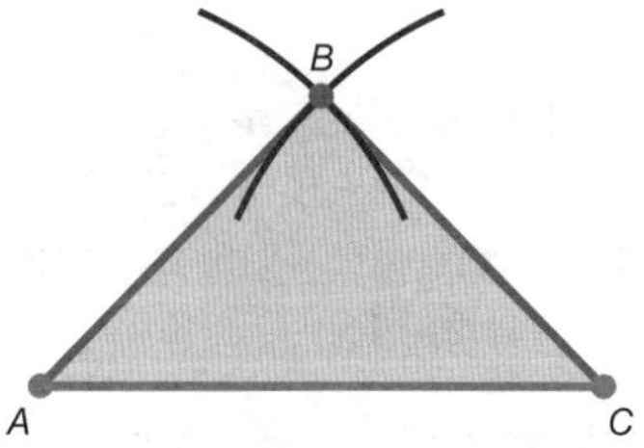

Triangle ABC was constructed from $\overline{AC}$ as follows:

1) Set the compass to an arbitrary distance less than AC.

2) Place the tip of the compass at point A and construct an arc above $\overline{AC}$.

3) Place the tip of the compass at point C and construct an arc above $\overline{AC}$.

4) Label the point of intersection of the arcs as point B.

5) Use a straightedge to draw line $\overline{AB}$ and $\overline{CB}$.

Choose the statement that is true based on this construction.

A. $\angle A$ and $\angle B$ are congruent.

B. The triangle's sides $\overline{AC}$ and $\overline{AB}$ are congruent.

C. The triangle's sides $\overline{AB}$ and $\overline{CB}$ are congruent.

D. The measure of $\angle B$ is $\frac{1}{2}$ the measure of $\angle C$.

Apply

Is the Logo Ready for Production?

The Tri-City Tigers soccer team wants a new logo for their jerseys. Watch the video to learn more.

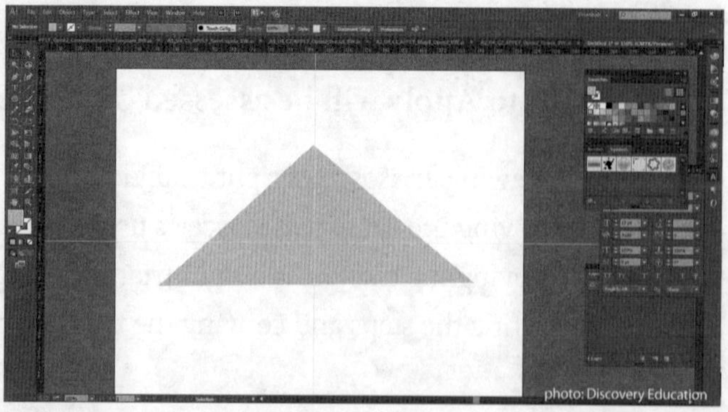

How can you design the best logo?

Jersey Logo: View this video segment to see how you can design the best logo.

Based on the fan feedback noted in the video, the triangle in the background of the logo needs to be a right isosceles triangle. To make it more visually interesting, the legs cannot be vertical or horizontal segments.

The jersey company provided a grid so that the logo design will be centered on the front of the jerseys appropriately. Use the Dynamic Geometry Tool or a printed copy of your logo to get started.

Dynamic Geometry Tool

- Upload a snapshot of your logo.
- Use coordinate geometry to prove that you have drawn a right isosceles triangle.
- Construct a copy of the triangle using the Straight Liner and Compass.
- Describe the steps you took to make the copy.

Paper

- Print a copy of your logo.
- Use coordinate geometry to prove that you have drawn a right isosceles triangle.
- Construct a copy of the triangle using a straightedge and compass.
- Describe the steps you took to make the copy.

Show what you've learned by completing the other performance tasks in the online Apply section.

Apply *(continued)*

Your answer to Apply will be assessed on the following criteria:

1. Drawing an isosceles right triangle in the Dynamic Geometry Tool
2. Proving the triangle is isosceles using coordinate geometry
3. Proving the triangle is a right triangle using coordinate geometry
4. Listing the steps and copying the triangle using a compass and straightedge

Scale / Criteria	4 Exceeds Criteria	3 Meets Criteria	2 Progressing to Criteria	1 Below Expectations	0 No Expectation
Triangle in Logo	Constructs a right isosceles triangle with legs that are not horizontal or vertical.	Constructs a right isosceles triangle, but has legs that are vertical or horizontal.	Constructs a right triangle or an isosceles triangle, but it is not both.	Constructs a triangle that is neither a right triangle nor an isosceles triangle.	Does not construct a triangle.
Isosceles Triangle	Proves the triangle is isosceles using coordinates to find all the necessary measurements.	Proves the triangle is isosceles using coordinates, but makes a minor calculation error.	Attempts to prove the triangle is isosceles using coordinates, but makes a major calculation error.	Attempts to prove the triangle is isosceles without using coordinates.	Does not attempt to prove the triangle is isosceles.
Right Triangle	Proves the triangle is right using coordinates to find all the necessary measurements.	Proves the triangle is right using coordinates, but makes a minor calculation error.	Attempts to prove the triangle is right using coordinates, but makes a major calculation error.	Attempts to prove the triangle is right without using coordinates.	Does not attempt to prove the triangle is right.
Copy	Correctly lists the steps and copies the triangle using only a compass and a straightedge.	Lists the steps and copies the triangle using only a compass and a straightedge, but makes one error.	Lists the steps and copies the triangle using only a compass and a straightedge, but makes two or three errors.	Lists the steps and copies the triangle, but the list may be incomplete or there is evidence that other tools were used to make the copy.	Does not list the steps or copy the triangle.

photo: Discovery Education

UNIT 2: Geometric Transformations

2.1 Explore Transformations

Lesson Objectives

- Define and identify rigid transformations of points, line segments, and polygons informally and with coordinates.
- Use a variety of tools to investigate and draw transformations of segments, angles, and polygons.
- Define, identify, and draw line and rotational symmetry, including tessellations informally and with coordinates.

Essential Questions

- What distinct features identify transformations as a family of functions?
- What effects do rigid motion transformations have on the properties of a figure?
- How are rigid motion transformations used to determine the type and number of symmetries of a figure?

Investigations

Transformations and Animations

Create an animation using a flip book. How does the figure move? Be precise!

Identifying Transformations in Animations

What are reflections, rotations, and transformations? There's function notation for that!

Transformations in the Plane

Perform transformations and watch what happens to side lengths and angle measures.

Animation and the Coordinate Plane

The coordinate plane adds precision to transformations. Function rules take them to the next level.

A Closer Look

Is that transformation function a reflection, rotation, or translation? Coordinates to the rescue!

Animation and Symmetry

How are symmetry, reflections, and rotations related?

Tessellations

What makes a shape pattern a tessellation? Look at examples and nonexamples to decide.

Key Vocabulary

isometry/congruence transformation, point symmetry, preimage, reflection, rigid motion, rotation, symmetry, transformation, translation

Discover

As you complete Engage and the investigations, record the most important ideas you've learned.

Engage

Investigation 1

Investigation 2

Investigation 3

Investigation 4

Investigation 5

Investigation 6

Name ____ Date ____________

Check for Understanding

Explore Transformations: Investigation 1

1. Write the transformations $T(A)$, $Ro(A)$, or $R(A)$ that would take A onto the two hang gliders shown.

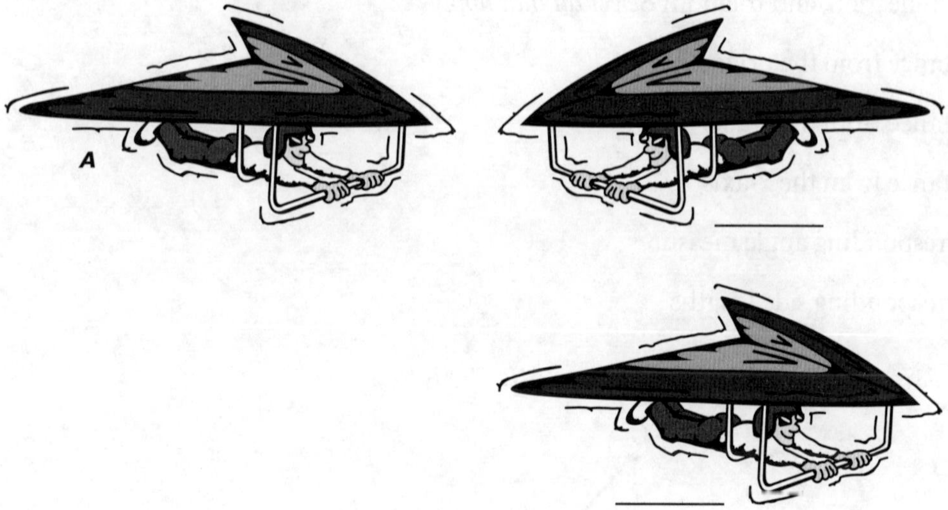

2. Circle to indicate whether the statement about the transformation of $\overline{AB}$ is always true, sometimes true, or never true.

$T(\overline{AB}) \parallel \overline{AB}$	[Always / Sometimes / Never]
$R(\overline{AB}) \parallel \overline{AB}$	[Always / Sometimes / Never]
$T(\overline{AB}) \perp \overline{AB}$	[Always / Sometimes / Never]
$Ro(\overline{AB}) \perp \overline{AB}$	[Always / Sometimes / Never]
$Ro(\overline{AB}) \cong T(\overline{AB})$	[Always / Sometimes / Never]
$R(\overline{AB}) \cong Ro(\overline{AB})$	[Always / Sometimes / Never]

3. The transformation $Ro(M)$ of $x°$ clockwise around the origin results in an image M'. Other than a counterclockwise rotation, what transformation would map M' back onto M?

Name ______________________________ Date ______________

Check for Understanding

Explore Transformations: Investigation 2

1. Which of the following remain constant between any pre-image and image for the transformation's translation, reflection, and rotation? *Select all that apply.*

   A. distance from the origin

   B. distance from the *y*-axis

   C. distance from the *x*-axis

   D. corresponding angle measures

   E. corresponding side lengths

2. Examine the transformation $Ro(ABCDEF) = A'B'C'D'E'F'$.

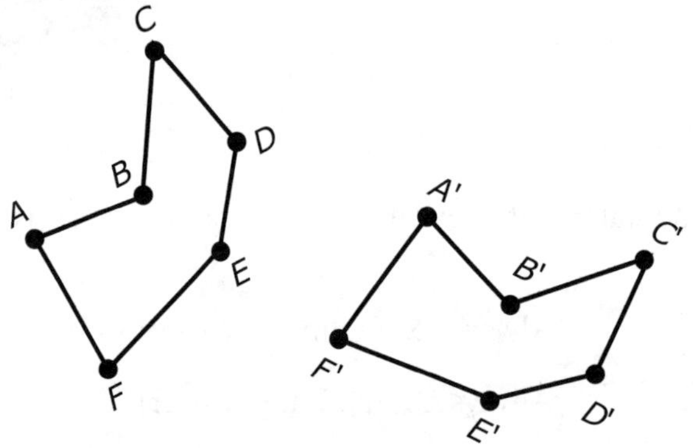

   The rotation is about unknown point *X*. Which relationship must be true and can help you locate point *X*?

   A. $\angle AXA' \cong \angle A'XA$

   B. $\angle AXA' \cong \angle FXF'$

   C. $\angle AFX \cong \angle XA'F'$

   D. $\angle XFA' \cong \angle XAF'$

3. Describe how to find the line of reflection for $R(\triangle XYZ) = \triangle X'Y'Z'$.

Name ________________________________ Date ____________

Check for Understanding

Explore Transformations: Investigation 3

1. Match the transformation of (−3, 4) with the resulting coordinates. *Not all resulting coordinates will be used, and some may be used more than once.*

Transformation	Resulting Coordinates
A. $T_{<7,-1>}(-3, 4)$	**I.** (−3, 4)
B. $T_{<6,0>}(-3, 4)$	**II.** (−4, −3)
C. $Ro_{180°}(-3, 4)$	**III.** (4, 3)
D. $Ro_{90°}(-3, 4)$	**IV.** (−3, −4)
E. $R_{x\text{-axis}}(-3, 4)$	**V.** (3, 4)
F. $R_{y\text{-axis}}(-3, 4)$	**VI.** (3, −4)
	VII. (4, −3)

2. What is the distance between the input and output points of $R_{x\text{-axis}}(5, 3)$?

A. 3

B. 5

C. 6

D. 10

3. $R(\overline{AB}) = \overline{A'B'}$ for points $A(1, 3)$, $B(4, -3)$, $A'(-5, 3)$, and $B'(-8, -3)$. How would you determine the line of reflection for this transformation?

Name ______________________________ Date ____________

Check for Understanding

Explore Transformations: Investigation 4

1. Which statement is always true about rotations?

   A. The segment connecting the center to a point of the pre-image is perpendicular to the segment connecting the center to the corresponding point of the image.

   B. The segments connecting the corresponding vertices of the pre-image and image are congruent, so the lengths are preserved, but the angles can change.

   C. The angle of rotation is formed by the intersection of the segments connecting all the points of the pre-image to the corresponding points on the image.

   D. The distance from the center to a point of the pre-image is equal to the distance from the center to the corresponding point of the image.

2. Point $Q(-13, 7)$ is in the domain and Point $R(11, 9)$ is in the range of a reflection function. Use function notation to describe this transformation. Include the line of reflection.

   A. $R_{y-12x+4}(x, y)$

   B. $R_{y-\frac{1}{12}x+4}(x, y)$

   C. $R_{y-12x-4}(x, y)$

   D. $R_{y-\frac{1}{12}x-4}(x, y)$

   E. $R_{y=12x+4}(x, y)$

   F. $R_{y=\frac{1}{12}x+4}(x, y)$

3. Triangle ABC has vertices $A(-10, 1)$, $B(-4, 5)$, and $C(-2, 2)$, and $\triangle RST$ has vertices $R(3, 8)$, $S(7, 2)$, and $T(4, 0)$. The transformation $Ro_{90°, P(x, y)}$ maps $\triangle ABC$ onto $\triangle RST$. Is Point $P(0, -2)$ the center of rotation of this rotation function? Use the coordinates and algebraically support your claim.

Name ______________________________ Date ____________

Check for Understanding

Explore Transformations: Investigation 5

1. A shape with an angle of rotational symmetry of 45° has rotational order ______.

2. For each shape, write the type of symmetry that applies. *Select all that apply.*

A. Rotational Symmetry (order $n > 1$)

B. Line Symmetry

C. No Symmetry

______ ______ ______ ______

3. For a given figure, there is an angle of measure $x°$ such that for every point A on the figure there exists a point A' on the figure such that $Ro_{x°}(A) = A'$. Which equation could be used to determine the order, n, of the rotational symmetry, in terms of x?

A. $\frac{180°}{x} = n$ **B.** $\frac{360°}{x} = n$ **C.** $180°x = n$ **D.** $360°x = n$

4. Examine rectangle $WXYZ$, for which $WX > XY$.

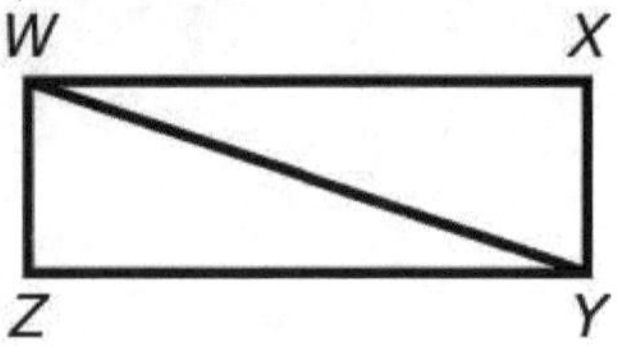

Explain, using transformations, why the diagonal, $\overline{WY}$, is not a line of symmetry.

Name ______________________________ Date ______________

Check for Understanding

Explore Transformations: Investigation 6

1. Choose True or False for each of the following statements about tessellations.

A tessellation must consist of only one polygon type.	True	False
A tessellation pattern must be a repeating pattern.	True	False
A tessellation may contain overlaps between polygons.	True	False
A tessellation may include irregular polygons.	True	False

2. The tessellation below is created with congruent triangles. Use the clues to determine the measure of each angle indicated.

- Clue 1: $m\angle 1 = m\angle 4$
- Clue 2: $m\angle 1 = (m\angle 2 - 15°)$
- Clue 3: $m\angle 2 = m\angle 3 = m\angle 5 = m\angle 6$

$m\angle 1 =$ ________ $m\angle 2 =$ ________

$m\angle 3 =$ ________ $m\angle 4 =$ ________

$m\angle 5 =$ ________ $m\angle 6 =$ ________

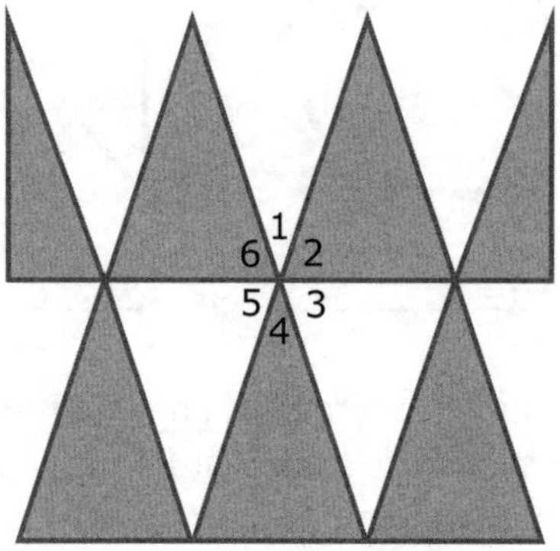

3. Is it possible to make a tessellation with trapezoids that do not possess line symmetry? Explain or demonstrate a way of doing this.

Summary

Before you attempt the Practice Exercises, review what you've learned.

A **transformation** is a function that maps each point in the pre-image to a unique point in the image. Transformations can be used to describe symmetry and create tessellations.

Rigid Transformations

A **translation** T of point P along $\vec{v}$ is a transformation $T(P) = P'$ with the following properties:

a. PP' is equal to the magnitude of $\vec{v}$.

b. $\overline{PP'}$ is parallel to $\vec{v}$.

EXAMPLE: Translation

$T(CDEF) = C'D'E'F'$

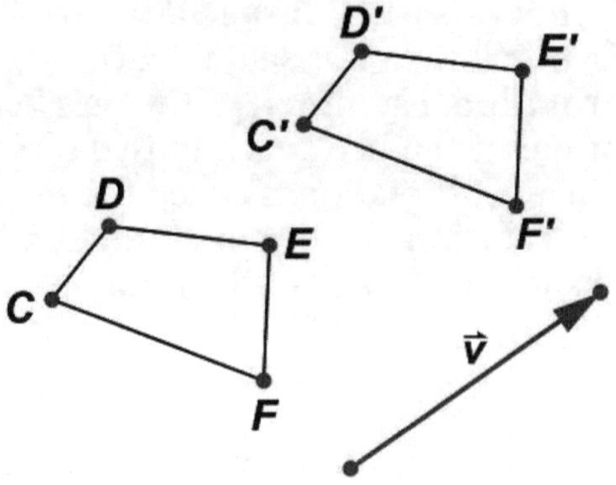

A **rotation** Ro of point P around a given point C, by t degrees with $-360° < t < 360°$, is a transformation $Ro(P) = P'$ with the following properties:

a. $\overline{CP} \cong \overline{CP'}$

b. $m\angle PCP' = t°$

EXAMPLE: Rotation

$Ro(ABC) = A'B'C'$

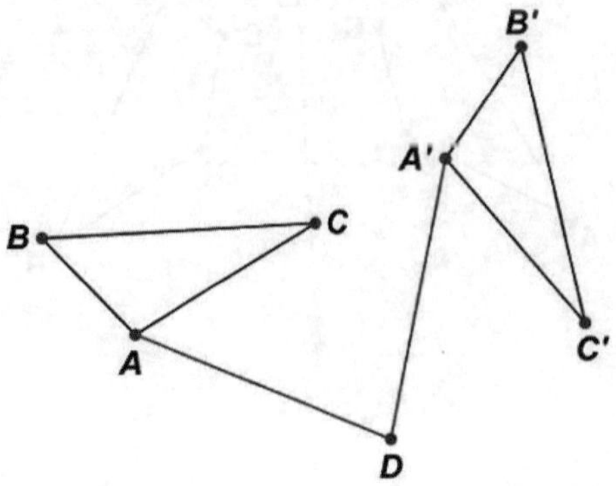

My Notes

Summary *(continued)*

A **reflection** R of point P across a line of reflection l is a transformation $R(P) = P'$ with the following property:

a. l is the perpendicular bisector of $\overline{PP'}$.

EXAMPLE: Reflection

$R(ABCD) = A'B'C'D'$

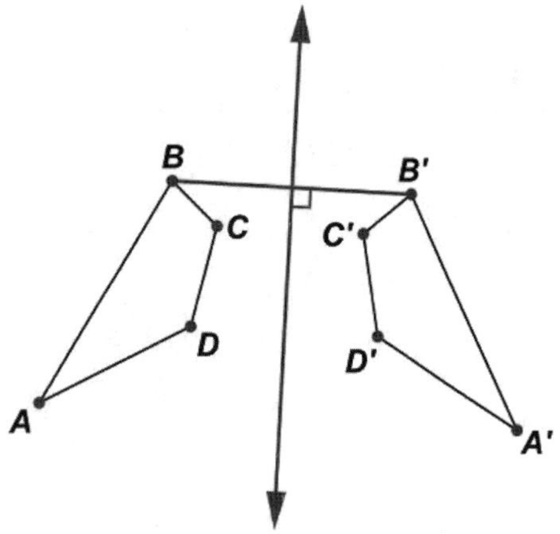

My Notes

Symmetry

A figure has **symmetry** if there is a transformation that maps the figure onto itself.

A figure has **line symmetry** if there exists a reflection such that the image coincides with the pre-image. The line is called a **line of symmetry**. A figure can have more than one line of symmetry.

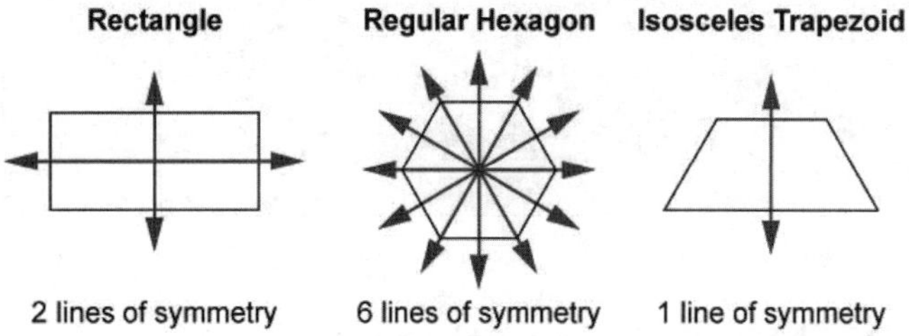

If a figure is superimposed on itself while it is being turned around a point, then it is said to have **rotational symmetry**. The number of times it can be superimposed on itself in a 360° rotation is called the **order of rotation**. The point of rotation is called the **center** of the rotation.

A figure has **point symmetry** if the figure's appearance is unchanged by a 180° rotation.

Tessellations

A **tessellation** is a repeating pattern of one or more shapes that match up along their edges and cover the plane with no gaps or overlaps.

Practice Exercises

Review what you've learned using these practice problems. For practice problems with feedback, try the Coach and Play items in the Practice section online.

1. Study the graph of $\angle B$ and $\angle B'$.

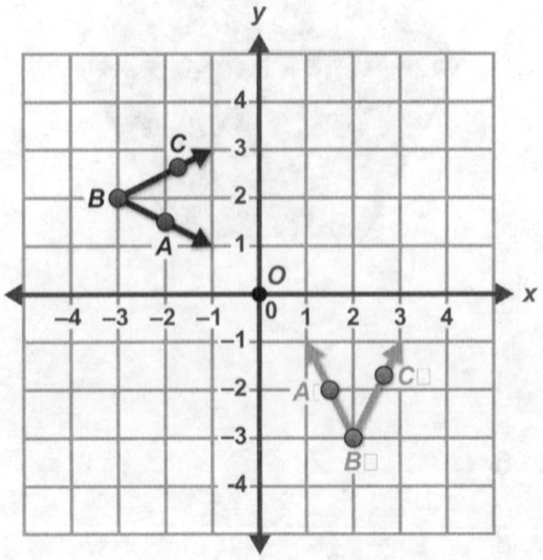

Which statements are true about the transformation that maps $\angle B$ onto $\angle B'$? *Select all that apply.*

A. The transformation is a rotation 180° clockwise about the origin.

B. The transformation is a reflection over the line $y = x$.

C. The transformation is a translation 5 units right and 3 units down.

D. The distance between A and C is equal to the distance between A' and C'.

E. The measure of $\angle B$ is equal to the measure of $\angle B'$.

2. The coordinates of the endpoints of $\overline{CD}$ are $C(3, 1)$ and $D(2, 5)$.

If $\overline{C'D'} = R_{y=3x-1}(\overline{CD})$, which statements are true? *Select all that apply.*

A. The coordinates of C' are (–3, –1).

B. The coordinates of D' are (2, 5).

C. The slope of $\overline{CC'}$ is $-\frac{1}{3}$.

D. $\overline{CD} \parallel \overline{C'D'}$.

E. $\overline{C'D'} = 4.12$ units.

3. A line segment has endpoints $A(2, -1)$ and $B(3, 5)$.

What is the output of the transformation $T_{\langle -4,0 \rangle}(\overline{AB})$?

A. a line segment with endpoints $A'(-2, -5)$ and $B'(-1, 1)$

B. a line segment with endpoints $A'(-2, -1)$ and $B'(-1, 5)$

C. a line segment with endpoints $A'(6, -1)$ and $B'(7, 5)$

D. a line segment with endpoints $A'(6, 3)$ and $B'(7, 9)$

4. Quadrilateral $ABCD$ is rotated 180° clockwise about $E(1, 0)$.

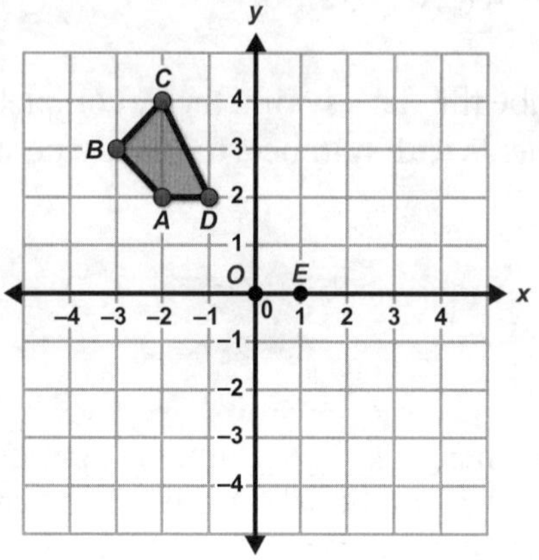

What are the coordinates of the vertices of $A'B'C'D' = Ro_{180°, E}(ABCD)$?

A'(__________, __________)

B'(__________, __________)

C'(__________, __________)

D'(__________, __________)

Practice Exercises *(continued)*

5. Stewart has to make a tessellation poster for his art class. His parallelogram *ABCD* is *A*(2, 3), *B*(3, 5), *C*(8, 5), *D*(7, 3) is reflected across the *y*-axis. What will the new coordinates be on his project?

A. $A'(-2, 3), B'(-3, 5), C'(-8, 5), D'(-7, 3)$

B. $A'(2, -3), B'(3, -5), C'(8, -5), D'(7, -3)$

C. $A'(2, 3), B'(3, 5), C'(8, 5), D'(7, 3)$

D. $A'(-2, -3), B'(-3, -5), C'(-8, -5), D'(-7, -3)$

6. Segment *AB* has endpoints *A*(1, 5) and *B*(3, 2). The segment is translated 2 units down and 2 units right. What are the endpoints of the translated segment *A′B′*?

A. $A'(3, 7), B'(5, 4)$

B. $A'(3, 3), B'(5, 0)$

C. $A'(-1, 3), B'(1, 0)$

D. $A'(2, 0), B'(2, -3)$

7. Describe the star's symmetry by completing the statements with numbers to make true statements.

The star has __________ lines of symmetry.
The star has rotational symmetry. The center of rotation is the center of the star. The angle of rotation is __________.

8. How many lines of symmetry does a regular hexagon have?

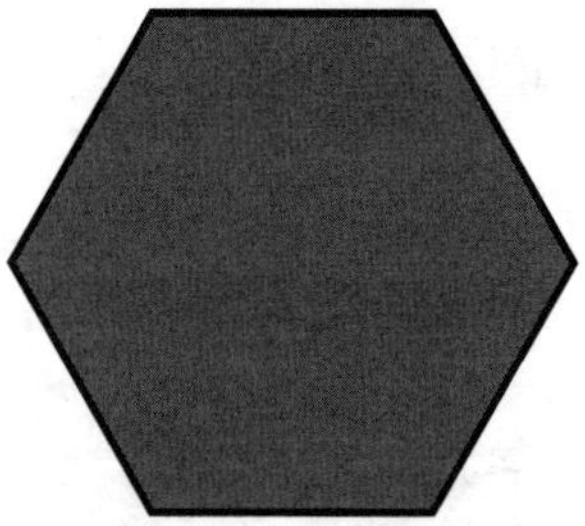

A. 2

B. 4

C. 6

D. 8

9. If a shape is reflected, rotated, and/or translated, what relationship is there between the original shape and the new one?

A. They are similar.

B. They are congruent.

C. They are similar but not congruent.

D. They are not congruent or similar.

10. Which of the letters below does not have point symmetry?

H L N S

A. H

B. L

C. N

D. S

Apply

What Path Should the Rover Take?

Suppose you're working as part of a team directing a rover on the surface of Mars. The map shows the current position of the rover, as well as the locations of three rocks that the team would like the rover to analyze. The map also shows the rover's ending destination after the analysis of the rocks is complete.

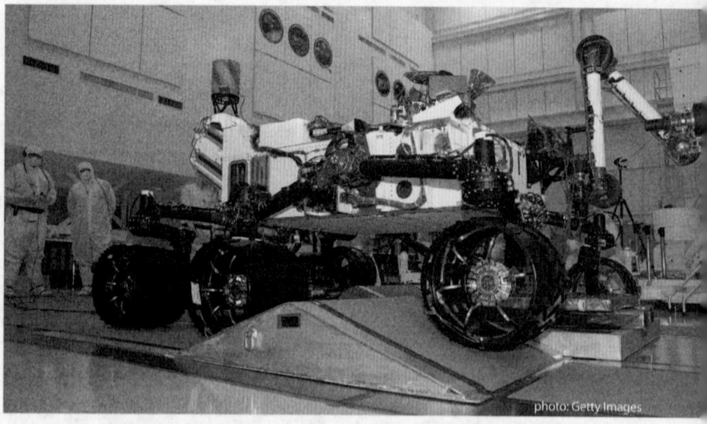

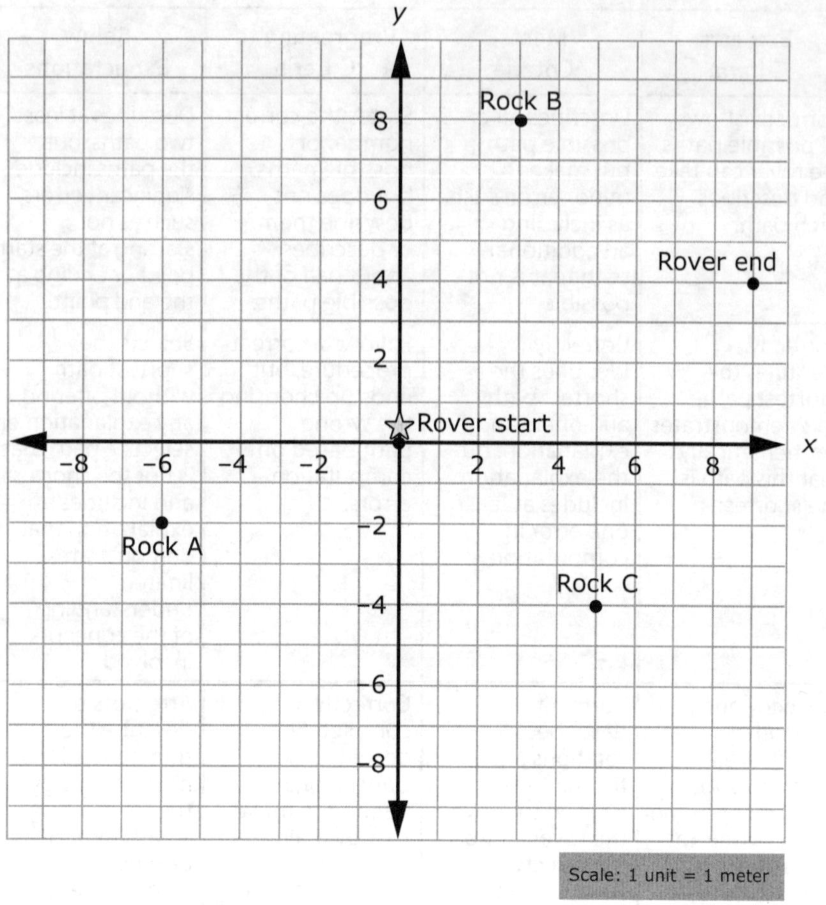

In order to conserve the rover's batteries, the team wants to determine the shortest path the rover can take to visit all three rocks and then travel to its ending point. Your job is to determine this shortest path, and then model it by using transformations.

Assume that the rover can rotate about its center and can only travel forward horizontally or vertically in the direction it is facing. The star on the map shows the front of the rover.

Show what you've learned by completing the other performance tasks in the online Apply section.

Apply *(continued)*

Your answer to Apply will be assessed on the following criteria:

1. Determining all the possible paths that the rover can take
2. Selecting a path for the rover and showing that the path you select is the shortest
3. Describing the translations and rotations the rover needs to make to follow the path
4. Showing that the translations you describe result in the rover reaching the correct coordinates at each stage

Scale / Criteria	4 Exceeds Criteria	3 Meets Criteria	2 Progressing to Criteria	1 Below Expectations	0 No Expectation
Determine Paths	Correctly shows all possible paths the rover can take and describes each path.	Describes all possible paths but makes a minor error, such as including an additional path that is not possible.	States the correct number of possible paths, but does not describe them, or describes about half of the possible paths.	Describes at least two paths, but the paths include significant errors, such as not starting at the start point or ending at the end point.	Response, if any, discusses only a single path.
Shortest Path	Correctly identifies the shortest path, and demonstrates mathematically that this path is the shortest.	Correctly identifies the shortest path and offers an explanation, but the explanation includes at least one error in computation.	Follows a correct procedure, but ends up choosing the wrong path based on computation errors.	Selects the shortest path without offering any explanation or selects a path that is not the shortest and includes an explanation that demonstrates limited understanding of the concepts involved.	Selects a path that is not the shortest and offers no explanation for this choice.
Descriptions	Correctly and completely describes all rotations and translations in the correct order to move the rover along the shortest path.	Correctly describes all rotations and translations needed to move the rover along the selected path, but omits minor details, such as whether a rotation should be clockwise or counterclockwise.	Correctly and completely describes the translations needed to move the rover along the selected path, but omits the rotations, or includes all needed transformations, but makes several errors in describing them.	Attempts to describe the transformations needed to move the rover along a path, but the description is incomplete, and it demonstrates only a limited understanding of transformations.	Does not attempt to describe any transformations.
Translations	Clearly and correctly demonstrates that each translation moves the rover to the correct coordinates.	Shows that the translations described are correct, but the explanation omits a few of the steps involved.	Explains how the translations affect the rover's position, but the explanation includes several errors in computation or reasoning.	Attempts to show at least one translation is correct, but the attempt demonstrates a very limited understanding of translations.	Does not attempt to show that any translations are correct.

UNIT 2: Geometric Transformations

2.2 Investigate and Apply Congruence Definitions

photo: Getty Images

Lesson Objectives

- Explain congruence of two objects in terms of rigid motion.
- Predict and create a sequence of translations, rotations, and reflections to carry a given figure onto another.
- Determine whether two objects are congruent.
- Use rigid motions to discover and explain triangle congruence theorems.
- Show that two triangles are congruent.

Essential Questions

- How can rigid motion transformations be used to show that two figures are congruent?
- What is the relationship between rigid motion transformations and triangle congruence criteria?

Investigations

Identity and Inverse Transformations

You used properties of identity and inverse in algebra. Do these properties apply to transformations?

Explore Combined Transformations

Combine transformations for more complex movement. What's the function notation for that?

Analyzing Combined Transformations

Simplify movement. When can two transformations be replaced by one?

Combined Transformations in the Coordinate Plane

Animators must be precise. Use function notation and coordinates to describe combined transformations.

Defining Congruence

How can you be sure animated figures remain congruent?

Triangle Congruence

Congruent triangles beyond transformations. What evidence do you need to prove congruence?

Key Vocabulary

isometry/congruence transformation, frieze pattern, rigid motion

Discover

As you complete Engage and the investigations, record the most important ideas you've learned.

Engage	**Investigation 1**
Investigation 2	**Investigation 3**
Investigation 4	**Investigation 5**

Name ______________________________ Date __________

Check for Understanding

Investigate and Apply Congruence Definitions: Investigation 1

1. In the combined transformation $R(Ro(T(KLMN)))$, which transformation should be performed first?

   A. dilation

   B. reflection

   C. rotation

   D. translation

2. Look at the pre-image $ABCD$ and image $PQRS$.

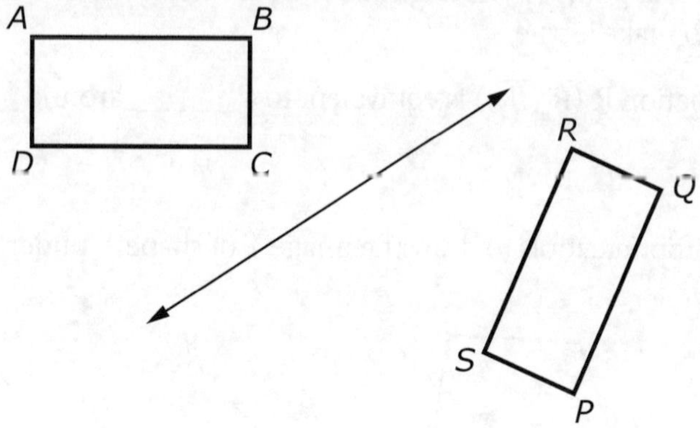

Which combined transformation could map $ABCD$ onto $PQRS$?

   A. $Ro(T(ABCD)) = PQRS$

   B. $T(T(ABCD)) = PQRS$

   C. $R(R(ABCD)) = PQRS$

   D. $T(R(ABCD)) = PQRS$

3. Are there any angles of rotation for which $Ro(T(\overline{AB})) = \overline{AB}$? If so, explain an angle that would work. If not, explain why it is impossible.

Name ______________________________ Date ____________

Check for Understanding

Investigate and Apply Congruence Definitions: Investigation 2

1. For, $R(T(\overline{AB})) = \overline{A'B'}$ which relationships are always, sometimes, or never true? *Circle* Always, Sometimes, *or* Never *for each option.*

A. $\overline{AB} \cong \overline{A'B'}$ [Always / Sometimes / Never]

B. $AB < A'B'$ [Always / Sometimes / Never]

C. $\overline{AB} \perp \overline{A'B'}$ [Always / Sometimes / Never]

D. $\overline{AB} \parallel \overline{A'B'}$ [Always / Sometimes / Never]

2. Two lines, *a* and *b*, intersect at point *P*. The angle measure formed by lines *a* and *b* at point *P* is 25°. Complete the statement to make it true.

The compound transformation $R_a(R_b(\overline{JK}))$ is equivalent to a ______ around ______ by an angle measure of ______.

3. Given that $m \parallel n$, use appropriate tools to draw the image *Y* of shape *X* under the combined transformation $R_n(R_m(X))$.

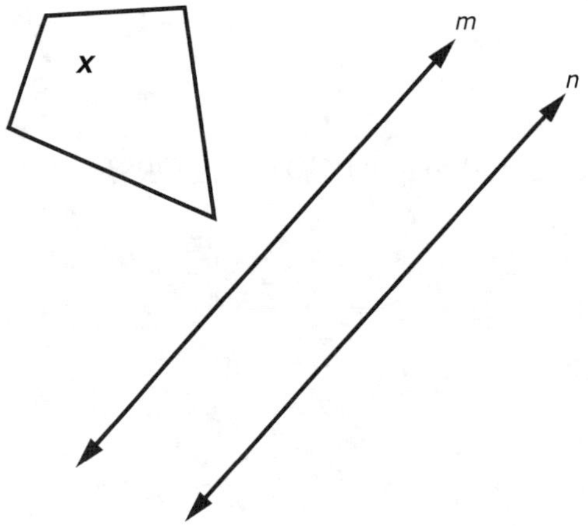

4. Take on the role of the teacher! Explain how you might use the distance between lines *m* and *n* to verify that the combined transformation $R_n(R_m(X))$ in problem 3 is drawn correctly.

Name ______________________________ Date ____________

Check for Understanding

Investigate and Apply Congruence Definitions: Investigation 3

1. Which point of quadrilateral *QRST* remains in the same location under $R_{x\text{-}axis}(QRST)$?

   A. $Q(-3, 5)$

   B. $R(0, 2)$

   C. $S(1, 0)$

   D. $T(3, 3)$

2. Write a rule for the coordinates of the combined transformations, in terms of *x* and *y*.

   $R_{y=x}\left(T_{\langle 1,-2\rangle}(x, y)\right) = (\_\_\_\_\_\_, \_\_\_\_\_\_)$

   $R_{y\text{-}axis}\left(T_{\langle 1,-2\rangle}(x, y)\right) = (\_\_\_\_\_\_, \_\_\_\_\_\_)$

   $R_{x\text{-}axis}\left(T_{\langle 1,-2\rangle}(x, y)\right) = (\_\_\_\_\_\_, \_\_\_\_\_\_)$

3. The transformations *R* and *T* would be commutative if $R(T(x, y)) = T(R(x, y))$ for all points. Are any of the combined transformations from problem 2 commutative? Explain your answer.

Name ____________________ Date ____________

Check for Understanding

Investigate and Apply Congruence Definitions: Investigation 4

Examine the two pentagons shown.

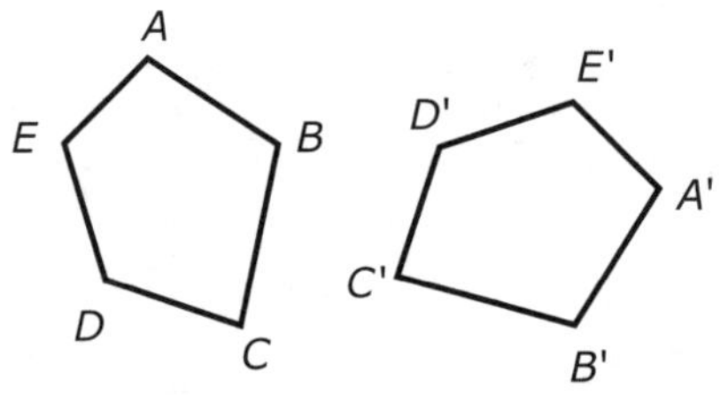

1. Which of the following combined transformations can be used to map $ABCDE$ onto $A'B'C'D'E'$?

A. $R(T(ABCDE))$

B. $R(Ro(ABCDE))$

C. $T(Ro(ABCDE))$

D. $R(T(Ro(ABCDE)))$

2. Which of the following relationships help prove that $ABCDE \cong A'B'C'D'E'$? *Select all that apply.*

A. $\overline{ED} \cong \overline{DC}$

B. $\overline{DC} \cong \overline{D'C'}$

C. $\angle ABC \cong \angle A'B'C'$

D. $\overline{ED} \cong \overline{E'D'}$

E. $\angle ABC \cong \angle A'C'B'$

F. $\angle ACB \cong \angle A'B'C'$

3. Examine figures ABC and $A'B'C'$.

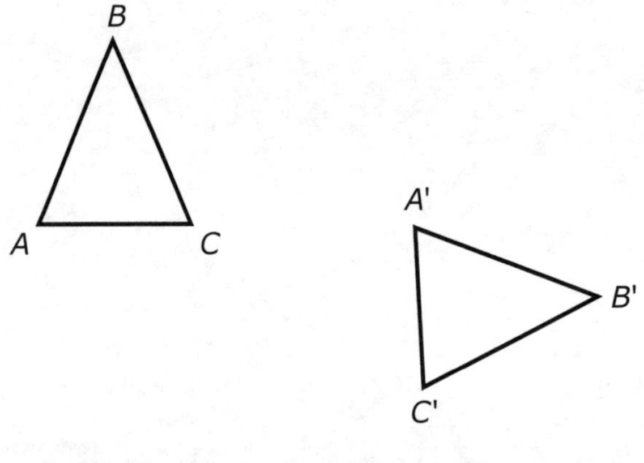

Use appropriate tools to draw the transformation that will map the pre-image ABC onto the image $A'B'C'$. Describe the transformation using function notation.

Name ______________________________ Date __________

Check for Understanding

Investigate and Apply Congruence Definitions: Investigation 5

1. Which acronyms represent the minimum conditions needed to conclude that two triangles are congruent? *Select all that apply.*

A. AAA **B.** AAS **C.** ASA

D. SAS **E.** SSA **F.** SSS

2. Given $\triangle ABC$ and $\triangle DEF$ so that $AB = FE$ and $m\angle A = m\angle F$ what additional information do you need to conclude that the triangles are congruent by the ASA congruence criteria?

A. $AD = CF$

B. $m\angle B = m\angle E$

C. $AC = DF$

D. $m\angle C = m\angle D$

3. Given $P(-2, 0)$, $Q(-1, 2)$, $R(0, -1)$ and $K(7, 2)$, $L(6, 4)$, $M(5, 1)$, draw $\triangle PQR$ and $\triangle KLM$ in the coordinate plane. Can the HL congruence criteria be used to show that $\triangle PQR \cong \triangle KLM$? Explain your reasoning.

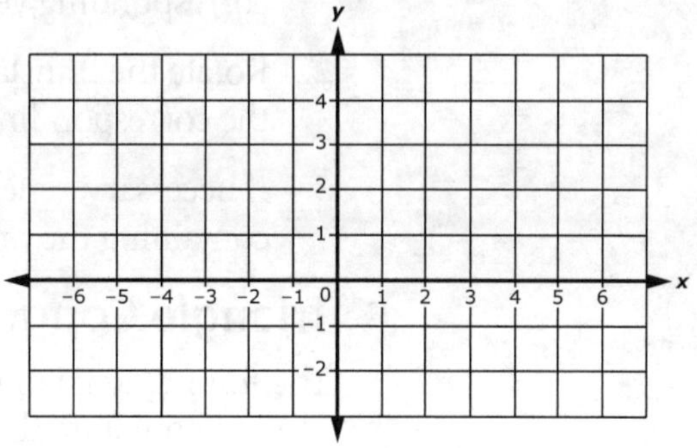

Summary

Before you attempt the Practice Exercises, review what you've learned.

Transformations can be performed in sequence as a combined transformation to represent complex motions in the plane.

Notation for Combined Transformations

Combined transformations can be described by using a **composition of functions**. For example, if a translation T is followed by a reflection T, the combined transformation is $R(T(\text{A}))$. Because there are two transformations, the final image point is written as A''.

Combined Reflections

A combined transformation across two parallel lines is equivalent to a translation. The translation vector is perpendicular to both parallel lines, and its length is twice the distance between the lines.

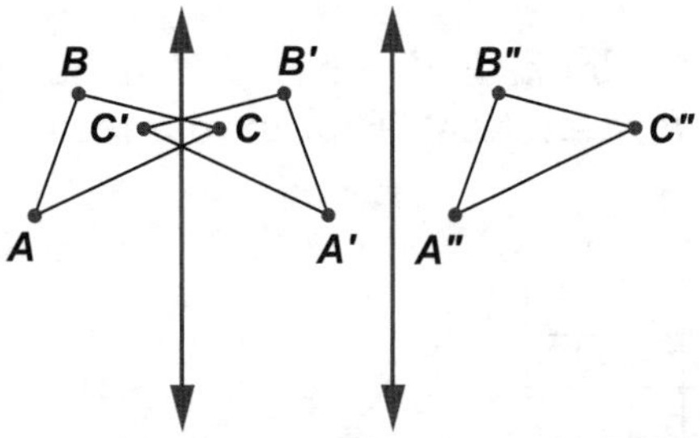

A combined transformation across two intersecting lines is equivalent to a rotation. The center of rotation is the intersection of the lines, and the measure of the angle or rotation is twice the measure of the angle formed by the lines.

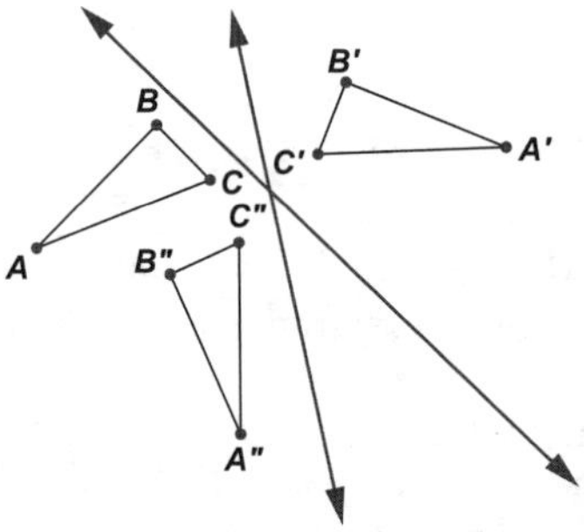

Congruence and Rigid Motions

If two figures are congruent, then there is a sequence of rigid motions that can map one figure onto the other. Here is one possible sequence of transformations:

1. Translate one of the figures so that two of the corresponding vertices match up.
2. Rotate the translated image so that (at least) two of the corresponding sides match up.
3. If necessary, reflect the rotated image across a line containing the matched sides.

Triangle Congruence Criteria

The following triangle congruence criteria can be used to conclude that two triangles are congruent:

- SSS – Three sides of one triangle are congruent to three sides of another triangle.
- SAS – Two sides and the angle included between them of one triangle are congruent to two sides and the angle included between them of another triangle.
- ASA – Two angles and the side included between them of one triangle are congruent to two angles and the side included between them of another triangle.
- AAS – Two angles and the side that is not included between them of one triangle are congruent to two angles and the side that is not included between them of another triangle.
- HL – The hypotenuse and one leg of one triangle are congruent to the hypotenuse and one leg of another triangle.

My Notes

Practice Exercises

Review what you've learned using these practice problems. For practice problems with feedback, try the Coach and Play items in the Practice section online.

1. You are constructing the A-frame swing set shown below.

So far, you have assembled the two A-shaped sides of the swing set, but you need to verify that these triangular sides are congruent before you attach the horizontal top bar and swings.

You have already measured and verified that the lengths of pieces $\overline{BC}$ and $\overline{EF}$ are the same.

Which other measurements of sides or angles could you show are equal in order to prove $\triangle ABC$□ $\triangle DEF$? *Select all that apply.*

A. Show that $DE = AB$ and $DF = AC$.

B. Show that $DF = AC$ and $m\angle C = m\angle F$.

C. Show that $DF = AC$ and $m\angle A = m\angle D$.

D. Show that $DE = AB$ and $m\angle C = m\angle F$.

E. Show $m\angle A = m\angle D$ and that $m\angle C = m\angle F$.

F. Show $m\angle B = m\angle E$ and $m\angle C = m\angle F$.

2. If triangle ABC $A(-1, 4)$, $B(2, 1)$, $C(0, -3)$ is translated 2 units up, what are the new coordinates for triangle $A'B'C'$?

A. $A(-1, 6)$, $B(2, 3)$, $C(0, -1)$

B. $A(-1, 2)$, $B(2, -1)$, $C(0, -5)$

C. $A(-3, 4)$, $B(0, 1)$, $C(-2, -3)$

D. $A(4, -1)$, $B(1, 2)$, $C(-3, 0)$

3. Triangle ABC has vertices located at $A(0,0)$, $B(3, 0)$, and $C(0, 4)$. The triangle undergoes a rotation 90° counterclockwise about the origin followed by a translation, so that the coordinates of its image are located at $A''(5, -5)$ and $B''(5, -2)$.

The coordinates of vertex C'' are (__________, __________).

4. The coordinates for $\triangle A'B'C'$, the image of $\triangle ABC$ under transformation, are given in the table.

Pre-image	Image
$A(3, 4)$	$A'(-3, -4)$
$B(0, 0)$	$B'(0, 0)$
$C(3, -4)$	$C'(-3, 4)$

Select every transformation (or sequence of transformations) that would take each input point from $\triangle ABC$ and map it onto the corresponding output point of $\triangle A'B'C'$.

A. $R_{y=x}(x, y)$

B. $R_{x\text{-axis}}(x, y)$ followed by $R_{y\text{-axis}}(x, y)$

C. $Ro_{180°}(x, y)$

D. $Ro_{180°}(x, y)$ followed by $R_{x\text{-axis}}(x, y)$

E. $T_{\langle 0, 6\rangle}(x, y)$

F. $T_{\langle 6, 0\rangle}(x, y)$ followed by $T_{\langle 0, 8\rangle}(x, y)$

5. In $\triangle MPB$ and $\triangle TAK$, $MB = 7$ inches, $TA = 5$ inches, and $KA = 10$ inches.

Which of the following statements would show $\triangle MPB \cong \triangle TAK$ by the SSS triangle congruence criterion? *Select three.*

A. $\overline{MB} \cong \overline{AT}$

B. $MP = 5$ in.

C. $\overline{MB} \cong \overline{TK}$

D. $TK = 5$ in.

E. $PB = 10$ in.

Practice Exercises *(continued)*

6. In the figure shown, $\triangle ALJ \cong \triangle PTB$

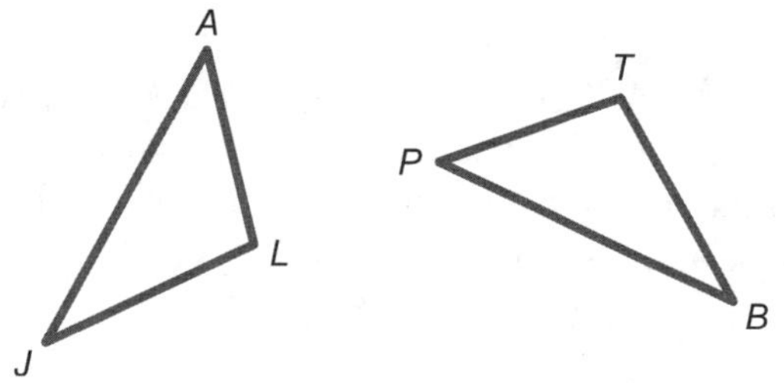

Which statements correctly relate the corresponding parts of these triangles?

Select all that apply.

A. $\overline{AL} \cong \overline{PT}$

B. $\overline{LJ} \cong \overline{PB}$

C. $\angle J \cong \angle B$

D. $\overline{AJ} \cong \overline{PB}$

E. $\angle A \cong \angle B$

7. Which two transformations map parallelogram *ABCD* onto parallelogram *A″B″C″D″*?

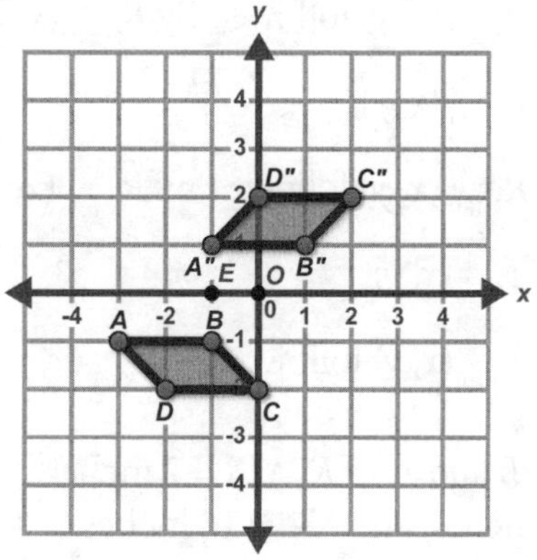

A. a reflection over the *y*-axis and then a translation 2 units right

B. a rotation 180° counterclockwise about *E*(–1, 0) and then a reflection over the *y*-axis

C. a reflection over the *x*-axis and then a reflection over the *y*-axis

D. a rotation 180° counterclockwise about *O*(0, 0) and then a reflection over the *y*-axis

8. Quadrilateral *WXYZ* is translated 2 units up and 3 units right and then reflected across the *y*-axis.

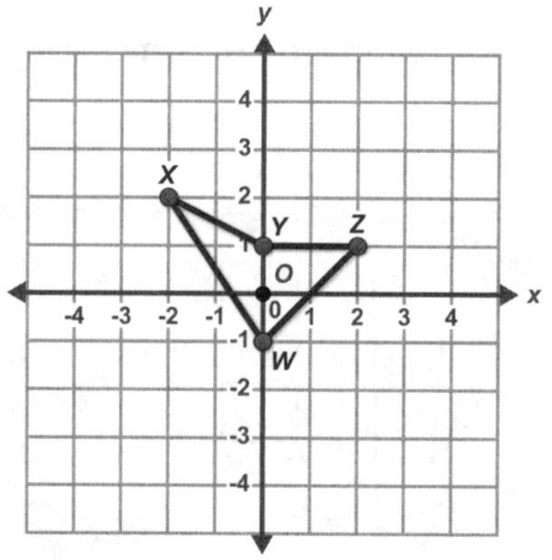

What are the coordinates of the vertices of the new figure after both transformations?

W″ (______, ______)

X″ (______, ______)

Y″ (______, ______)

Z″ (______, ______)

9. When *V*(5, 3), *H*(6, –1), and *K*(2, –1) are reflected across the line $y = x$, what are the coordinates for *V′*, *H′*, and *K′*?

A. *V′*(3, –5), *H′*(1, 6), and *K′*(–1, –2)

B. *V′*(3, 5), *H′*(6, 1), and *K′*(–1, –2)

C. *V′*(3, 5), *H′*(–1, 6), and *K′*(–1, 2)

D. *V′*(–3, –5), *H′*(–6, –1), and *K′*(–1, –2)

10. Tina is given a diagram containing $\triangle ABC$ with *A*(–1, 1), *B*(–2, 1), and *C*(–2, 4).

She compares it to a diagram containing $\triangle PQR$ with *P*(1, 1), *Q*(1, 2), and *R*(4, 2).

Are Tina's two triangles congruent? ______

Apply

How Are Geometric Transformations Used to Design Textile Patterns?

Photo: Getty Images

Combined transformations often are used in fashion design.

Use the Internet to research fabric patterns. Choose a design, and describe any transformations you find.

Make a fabric pattern using combined transformations. Focus on the use of rigid motions to make a pattern that can be used to design an article of clothing or accessory. Use the Dynamic Geometry Tool to help construct the finished article or draw the patterns below. Design one picture on a grid and one picture not on a grid.

One way to formally describe the pattern you created is in terms of the transformations (rigid motions) it uses. Overlay your created pattern on a coordinate plane. Use function notation and composition of functions to describe the transformations (rigid motion) in the pattern you designed.

Show what you've learned by completing the other performance tasks in the online Apply section.

Apply *(continued)*

Your answer to Apply will be assessed on the following criteria:

1. Researching fabrics and clothing patterns for designing your own fabric pattern
2. Using multiple geometric figures and combined transformations to design a complex fabric pattern
3. Explaining how transformations are used in the fabric pattern
4. Describing the transformations in your design using function notation

Scale / Criteria	4	3	2	1	0
	Exceeds Criteria	Meets Criteria	Progressing to Criteria	Below Expectations	No Expectation
Research	Conducts thorough research using reliable sources to design a fabric pattern using combined transformations.	Conducts incomplete research using reliable sources to design a fabric pattern using combined transformations.	Conducts incomplete research using reliable sources which limits fabric patterns to single transformations.	Conducts incomplete research using unreliable sources which limits fabric patterns to single transformations.	Does not attempt the task.
Pattern	Makes a complex pattern using combined transformations, uses rigid motions, and shows a drawing of the finished article of clothing.	Makes a medium-complexity pattern using combined transformations, uses rigid motions, and shows a drawing of the finished article of clothing.	Makes a medium-complexity pattern using combined transformations and uses rigid motions, but does not show a drawing of the finished article of clothing.	Makes a simple pattern using combined transformations and uses rigid motions, but does not show a drawing of the finished article of clothing.	Does not attempt the task.
Explanation	Explains accurately and completely how transformations are used in the pattern.	Partially but accurately explains how transformations are used in the pattern.	Partially explains how transformations are used in the pattern but the descriptions are not accurate.	Provides an unreasonable explanation of the transformations used in the pattern.	Does not attempt the task.
Function Notation	Correctly and completely describes the rigid motions in the pattern using correct function notation.	Correctly but incompletely describes the rigid motions in the pattern using correct function notation.	Incorrectly describes the rigid motions in the pattern using correct function notation.	Incorrectly describes the rigid motions in the pattern and does not use correct function notation.	Does not attempt the task.

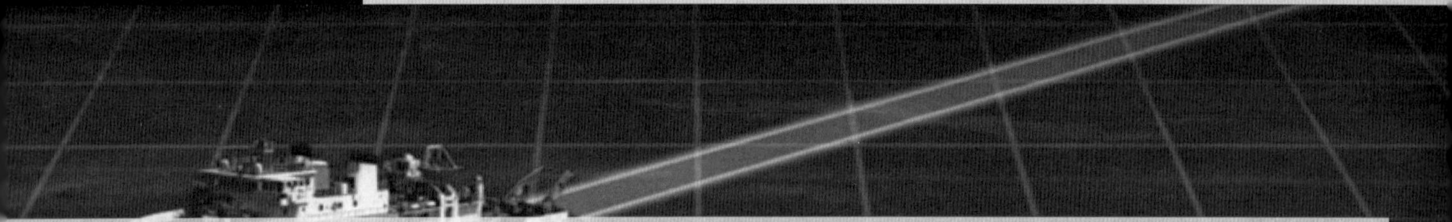

UNIT 3: Intersecting Lines

3.1 Explore Parallel and Perpendicular Lines

photo: Discovery Education

Lesson Objectives

- Construct parallel and perpendicular lines manually and with technology.
- Construct a line through a point that is perpendicular or parallel to a given line.
- Write equations for lines parallel and perpendicular to a given line.

Essential Question

- How can you use geometric transformations and algebra to determine whether two lines are parallel or perpendicular?

photo: Getty Images

Investigations

ROV Paths

Sometimes vehicles must stay on parallel paths. What requirements are necessary for paths to be parallel?

Keeping ROV Paths Parallel

Create parallel and perpendicular paths for the ROVs. How can you be sure the paths are precise?

Paper Folding

Use paper-folding techniques and transformations to construct parallel and perpendicular lines.

Constructing Lines

Use the Dynamic Geometry Tool and transformations to construct parallel and perpendicular lines.

Lines in the Coordinate Plane

Use the coordinate plane to set precise locations of perpendicular and parallel lines.

Key Vocabulary

slope

Discover

As you complete Engage and the investigations, record the most important ideas you've learned.

<table>
<tr><td colspan="2">Engage</td></tr>
<tr><td>Investigation 1</td><td>Investigation 2</td></tr>
<tr><td>Investigation 3</td><td>Investigation 4</td></tr>
</table>

Name ______________________________ Date ____________

Check for Understanding

Explore Parallel and Perpendicular Lines: Investigation 1

1. Suppose two lines intersect at a single point. Which statement below is always true?

A. The lines are parallel. **B.** The lines are perpendicular.

C. The lines are not parallel. **D.** The lines are not perpendicular.

2. In the figure, line *m* is perpendicular to line *n*. *Circle the correct words in the sentences.*

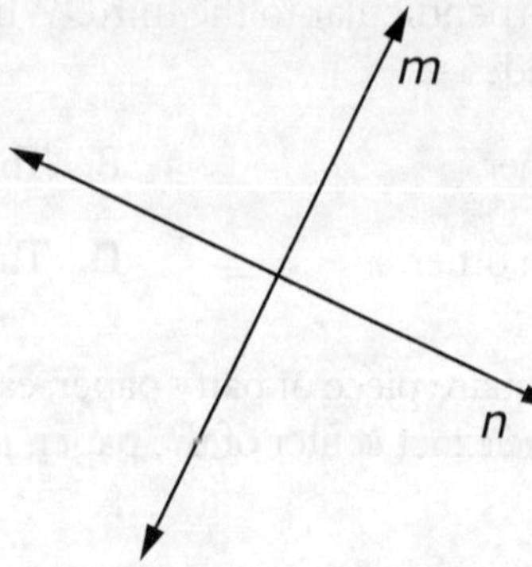

Reflecting line *m* over line *n* [would / would not] result in the pre-image and image of line *m* coinciding.

Reflecting line *n* over line *m* [would / would not] result in the pre-image and image of line *n* coinciding.

3. Pablo and Jean each drew a line. Pablo's line has a slope of $\frac{3}{5}$. *Fill in the blanks below.*

If Jean's line is parallel to Pablo's line, then it has a slope of ________, and if it is perpendicular to Pablo's line, then it has a slope of ________.

4. In a city's business district, Arch Street and Market Street are intersected by 1st Avenue and 2nd Avenue, and all four streets are straight. The intersection of Arch Street and 1st Avenue is 250 feet from that of Market Street and 1st Avenue. Also, the intersection of Arch Street and 2nd Avenue is 250 feet from that of Market Street and 2nd Avenue. Are Arch Street and Market Street parallel to each other? Explain your answer.

Name ______________________________ Date ____________

Check for Understanding

Explore Parallel and Perpendicular Lines: Investigation 2

1. On a square piece of patty paper, Lois constructed a line segment that is parallel to the top edge of the paper. *Circle the correct words in the sentence below.*

To construct a new line segment that is perpendicular to the existing line segment, Lois could fold the paper so that the top left corner aligns with the [bottom left / bottom right / top right] corner.

2. Juan used paper folding to construct three line segments on a square piece of patty paper. Each of the first two line segments he constructed is perpendicular to the third. Which statement must be correct about the first two line segments Juan constructed?

A. They are congruent to each other.

B. They are parallel to each other.

C. They are perpendicular to each other.

D. They bisect each other.

3. To construct four line segments on a square piece of patty paper, each of the corners of the paper were folded inward so that they all met in the exact center of the paper. *Fill in the blanks with the correct numbers to complete the sentence.*

Each line segment constructed is parallel to ________ of the other line segments and perpendicular to ________ of the other line segments.

4. A line segment constructed by folding a square piece of patty paper is shown below.

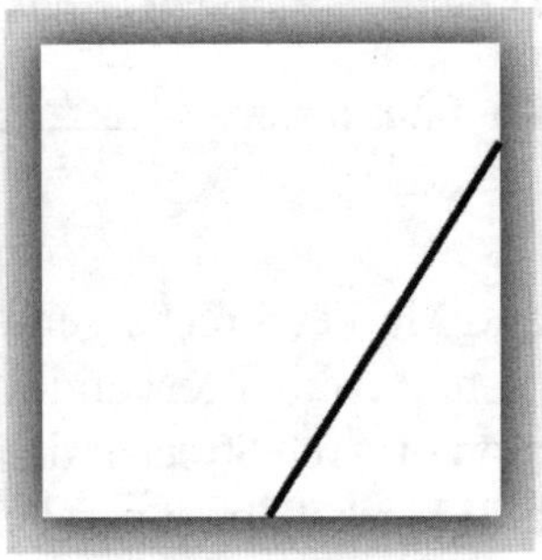

Suppose another line segment was constructed by folding the same piece of patty paper along the diagonal going from the top left corner to the bottom right corner. If the existing line segment was reflected over this new line segment, would the pre-image and image of the existing line segment coincide? Explain your answer.

Name ______________________________ Date ____________

Check for Understanding

Explore Parallel and Perpendicular Lines: Investigation 3

1. Maggie wants to apply a single rigid motion transformation to a line that will result in the line's image being perpendicular to its pre-image. Which could she apply?

A. a rotation but not a translation

B. a rotation or a translation up

C. a rotation or a translation down

D. a rotation or a translation left or right

2. Suppose your teacher has asked you to translate a line so that the line's image is parallel to its pre-image. *Circle the correct words in the following sentence.*

It [does / does not] matter which translation you choose, because any point on the line [will / will not] always move the same distance and direction as any other point.

3. Chet drew lines *m* and *n*. If he reflects line *m* over line *n*, which of these pairs of slopes would result in the pre-image and image of line *m* being parallel? *Select all that apply.*

A. line *m*: 5; line *n*: –5

B. line *m*: –1; line *n*: –5

C. line *m*: –1; line *n*: –1

D. line *m*: –1; line *n*: 1

E. line *m*: 5; line *n*: 5

F. line *m*: 5; line *n*: 1

4. The line shown in the figure below is vertical. Point *A* lies on the line.

Give two different counterclockwise rotations to the line that would result in the line's image being perpendicular to its pre-image. Assuming that the center of rotation is a point on the line, for each rotation, describe the condition(s) that would result in:

- point *A*'s image being to the left and either above or below its pre-image
- point *A*'s image being to the right and either above or below its pre-image
- point *A*'s image being in the same place as its pre-image

Name ______________________________ Date ____________

Check for Understanding

Explore Parallel and Perpendicular Lines: Investigation 4

1. A line in a coordinate plane passes through (−8, −7) and (12, 8), while another line passes through (−16, −8) and (−4, 1). Are the two lines perpendicular?

   A. Yes, because their slopes are the same.

   B. No, because their slopes are not the same.

   C. Yes, because their slopes are opposite reciprocals.

   D. No, because their slopes are not opposite reciprocals.

2. Consider what it means for two lines to be parallel. *Fill in the blanks below.*

   A line passing through (−2, 9) and (1, −3) in a coordinate plane has a slope of ______, so a parallel line passing through (−3, 7) must pass through (2, ______).

3. Ashton reflected $\overleftrightarrow{AB}$ over the y-axis to get $\overleftrightarrow{A'B'}$ as shown in the following figure.

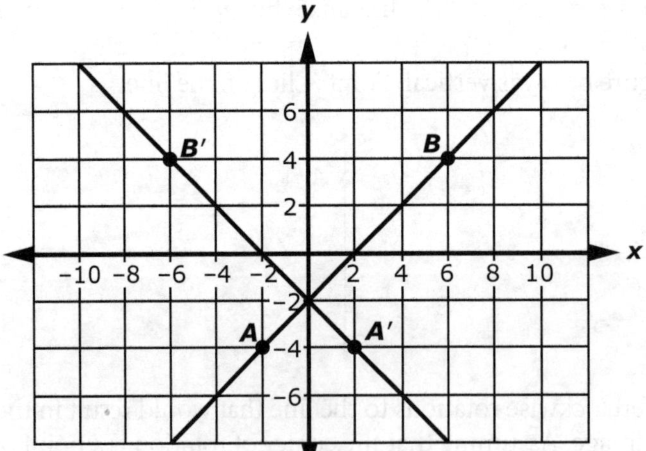

According to his calculations, $\overleftrightarrow{AB}$ is perpendicular to $\overleftrightarrow{A'B'}$, so he concluded that when any line is reflected over the y-axis in a coordinate plane, its pre-image and image form a 90° angle. Is Ashton correct in his calculations and reasoning? Explain your answer.

Summary

Before you attempt the Practice Exercises, review what you've learned.

In this lesson, you explored how transformations can be used to construct **parallel lines** and **perpendicular lines**. Properties of parallel and perpendicular lines are important in geometry because many geometric figures are constructed with parallel or perpendicular line segments.

Parallel Lines

Lines are parallel if they lie in the same plane and are the same distance apart over their entire lengths. Parallel lines in a plane never intersect. The symbol || is used to name two parallel lines. Arrowheads (>) may be used in a figure to show that lines are parallel. In a coordinate plane, parallel lines have the same slope.

Perpendicular Lines

Lines are perpendicular if they intersect to form 90° angles. If two lines are perpendicular, each line is a 90° rotation of the other, and each is a line of symmetry of the other. The symbol ⊥ is used to name two perpendicular lines. A small box (¬) may be used in a figure to show that lines are perpendicular. In a coordinate plane, the slopes of perpendicular lines are opposite reciprocals.

EXAMPLE: Determine Whether Lines Are Parallel or Perpendicular in a Coordinate Plane

Are any of the lines in the coordinate plane parallel? Are any lines perpendicular?

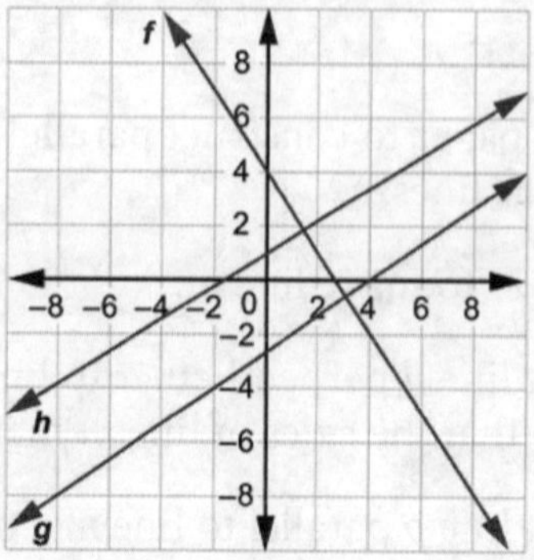

Compare the slopes of the lines.

The slope of line f is $-\frac{3}{2}$.

The slope of line g is $\frac{2}{3}$.

The slope of line h is $\frac{3}{5}$.

None of the lines have the same slope, so there are no parallel lines in the coordinate plane. The slopes of lines f and g are opposite reciprocals, so line $f \perp$ line g.

My Notes

Summary *(continued)*

Constructing Parallel and Perpendicular Lines

A **construction** is a mathematically precise drawing. The geometric properties of a figure are used in a construction, instead of measurement tools.

To use patty paper to construct parallel and perpendicular lines:

Fold the paper to make line m.

To construct line n perpendicular to line m, fold the paper so that the sides of line m match up.

To construct line p parallel to line m, construct a third line that is perpendicular to line n.

To use geometry software to construct parallel and perpendicular lines:

Draw a line. To construct a parallel line, you can translate the figure by a vector that is not parallel to the line. You may also use the Parallel Line command on the Construct menu.

To construct a perpendicular line, you can rotate the line by 90° around a point, or use the Perpendicular Line command on the Construct menu.

My Notes

Practice Exercises

Review what you've learned using these practice problems. For practice problems with feedback, try the Coach and Play items in the Practice section online.

1. The graph shows a line through the points (−2, −2) and (2, 0). Consider the line parallel to this line through the point (4, 4).

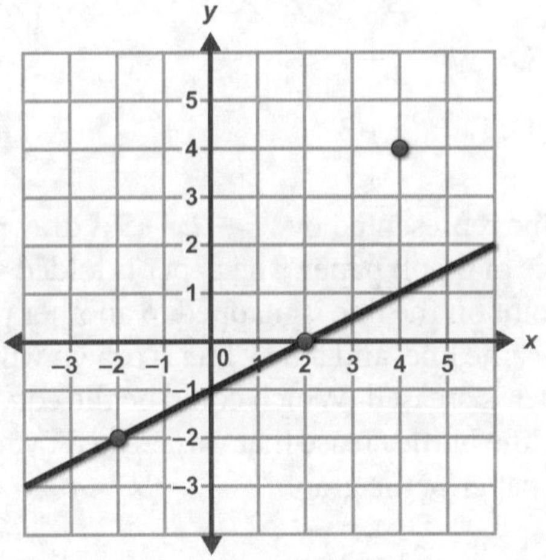

Where will the parallel line cross the y-axis (its y-intercept)?

A. (0, 1)

B. (0, 1.5)

C. (0, 2)

D. (0, 2.5)

E. (0, 3)

2. Which of the following lines are parallel to the line $y = -\frac{1}{2}x + 5$? *Select all that apply.*

A. $y = -2x - 5$

B. $y = -\frac{1}{2}x - 5$

C. $y = -\frac{1}{2}x - \frac{1}{5}$

D. $y = \frac{1}{2}x - 5$

E. $y = 2x - 5$

F. $y = 2x - \frac{1}{5}$

3. The graph shows a line formed by the points (−2, 1) and (2, 3). Consider the line perpendicular to this line through the point (1, 1).

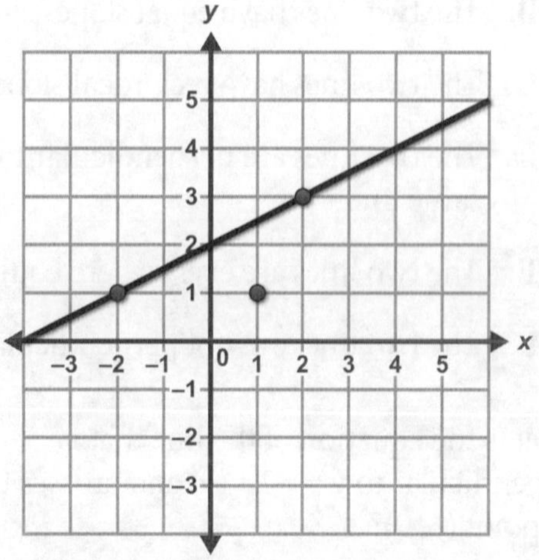

Where will this perpendicular line cross the y-axis (its y-intercept)?

A. (0, 3.5) **B.** (0, 3)

C. (0, 2.5) **D.** (0, 2)

E. (0, 1.5) **F.** (0, 1)

4. In the spaces provided below, enter values of m and b so that the line $y = mx + b$ is perpendicular to the line $y = -4x + 3$ and crosses the y-axis at the point (0, −3).

m = ________

b = ________

5. What is the equation of the line that is parallel to $y = \frac{2}{3}x + 6$ and passes through the point (12, 8)?

A. $y = \frac{3}{2}x$

B. $y = \frac{2}{3}x$

C. $y = \frac{2}{3} + 16$

D. $y = -\frac{3}{2}x + 26$

Practice Exercises *(continued)*

6. Two different lines lie in the same plane. Which of the following criteria would be able to show—without any other information provided—that the two lines are parallel? *Select all that apply.*

A. The two lines have opposite slopes.

B. The two lines have equal slopes.

C. The two lines have reciprocal slopes.

D. The two lines are perpendicular to the same line.

E. The two lines are everywhere equidistant.

F. The two lines are not perpendicular.

7. What is the equation of the line that is perpendicular to $y = -2x + 4$ and crosses through the point (6, 9)?

A. $y = -2x + 21$

B. $y = -\frac{1}{2}x + 12$

C. $y = \frac{1}{2}x + 6$

D. $y = 2x - 3$

8. The length of an official tennis court is 78 feet. The 3.5 foot high net is exactly between the two baselines and divides the court into equal playing regions.

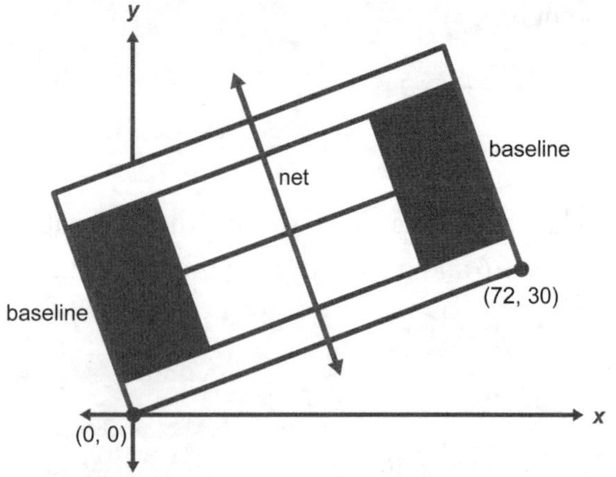

Use the corner coordinates in the given figure to write the equation of the line containing the net.

Write the entire equation below.

$y =$ ______ $x +$ ______

9. What is the equation of the line that is perpendicular to $y = \frac{3}{4}x + 8$ and passes through the point (−12, −6)?

A. $y = -\frac{4}{3}x - 22$

B. $y = -\frac{4}{3}x + 10$

C. $y = \frac{3}{4}x + 15$

D. $y = \frac{3}{4}x + 3$

10. A line represented by $y = -2x - 5$ is drawn on a piece of graph paper. The paper is folded so that a point on the line is mapped to another point on the same line, and a new line is drawn where the paper is creased. Which could be the equation of the line of the crease that was created by folding the paper of the graph?

A. $y = 2x + 5$

B. $y = -2x - 1$

C. $y = \frac{1}{2}x + 3$

D. $y = -\frac{1}{2}x - 5$

Apply

Where Are the Parallel and Perpendicular Folds in an Origami Design?

photo: Getty Images

Origami is the Japanese art of paper folding. In the most common type of origami, a single square sheet of paper is folded to make a three-dimensional sculpture of an object, such as an animal or flower. The folding patterns used in origami often result in parallel or perpendicular lines.

For this problem, you will need two square sheets of paper. Try to use paper that is white on one side and a different color on the other side. If this type of paper is not available, start with plain white paper, but color one side with a colored pencil.

Hands-On Activity: Origami Instructions. Use these instructions to make an origami cicada.

Start by finding a set of instructions for a simple origami design online. Follow the set of instructions using your first sheet of paper. Then, follow the instructions again using the second piece of paper. This time, unfold the paper after you are finished. Mark all fold lines that are perpendicular and all fold lines that are parallel on the unfolded sheet.

Origami

Explain how you know that the lines are perpendicular or parallel by using what you have learned about constructions. Write your explanations for this problem in the space below. Be sure to submit your origami design and your unfolded sheet of paper directly to your teacher.

Show what you've learned by completing the other performance tasks in the online Apply section.

Apply *(continued)*

Your answer to Apply will be assessed on the following criteria:

1. Following a set of instructions to fold an origami design
2. Marking the parallel and perpendicular fold lines with the appropriate notation
3. Using constructions to explain how you know that the perpendicular folds are perpendicular
4. Using constructions to explain how you know that the parallel folds are parallel

Scale / Criteria	4 Exceeds Criteria	3 Meets Criteria	2 Progressing to Criteria	1 Below Expectations	0 No Expectation
Follows Instructions	Correctly follows the selected set of instructions to fold an origami figure.	Correctly follows almost all of the instructions, but makes a minor mistake in one of the folds.	Correctly follows the first few steps of the instructions, but makes some errors in the later folds.	Attempts to fold an origami figure, but shows little understanding of the instructions.	Does not submit a folded origami figure.
Marks Lines	Correctly marks all obviously parallel and perpendicular folds using the appropriate notation.	Marks most of the obviously parallel and perpendicular folds, but omits one or two.	Correctly marks some of the obviously parallel and perpendicular folds, but omits several or marks some incorrectly.	Makes some attempt to mark the perpendicular and parallel folds, but marks indicate little understanding of parallel or perpendicular lines.	Does not show any marks on the unfolded paper, or does not submit an unfolded paper.
Justifies Perpendicular Folds	Uses paper folding constructions to prove that all obviously perpendicular folds are indeed perpendicular. Explanation is clear and complete.	Uses paper folding constructions to show that all obviously perpendicular folds are perpendicular, but the explanations may include minor errors.	Uses paper folding constructions to show that some pairs of perpendicular folds are perpendicular, but the explanations for other pairs of perpendicular folds are omitted or contain errors.	Attempts to demonstrate that folds are perpendicular, but the explanation shows little understanding of paper folding constructions.	Does not attempt to prove that folds are perpendicular.
Justifies Parallel Folds	Uses paper folding constructions to prove that all obviously parallel folds are indeed parallel. Explanation is clear and complete.	Uses paper folding constructions to show that all obviously parallel folds are parallel, but the explanations may include minor errors.	Uses paper folding constructions to show that some pairs of parallel folds are parallel, but the explanations for other pairs of parallel folds are omitted or contain errors.	Attempts to demonstrate that folds are parallel, but the explanation shows little understanding of paper folding constructions.	Does not attempt to prove that folds are parallel.

UNIT 3: Intersecting Lines

3.2 Prove Theorems about Lines and Angles

photo: Getty Images

Lesson Objectives

- Use a variety of methods to prove theorems about angle pairs, parallel lines, and perpendicular lines.
- Apply theorems about angle pairs, parallel lines, and perpendicular lines.

Essential Question

- What are some important theorems about lines and angles, and how can these theorems be proved?

photo: Getty Images

Investigations

Lines and Angles

Angles are formed when parallel lines are cut by a transversal. How are these angles related?

The Vertical Angles Theorem

Prove it! Consider different ways to prove a theorem. Choose a format to prove the vertical angles theorem.

Proving Relationships of Angle Pairs

Prove angle relationships with three proof formats. Then construct a proof for consecutive angles.

Are the Lines Parallel?

Turn it around and prove two lines are parallel.

Analyze Proofs

Find the holes in the arguments and patch them.

Lines and Slope

Take it to the plane. Prove lines are parallel or perpendicular on the coordinate plane.

Key Vocabulary

corresponding angles, linear pair, opposite reciprocal, slope, transversal, vertical angles

Discover

As you complete Engage and the investigations, record the most important ideas you've learned.

Engage	**Investigation 1**
Investigation 2	**Investigation 3**
Investigation 4	**Investigation 5**

Name ______________________________ Date __________

Check for Understanding

Prove Theorems about Lines and Angles: Investigation 1

1. Victor and Ophelia each wrote a correct paragraph proof of the congruent complements theorem, which states that if two angles are complementary to the same angle, then the angles are congruent. *Circle the correct words in the sentence below.*

   It [is / is not] possible that the two proofs contain a different number of sentences, because [all / not all] correct paragraph proofs of a certain theorem are the same.

2. A partial two-column proof of the linear pair theorem is shown. *Fill in the blanks.*

Statements	Reasons
1. Linear pair: $\angle ABC$ and $\angle CBD$	1. Given
2. $\angle ABD$ is a straight angle.	2. Definition of a linear pair
3. $m\angle ABD =$ ______	3. Straight angle postulate
4. $m\angle ABC + m\angle CBD = m\angle ABD$	4. Angle addition postulate
5. ______ + ______ = ______	5. Substitution property of equality
6. $\angle ABC$ and $\angle CBD$ are supplementary.	6. Definition of supplementary angles

3. Kathy wants to prove the straight angle congruence theorem. That is, given that $\angle 1$ and $\angle 2$ are straight angles, she wants to prove that $\angle 1 \cong \angle 2$. She is considering a flowchart proof with the following structure. Each of the seven shapes is called a node.

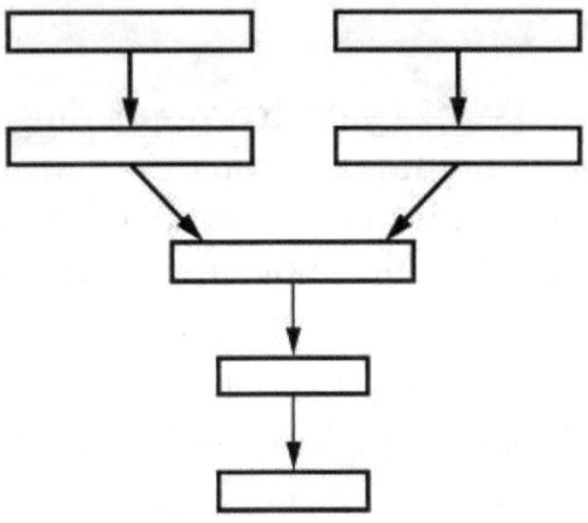

   Is the structure that Kathy is considering correct, or should it be modified? Explain what she should write in each node, and which node(s) she should delete, if necessary.

Name ______________________________ **Date** ____________

Check for Understanding

Prove Theorems about Lines and Angles: Investigation 2

1. Connor is creating a paragraph proof of the same-side exterior angle theorem, which states that if two parallel lines are cut by a transversal, then same-side exterior angles are supplementary. He is using the figure below, in which it is given that line m is parallel to line n, and he is attempting to prove that $\angle 1$ is supplementary to $\angle 7$.

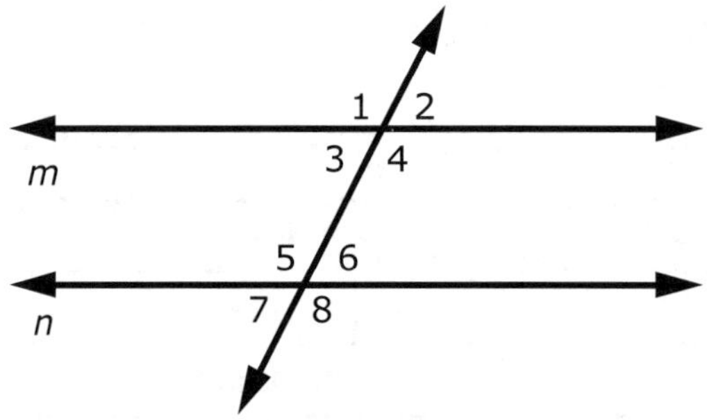

He began by stating that $\angle 2 \cong \angle 7$ by the alternate exterior angles theorem, and that $m\angle 2 = m\angle 7$ by the definition of congruent angles. To complete the proof in the fewest steps possible, what should be the reason for his next step?

A. the corresponding angles theorem

B. the linear pair theorem

C. the transitive property of equality

D. the vertical angles theorem

2. Another way that Connor could have begun his proof in question 1 is to state that $\angle 1$ and $\angle 3$ are supplementary by the linear pair property, and that $m\angle 1 + m\angle 3 = 180°$ by the definition of supplementary angles. *Circle the correct words below.*

Connor then could have stated that $\angle 3 \cong \angle 7$ by the [corresponding angles theorem / vertical angles theorem], and that $m\angle 3 = m\angle 7$ by the definition of congruent angles. Then, he could have stated that $m\angle 1 + m\angle 7 = 180°$ by the [substitution property of equality / transitive property of equality], and that $\angle 1$ is supplementary to $\angle 7$ by the definition of supplementary angles.

3. Suppose that when attempting to prove that $\angle 1$ is supplementary to $\angle 7$ in question 1, Connor had begun by stating that $\angle 1 \cong \angle 4$ by the vertical angles theorem, and that $m\angle 1 = m\angle 4$ by the definition of congruent angles. In paragraph form, explain one of the ways he could have completed the proof of the same-side exterior angle theorem.

Name ______________________ Date ____________

Check for Understanding

Prove Theorems about Lines and Angles: Investigation 3

1. In the figure shown below, line *m* and line *n* are intersected by transversal *t*.

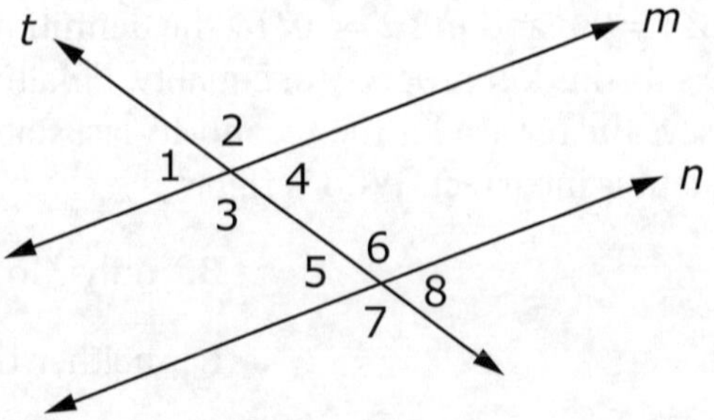

When given which of these pairs of angle measurements would you be able to prove that line *m* and line *n* are parallel to each other? *Select all that apply.*

A. $m\angle 3 = 119.9°$ and $m\angle 6 = 120.1°$

B. $m\angle 4 = 59.9°$ and $m\angle 6 = 120.1°$

C. $m\angle 1 = 59.9°$ and $m\angle 8 = 59.9°$

D. $m\angle 3 = 120.1°$ and $m\angle 5 = 60.1°$

E. $m\angle 1 = 59.9°$ and $m\angle 5 = 60.1°$

F. $m\angle 4 = 59.9°$ and $m\angle 8 = 60.1°$

2. The same-side exterior angle theorem states that if two parallel lines are cut by a transversal, then same-side exterior angles are supplementary. What is the converse?

A. If two parallel lines are cut by a transversal, then same-side exterior angles are not supplementary.

B. If two non-parallel lines are cut by a transversal, then same-side exterior angles are supplementary.

C. If two lines are cut by a transversal and same-side exterior angles are supplementary, then the lines are parallel.

D. If two lines are cut by a transversal and same-side exterior angles are not supplementary, then the lines are parallel.

3. Given that $\angle 2$ and $\angle 8$ in the figure in question 1 are supplementary, can you prove that $m \parallel n$ by using the converse of the corresponding angles theorem? Explain.

Name ______________________________ Date ____________

Check for Understanding

Prove Theorems about Lines and Angles: Investigation 4

1. Louis and Morgan analyzed a proof that attempted to show that if a transversal is perpendicular to two parallel lines, then consecutive interior angles are congruent. The consecutive interior angles used were $\angle 1$ and $\angle 2$. The proof stated that $m\angle 1 = 90°$ and $m\angle 2 = 90°$ by the definition of perpendicular lines. Then it stated that $m\angle 1 = m\angle 2$ by the substitution property of equality. Finally, it stated that $\angle 1 \cong \angle 2$ by the definition of right angles. Louis says the reason for the second-to-last statement is incorrect, while Morgan says the reason for the last statement is incorrect. Who is right?

A. only Louis

B. only Morgan

C. both Louis and Morgan

D. neither Louis nor Morgan

2. The congruent complements theorem states that if two angles are complementary to the same angle, then the angles are congruent. Suppose that $\angle 1$ is complementary to $\angle 3$, and $\angle 2$ is complementary to $\angle 3$. Below is a proof of the theorem showing that $\angle 1 \cong \angle 2$, but it has an error in it. *Circle the correct words in the sentence that follows.*

Statements	Reasons
1. $\angle 1$ is complementary to $\angle 3$; $\angle 2$ is complementary to $\angle 3$	1. Given
2. $m\angle 1 + m\angle 3 = 90°$; $m\angle 2 + m\angle 3 = 90°$	2. Definition of complementary angles
3. $m\angle 1 + m\angle 3 = m\angle 2 + m\angle 3$	3. Linear pair property
4. $m\angle 1 = m\angle 2$	4. Subtraction property of equality
5. $\angle 1 \cong \angle 2$	5. Definition of congruent angles

The error is in the [statement / reason] of the [second / third / fourth / fifth] step.

3. Quill attempted to prove the converse of the congruent supplements theorem. That is, given that $\angle 1 \cong \angle 2$, he tried to prove that $\angle 1$ is supplementary to $\angle 3$, and that $\angle 2$ is supplementary to $\angle 3$. First, he stated that $m\angle 1 = m\angle 2$ by the definition of congruent angles. Then, he stated that $m\angle 1 + m\angle 3 = m\angle 2 + m\angle 3$ by the addition property of equality. Next, he stated that $m\angle 1 + m\angle 3 = 180°$ and $m\angle 2 + m\angle 3 = 180°$ by the definition of straight angles. Finally, he stated that both $\angle 1$ and $\angle 2$ are supplementary to $\angle 3$ by the definition of supplementary angles. Did Quill make a mistake? Explain.

Name ________________________________ Date __________

Check for Understanding

Prove Theorems about Lines and Angles: Investigation 5

1. Points *A* and *B* lie on a line with a slope of 3. The coordinates of *A* are (x_1, y_1), while the coordinates of *B* are (x_2, y_2). Suppose you want to prove that $\overleftrightarrow{AB} \parallel \overleftrightarrow{CD}$. Which of these coordinates for points *C* and *D* would allow you to do so? *Select all that apply.*

   A. $C(x_1 - 1, y_1 - 1)$ and $D(x_2 + 5, y_2 + 5)$

   B. $C(x_1 + 5, y_1 - 1)$ and $D(x_2 + 5, y_2 - 1)$

   C. $C(x_1 - 5, y_1 + 5)$ and $D(x_2 - 1, y_2 + 1)$

   D. $C(x_1 - 1, y_1 + 5)$ and $D(x_2 + 1, y_2 - 5)$

   E. $C(x_1 - 1, y_1 + 5)$ and $D(x_2 - 1, y_2 + 5)$

   F. $C(x_1 + 5, y_1 + 5)$ and $D(x_2 - 1, y_2 - 1)$

2. A line with the equation $y = -\frac{7}{8}x + 4$ undergoes a transformation described by the translation vector $\langle p, q \rangle$. Which statement best explains whether or not it is possible to prove that the line's image is parallel to its pre-image?

   A. It is possible to prove for any values of *p* or *q*.

   B. It is only possible to prove if the value of *p* is $-\frac{7}{8}$.

   C. It is only possible to prove if the value of *q* is $-\frac{7}{8}$.

   D. It is not possible to prove for any values of *p* or *q*.

3. Stella's teacher told her that if $\overleftrightarrow{FG}$ and its image $\overleftrightarrow{F'G'}$ are perpendicular, with the coordinates of *F* being (x_1, y_1), and the coordinates of *G* being (x_2, y_2), then the coordinates of *F′* are $(-y_1, x_1)$, and the coordinates of *G′* are $(-y_2, x_2)$. Stella disagrees and says that the coordinates of *F′* are $(y_1, -x_1)$, and the coordinates of *G′* are $(y_2, -x_2)$. Who is correct? Explain your answer by citing the slope criteria for perpendicular lines.

Summary

Before you attempt the Practice Exercises, review what you've learned.

A variety of angle relationships can be found in intersections of two or more lines. These relationships can be used to prove properties of the angles or the lines.

Vertical Angles Theorem

Vertical angles theorem: If two angles form a pair of vertical angles, then they are congruent.

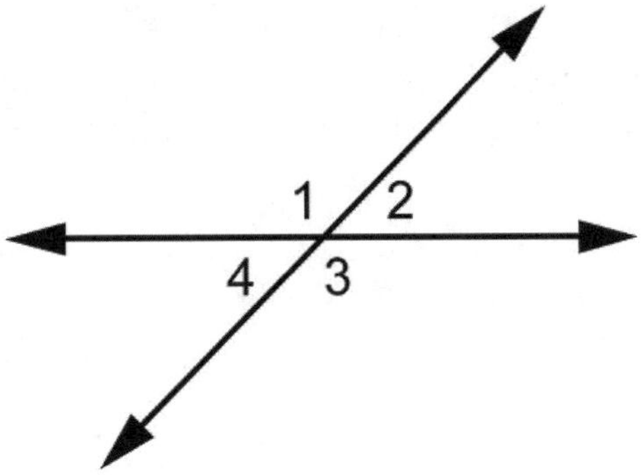

$\angle 1 \cong \angle 3, \angle 2 \cong \angle 4$

Angles Formed by Parallel Lines and a Transversal

Corresponding angles theorem: If two parallel lines are intersected by a transversal, then the corresponding angles are congruent.

Alternate exterior angles theorem: If two parallel lines are intersected by a transversal, then the alternate exterior angles are congruent.

Alternate interior angles theorem: If two parallel lines are intersected by a transversal, then the alternate interior angles are congruent.

Consecutive angles theorem: If two parallel lines are intersected by a transversal, then the consecutive interior angles are supplementary.

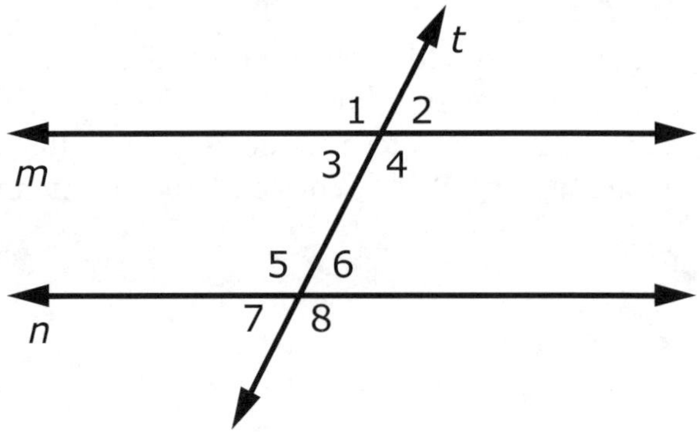

If lines *m* and *n* are parallel, the following relationships exist.

Corresponding angles: $\angle 1 \cong \angle 5, \angle 2 \cong \angle 6, \angle 3 \cong \angle 7, \angle 4 \cong \angle 8$

Alternate Exterior angles: $\angle 1 \cong \angle 8, \angle 2 \cong \angle 7$

Alternate Interior angles: $\angle 3 \cong \angle 6, \angle 4 \cong \angle 5$

Consecutive angles: $\angle 3$ and $\angle 5$ are supplementary; $\angle 4$ and $\angle 6$ are supplementary

My Notes

Summary *(continued)*

Proving Lines Parallel

Converse of the corresponding angles theorem: If two lines are intersected by a transversal so that corresponding angles are congruent, then the lines are parallel.

Converse of the alternate exterior angles theorem: If two lines are intersected by a transversal so that alternate exterior angles are congruent, then the lines are parallel.

Converse of the alternate interior angles theorem: If two lines are intersected by a transversal so that alternate interior angles are congruent, then the lines are parallel.

Converse of the consecutive angles theorem: If two lines are intersected by a transversal so that consecutive angles are supplementary, then the lines are parallel.

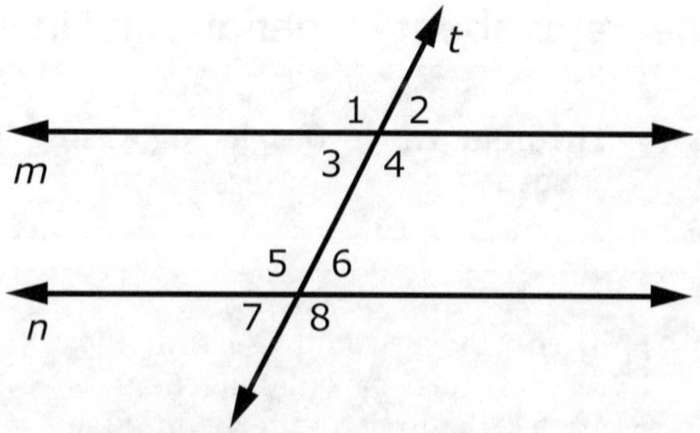

If $\angle 1 \cong \angle 5$, then $m \parallel n$.

If $\angle 1 \cong \angle 8$, then $m \parallel n$.

If $\angle 4 \cong \angle 5$, then $m \parallel n$.

If $\angle 3$ and $\angle 5$ are supplementary, then $m \parallel n$.

My Notes

Summary *(continued)*

Theorems about Perpendicular Lines

Theorem: Any two right angles are congruent.

Theorem: If two intersecting lines form a linear pair of congruent angles, the lines are perpendicular.

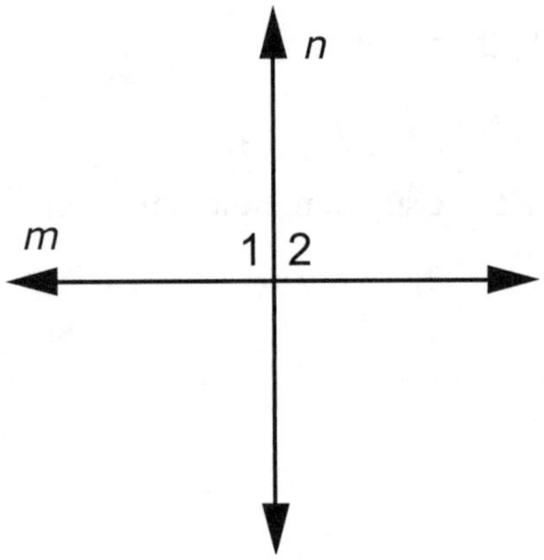

If $\angle 1 \cong \angle 2$, then $m \perp n$.

Theorem: If a transversal is perpendicular to one of two parallel lines, it is perpendicular to the other one.

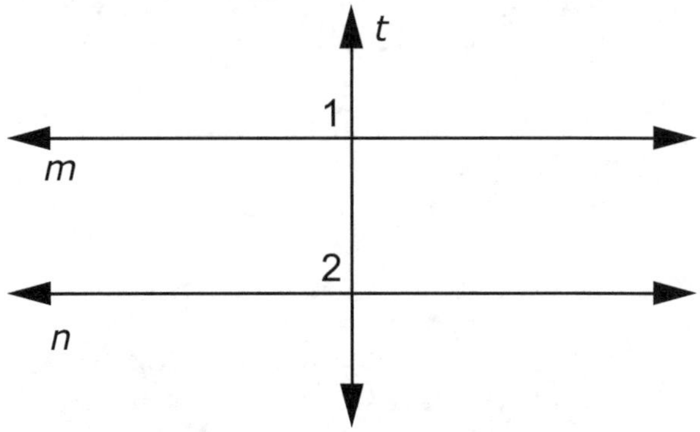

If $m \parallel n$ and $m \perp t$, then $n \perp t$.

Slope Criteria

Parallel Lines

If two lines are parallel, then they have the same slope.

Perpendicular Lines

If two lines are perpendicular, then their slopes are opposite reciprocals.

My Notes

Practice Exercises

Review what you've learned using these practice problems. For practice problems with feedback, try the Coach and Play items in the Practice section online.

1. Veronica is creating a modern design for a bookcase that has some shelves and panels that are parallel and some that are not. For example, in her diagram on the right, she has the shelves $l \parallel m$ and $l \parallel p$, but $l \nparallel n$.

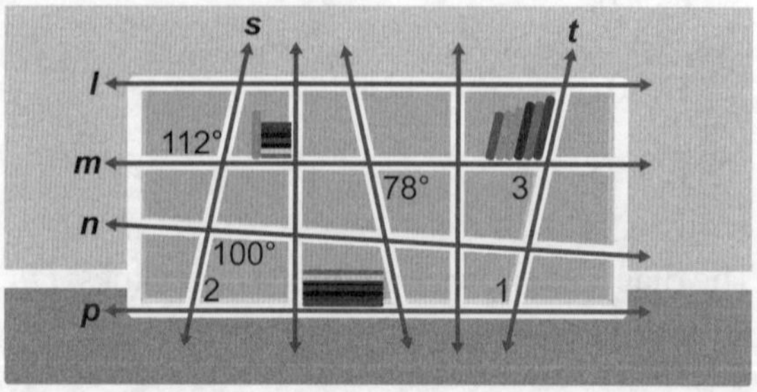

Based on the measures of the angles given in the diagram, what must be the measures of ∠1, ∠2, and ∠3 so that her design has parallel panels $s \parallel t$?

$m\angle1$ = __________

$m\angle2$ = __________

$m\angle3$ = __________

2. The diagram shows two parallel lines cut by a transversal.

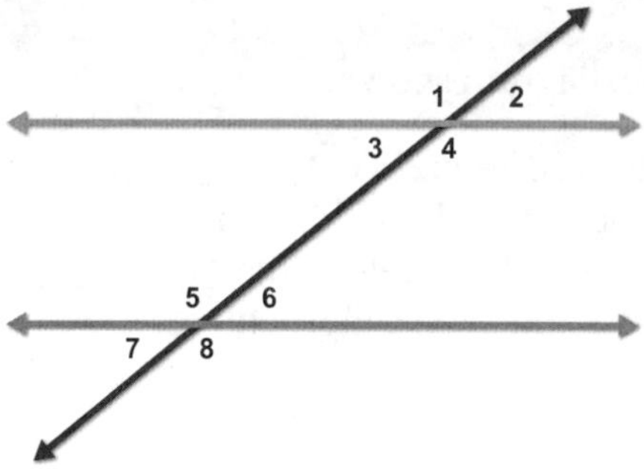

Fill in the blanks to complete the statements.

In the diagram, there are two parallel lines intersected by a transversal.

Angles 3 and __________ are consecutive interior angles.

Because ∠7 is a corresponding angle to ∠ __________ and because ∠7 forms a linear pair with ∠5 we know $m\angle5 + m\angle3 =$ __________°.

3. Two parallel lines are cut by a transversal such that angle 2 and angle 5 are consecutive interior angles. If the $m\angle2 = (2x - 17)°$ and $m\angle5 = (6x + 13)°$, then the $m\angle2 =$ __________

4. A given line passes through points $A(-2, 4)$ and $B(4, 1)$.

Suppose a new line will be drawn perpendicular to $\overleftrightarrow{AB}$ passing through the point (2, 2).

The new line will form angles with the existing line measuring __________. In the slope-intercept form of the equation for the new line, the slope will be

$m =$ __________ and the y-intercept will be

$b =$ __________.

5. The figure contains lines m and n and transversal line k.

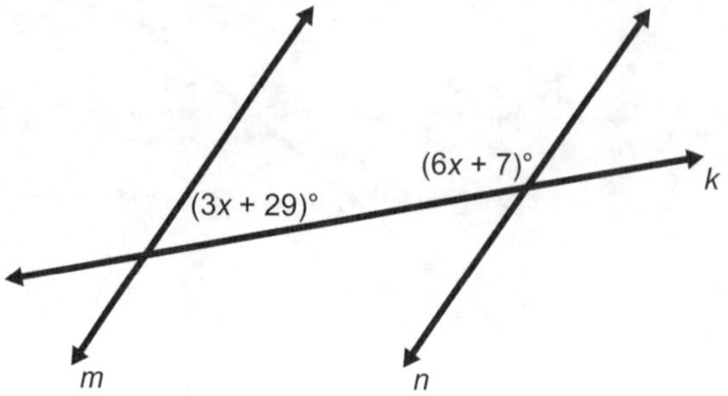

Determine the value of x that would make the lines m and n parallel.

If $x =$ __________, then $m \parallel n$.

6. Two parallel lines are cut by a transversal such that angle 2 and angle 3 are alternate interior angles. If the $m\angle2 = (9x - 21)°$ and $m\angle3 = (6x + 6)°$, what is the measure of angle 3?

   A. 60°

   B. 84°

   C. 96°

   D. 120°

Practice Exercises *(continued)*

7. Nicole constructed two lines cut by a transversal. She wants to verify that $\overline{CD}$ and $\overline{EG}$ are parallel. Which statement would prove that $\overline{CD}$ and $\overline{EG}$ are parallel?

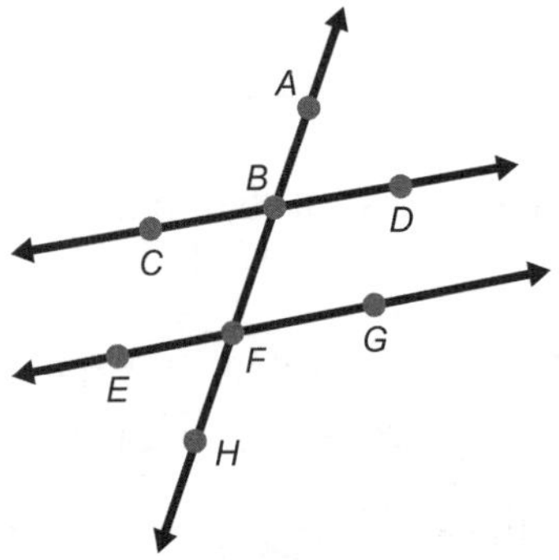

A. $\angle CBF$ and $\angle FBD$ are supplementary.

B. $\angle ABD$ and $\angle BFE$ are congruent.

C. $\angle CBF$ and $\angle BFG$ are congruent.

D. $\angle CBF$ and $\angle ABD$ are congruent.

8. The diagram shows two parallel lines cut by a transversal.

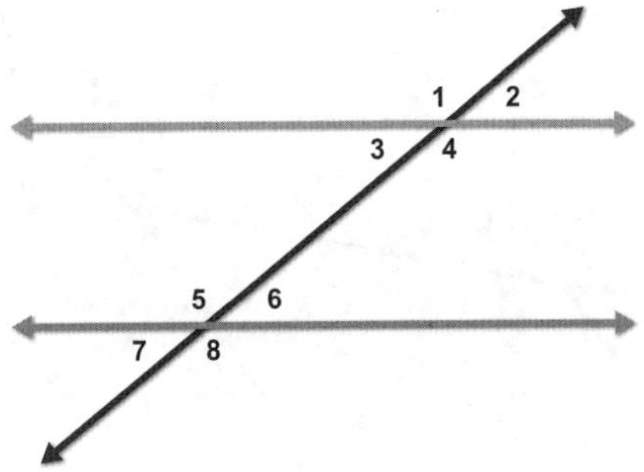

Fill in the blanks to complete the statements.

In the diagram, $\angle 3$ and $\angle 6$ are __________ interior angles.

Because $\angle 6$ is a corresponding angle with $\angle$ __________ and because $\angle 3$ is a vertical angle with $\angle$ __________, you can use the transitive property to show that $\angle 3$ and $\angle 6$ are __________.

9. Two parallel lines are cut by a transversal such that angle 1 and angle 8 are alternate exterior angles. If the $m\angle 1 = (3x + 27)°$ and $m\angle 8 = (5x - 7)°$, then what is the measure of angle 8?

A. 17°

B. 20°

C. 78°

D. 87°

E. 93°

10. The diagram shows two intersecting lines. *Fill in the blanks to complete the statements.*

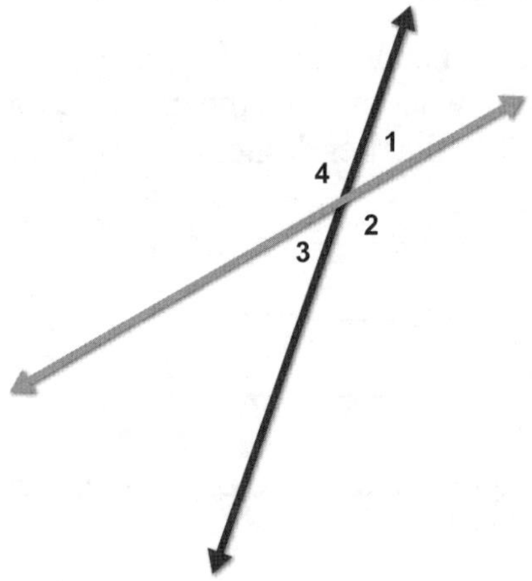

In the diagram shown,

$m\angle 2 = 180° - m\angle 1 =$ __________ $- m\angle 3$ because both $\angle 1$ and $\angle 3$ form a __________ pair with $\angle 2$. Therefore, the __________ angles 1 and 3 are congruent.

Apply

How Are Optical Illusions Created?

photo: Discovery Education

An optical illusion occurs when what you think you are seeing is not really what is there. Let's look at some optical illusions.

The first illusion is called the Café Wall Illusion.

Do the horizontal lines appear parallel or not? Measure the distance between the horizontal lines to see if you are correct, and describe what you discovered. How could you actually prove the lines are parallel?

The second illusion is called the Bent Lines Illusion.

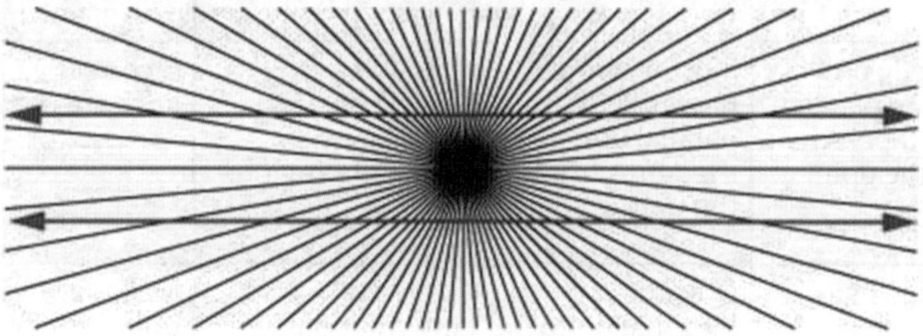

Your Optical Illusion

Research optical illusions, and describe how an illusion works.

Use the Dynamic Geometry Tool or patty paper to construct your own optical illusion involving parallel lines. Then, use theorems you learned in this unit to prove the lines are parallel.

Show what you've learned by completing the other performance tasks in the online Apply section.

Apply *(continued)*

Your answer to Apply will be assessed on the following criteria:

1. Researching information about optical illusions and presenting the information in an organized way
2. Constructing an optical illusion involving parallel lines using precision in geometric constructions
3. Using theorems to prove the lines in your illusion are parallel
4. Presenting a convincing argument about why the illusion works

Scale / Criteria	4	3	2	1	0
	Exceeds Criteria	**Meets Criteria**	**Progressing to Criteria**	**Below Expectations**	**No Expectation**
Research	Performs correct and useful Internet research to gather valid and usable information.	Performs Internet research to gather mostly valid and usable information.	Performs Internet research to gather sometimes valid and usable information.	Performs Internet research to gather mostly invalid and unusable information.	Does not perform research.
Constructs an Illusion	Creates an accurate optical illusion involving parallel lines using precision in constructions.	Creates a mostly accurate optical illusion involving parallel lines using precision in constructions.	Creates an accurate optical illusion without using parallel lines or geometric constructions.	Creates an inaccurate optical illusion involving parallel lines without using geometric constructions.	Does not create an optical illusion.
Proves Lines are Parallel	Correctly proves lines are parallel using theorems from the unit.	Mostly correctly proves lines are parallel using theorems from the unit.	Shows lines are parallel by measurement without proving.	Incorrectly proves lines are parallel.	Does not try to prove lines are parallel.
Conclusions with Justifications	Uses the information to justify a reasonable conclusion about why the illusions work.	Uses the information to justify a mostly reasonable conclusion about why the illusions work.	Uses the information to justify a somewhat reasonable conclusion about why the illusions work.	Makes an unreasonable conclusion.	Does not state a conclusion.

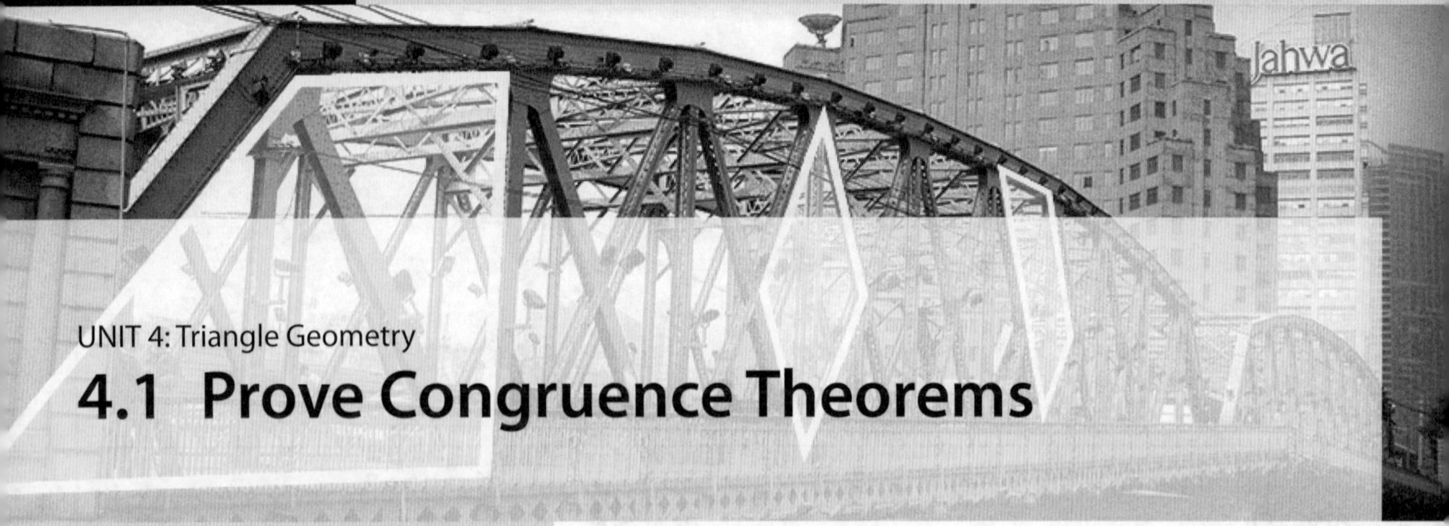

UNIT 4: Triangle Geometry

4.1 Prove Congruence Theorems

photo: Getty Images

Lesson Objectives

- Prove theorems about triangle angle measures.
- Prove theorems about triangle congruence.
- Use theorems about triangle congruence and angle measures to solve problems.

Essential Question

- Which triangle congruence statements and relationships in triangles are valid for all triangles?

Investigations

Geometric Structures

Keep it stable! What shapes are most often used in construction? What facts are needed to verify shapes?

Explore Properties of Triangles

Inspect triangles used in structures. Make plans and build proofs regarding the angles of triangles.

In Search of Evidence

What can you prove with transformations?

Explore the Louvre Pyramid

How are sides of triangles opposite congruent angles related? Can you prove it?

The SSS Congruence Theorem

Perpendicular bisectors are the key for building support and triangle congruence.

photo: Getty Images

Key Vocabulary

congruent, converse, exterior angle, isosceles triangle, perpendicular bisector, polygon, regular polygon, remote interior angle

Discover

As you complete Engage and the investigations, record the most important ideas you've learned.

Engage	
Investigation 1	**Investigation 2**
Investigation 3	**Investigation 4**

Name ______________________________ Date ____________

Check for Understanding

Prove Congruence Theorems: Investigation 1

1. Jack and Sigourney each were given a figure showing $\triangle ABC$. In Jack's figure, $\overleftrightarrow{DE} \parallel \overline{AB}$, with $\overleftrightarrow{DE}$ passing through point C. In Sigourney's figure, $\overleftrightarrow{DE} \parallel \overline{BC}$, with $\overleftrightarrow{DE}$ passing through point A. Who has enough information to prove the triangle sum theorem?

   A. only Jack
   B. only Sigourney
   C. both Jack and Sigourney
   D. neither Jack nor Sigourney

2. Suppose that $\angle FHK$ is an exterior angle of $\triangle FGH$, where $\angle HFG$ and $\angle FGH$ are the remote interior angles $m\angle FHK = 127°$. *Fill in the blanks in the following sentences with the correct angle measures in degrees.*

   Given the relationship among these angles it can be proven, if $m\angle HFG = 39°$, then $m\angle FGH =$ __________. However, if $m\angle FGH = 71°$, then $m\angle HFG =$ __________.

3. Given $\triangle PQR$ below, Ed tried to prove that $m\angle 1 + m\angle 2 + m\angle 3 + m\angle 4 + m\angle 5 + m\angle 6 = 360°$. Based on the results of his proof, he concluded that when a larger triangle is divided into two smaller triangles, the sum of the interior angles of the larger triangle is 360°.

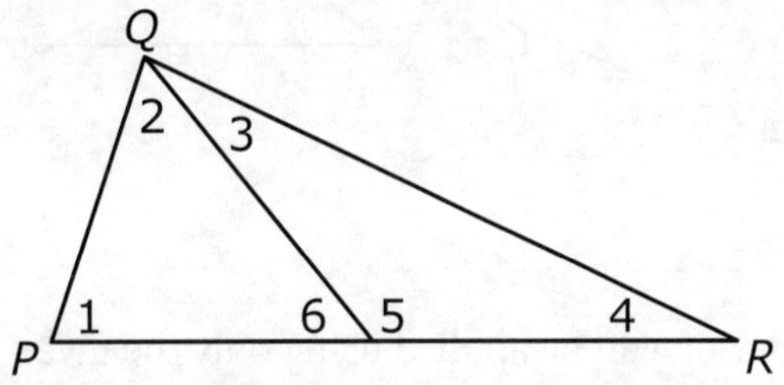

   First, he stated that $m\angle 1 + m\angle 2 + m\angle 6 = 180°$ and that $m\angle 3 + m\angle 4 + m\angle 5 = 180°$ by the triangle sum theorem. Then, by the addition property of equality, he stated that $m\angle 1 + m\angle 2 + m\angle 3 + m\angle 4 + m\angle 5 + m\angle 6 = 180° + m\angle 3 + m\angle 4 + m\angle 5$. Next, he stated that $m\angle 1 + m\angle 2 + m\angle 3 + m\angle 4 + m\angle 5 + m\angle 6 = 180° + 180°$ by the substitution property of equality. Finally, he simplified to $m\angle 1 + m\angle 2 + m\angle 3 + m\angle 4 + m\angle 5 + m\angle 6 = 360°$ by addition. Did Ed prove that the sum of the interior angles of $\triangle PQR$ is 360°? Explain.

Name ______________________________ Date ____________

Check for Understanding

Prove Congruence Theorems: Investigation 2

1. Given $\triangle ABC$ and $\triangle DEF$, so that $\overline{AB} \cong \overline{DE}$, $\overline{AC} \cong \overline{DF}$, and $\angle A \cong \angle D$, which composition of transformations cannot be used to prove the SAS congruence theorem?

   A. Translate A along vector $\overline{AD}$ so that vertex A maps onto vertex D, followed by a rotation with center D that maps the image of $\overline{AB}$ onto $\overline{DF}$, followed by a reflection over $\overline{DF}$.

   B. Translate $\triangle ABC$ along vector $\overline{BE}$ so that vertex B is mapped to vertex E, followed by a rotation with center E that maps the image of $\overline{BC}$ onto $\overline{EF}$, followed by a reflection over $\overline{EF}$.

   C. Translate A along vector $\overline{AD}$ so that vertex A maps onto vertex D, followed by a rotation with center D that maps the image of $\overline{AB}$ onto $\overline{DE}$, followed by a reflection over $\overline{DE}$.

2. Consider the two triangles shown.

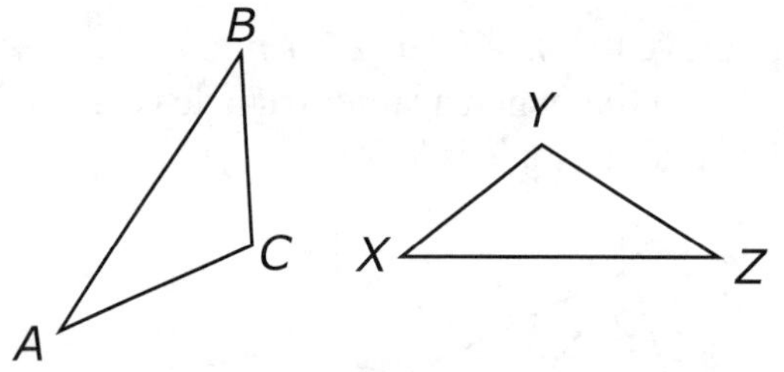

   Given: $\angle ACB \cong \angle ZYX$, $\overline{AC} \cong \overline{ZY}$, $\angle CBA \cong \angle YXZ$

   Which triangle congruence criterion can be applied using only the given information to prove that the two triangles are congruent?

   ____

3. Phillipe uses the HL triangle congruence theorem to establish $\triangle QRS \cong \triangle TUV$. Explain how the HL triangle congruence theorem could be considered a version of the SSS triangle congruence theorem.

Name ______________________ Date ____________

Check for Understanding

Prove Congruence Theorems: Investigation 3

1. You are given $\triangle ABC$, with $\overline{AB} \cong \overline{CB}$, and asked to prove the isosceles triangle theorem. *Fill in the blank in the following sentence with the appropriate angle.*

   If you add an auxiliary ray to the figure, it should bisect ____________.

2. Given the two triangles in the figure below, what can you prove with the isosceles triangle theorem? *Select all that apply.*

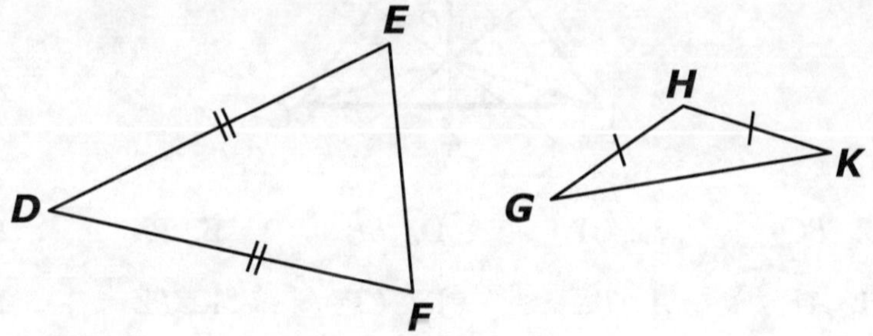

   A. $\angle D \cong \angle E$ B. $\angle G \cong \angle K$ C. $\angle D \cong \angle F$
   D. $\angle H \cong \angle K$ E. $\angle E \cong \angle F$ F. $\angle G \cong \angle H$

3. Trilby was given a triangle with one interior angle measuring 47° and another measuring 86°. Can she use it to prove the converse of the isosceles triangle theorem?

   A. Yes, because she has been given a triangle with two congruent angles.

   B. No, because she has not been given a triangle with two congruent angles.

   C. No, because she has been given a triangle with two congruent angles but not two congruent sides.

   D. Yes, because she has been given a triangle with two congruent sides but not two congruent angles.

4. Dale was given $\triangle LMN$, with $\overline{LM} \cong \overline{LN}$, while Felipe was given $\triangle PQR$, with $\overline{PQ} \cong \overline{PR} \cong \overline{QR}$. Can either one of them prove what the angles opposite the congruent sides in his triangle measure? If so, explain how it could be done.

Name ______________________________ Date ______________

Check for Understanding

Prove Congruence Theorems: Investigation 4

1. In the figure below, point P is the point of intersection of the perpendicular bisectors of the three sides of $\triangle ABC$. Which of these line segment lengths can be shown to be equal to AP by proving the perpendicular bisector theorem? *Select all that apply.*

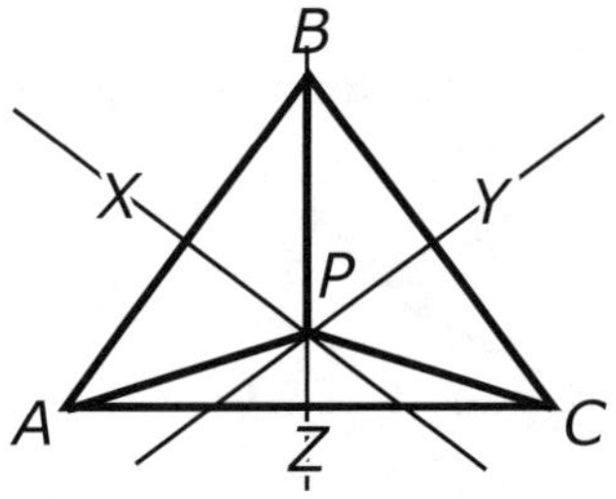

A. *AX*	**B.** *BC*	**C.** *CY*	**D.** *BP*	**E.** *PX*	**F.** *PY*
G. *BY*	**H.** *AB*	**I.** *BX*	**J.** *CP*	**K.** *PZ*	**L.** *AZ*

2. Conrad and Lawrence each want to create a proof of the converse of the perpendicular bisector theorem. Conrad was given $\triangle FGH$, with $GF = GH$, while Lawrence was given $\triangle LMN$, with $LM = LN$. Which statement is correct?

   A. Conrad should prove that point F is on the perpendicular bisector of $\overline{GH}$.

   B. Conrad should prove that point G is on the perpendicular bisector of $\overline{FH}$.

   C. Lawrence should prove that point N is on the perpendicular bisector of $\overline{LM}$.

   D. Lawrence should prove that point M is on the perpendicular bisector of $\overline{LN}$.

3. Genevieve's teacher proved the converse of the perpendicular bisector theorem and then used it to help prove the SSS triangle congruence theorem. Genevieve pointed out that now that the SSS Triangle congruence theorem has been proven, she can use it to help create an alternate proof of the converse of the perpendicular bisector theorem. Suppose Genevieve was given a diagram of $\triangle RST$, with $RS = TS$. Explain how the diagram and the SSS triangle congruence theorem will help her write her proof.

Summary

Before you attempt the Practice Exercises, review what you've learned.

Triangle Angle Measures

Triangle sum theorem: The sum of the interior angle measures of a triangle is 180°.

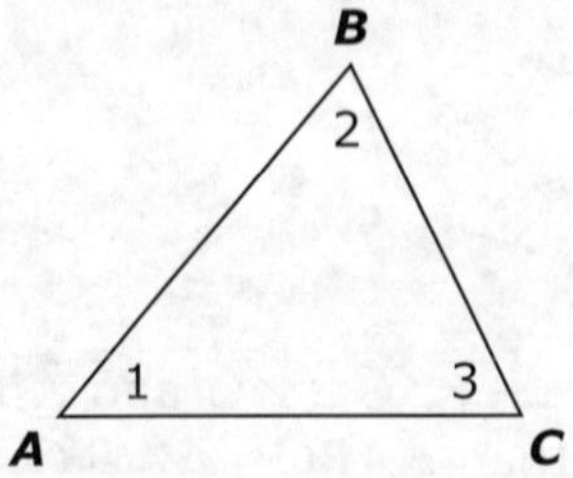

$m\angle 1 + m\angle 2 + m\angle 3 = 180°$

Exterior angle theorem: The measure of an exterior angle of a triangle is equal to the sum of the remote interior angle measures.

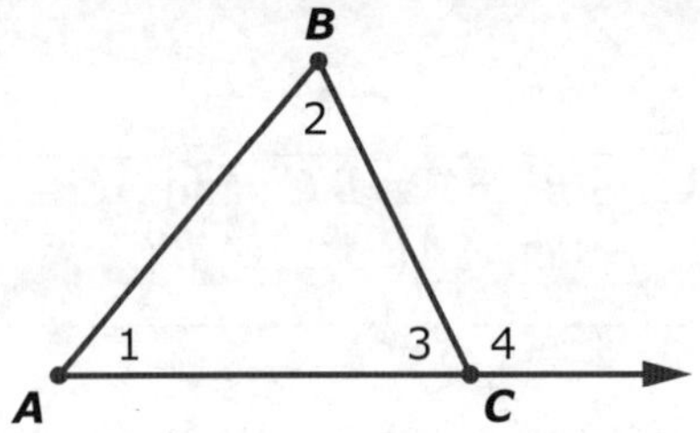

$m\angle 4 = m\angle 1 + m\angle 2$

Isosceles triangle theorem: The base angles of an isosceles triangle are congruent.

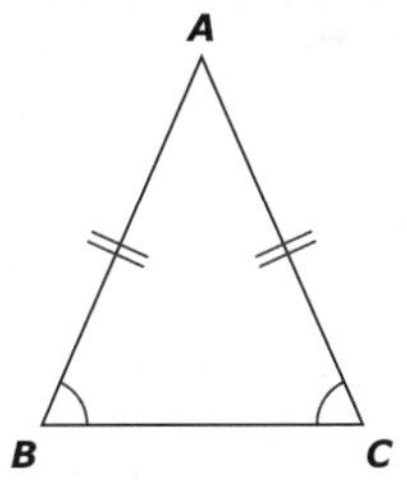

If $\overline{AB} \cong \overline{AC}$, then $\angle B \cong \angle C$

Equilateral triangle theorem: All three angles of an equilateral triangle are congruent.

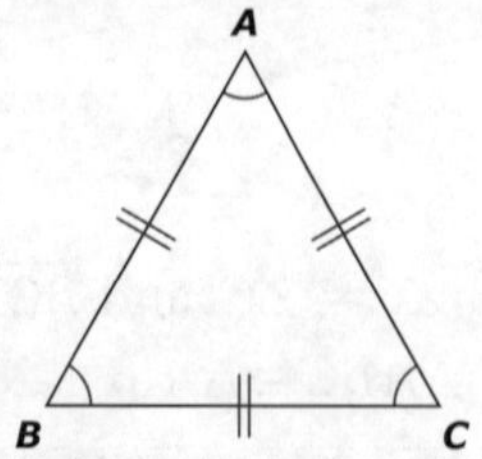

If $\overline{AB} \cong \overline{AC} \cong \overline{BC}$, then $\angle A \cong \angle B \cong \angle C$

Triangle Congruence

Two triangles are congruent if:

- all of their corresponding sides and angles are congruent;
- there exists a finite sequence of rigid motions that maps one triangle to the other; or
- one of the triangle congruence criteria described below applies.

My Notes

Summary *(continued)*

SAS triangle congruence theorem: If two triangles have two pairs of congruent sides, and the corresponding included angles are congruent, then the two triangles are congruent.

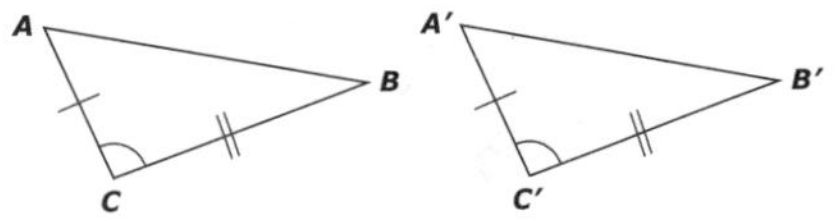

If $\overline{AC} \cong \overline{A'C'}$, $\angle C \cong \angle C'$ and $\overline{BC} \cong \overline{B'C'}$, then $\triangle ABC \cong \triangle A'B'C'$.

ASA triangle congruence theorem: If two triangles have two pairs of congruent angles, and the corresponding included sides of the angles are congruent, then the triangles are congruent.

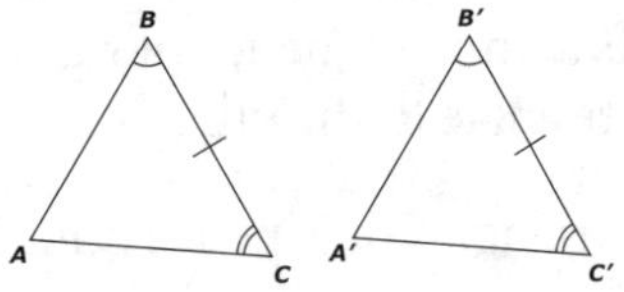

If $\angle B \cong \angle B'$, $\overline{BC} \cong \overline{B'C'}$, and $\angle C \cong \angle C'$, then $\triangle ABC \cong \triangle A'B'C'$.

AAS triangle congruence theorem: If two triangles have two pairs of congruent angles and a pair of nonincluded corresponding congruent sides, then the two triangles are congruent.

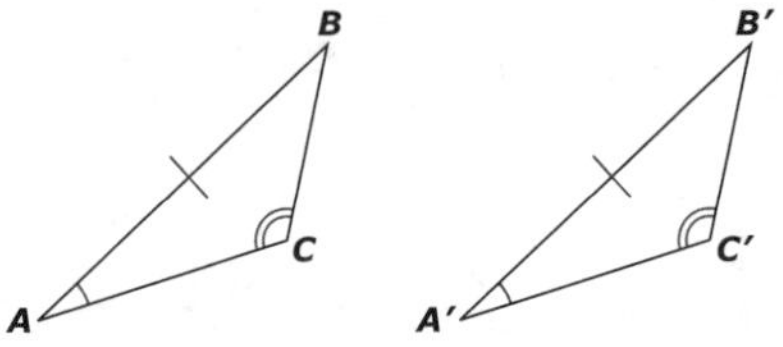

If $\angle A \cong \angle A'$, $\angle C \cong \angle C'$ and $\overline{AB} \cong \overline{A'B'}$, then $\triangle ABC \cong \triangle A'B'C'$.

SSS triangle congruence theorem: If two triangles have three pairs of congruent sides, then the two triangles are congruent.

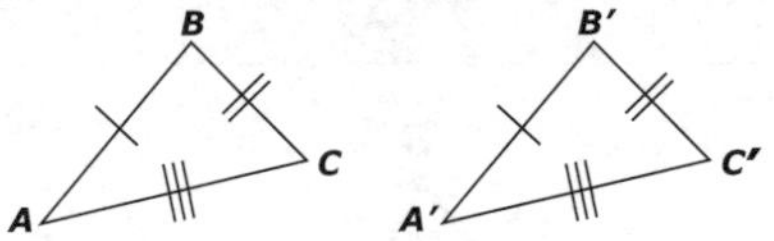

If $\overline{AB} \cong \overline{A'B'}$, $\overline{BC} \cong \overline{B'C'}$, and $\overline{AC} \cong \overline{A'C'}$, then $\triangle ABC \cong \triangle A'B'C'$.

HL triangle congruence theorem: Given two right triangles, if the hypotenuse and one leg of one triangle are congruent to the hypotenuse and one leg of another triangle, then the triangles are congruent.

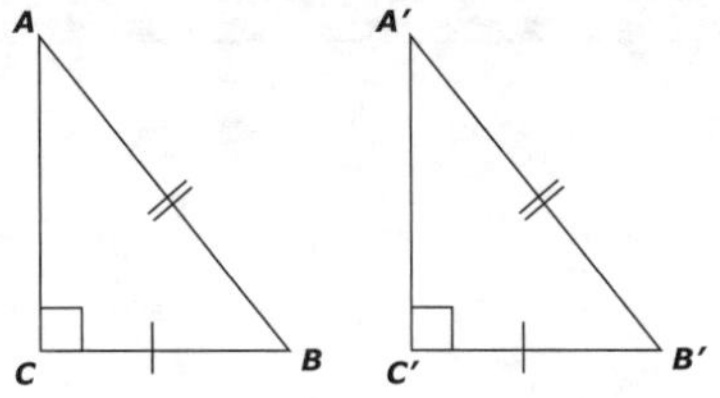

If $\triangle ABC$ and $\triangle A'B'C'$ are right triangles with $\overline{AB} \cong \overline{A'B'}$ and $\overline{BC} \cong \overline{B'C'}$, then $\triangle ABC \cong \triangle A'B'C'$.

My Notes

Practice Exercises

Review what you've learned using these practice problems. For practice problems with feedback, try the Coach and Play items in the Practice section online.

1. Jorge is improving the structural stability of a table by reinforcing the table's leg joints with triangular cross-braces. The cross-braces will connect the points where legs are attached to the tabletop to the midpoint of the bottom of the table's base.

   Notice in his sketch that the cross-braces will form a 53° angle with each other, and will form two angles that both measure $x°$ with the legs of the table.

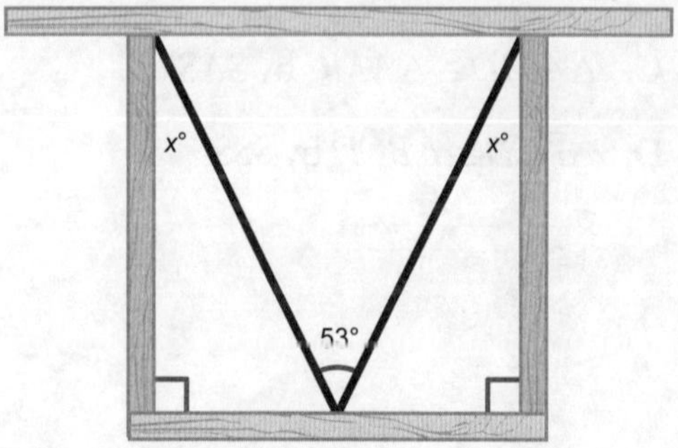

   What is the measure of each of the unknown angles in Jorge's sketch?

   $x° =$ ______________°

2. Use the information given in the diagram to determine $m\angle FED$.

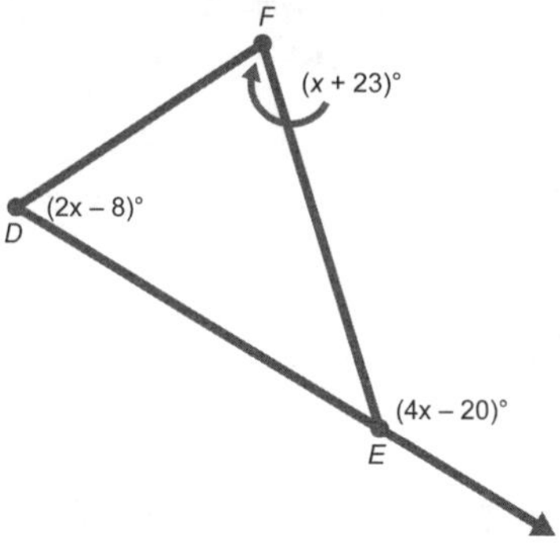

   $m\angle FED =$ ______________°

3. Consider two congruent triangles $\triangle ATN \cong \triangle JML$ such that $AN = 14$ m, $ML = 12$ m, and $AT = 7$ m.

   What is the length of side $\overline{JM}$?

   $JM =$ ______________ m

4. Joyce is making a quilt for her granddaughter. She made a sketch of one block on a square piece of fabric.

   The 4-pointed star is made of 8 congruent triangles, each having one 45° angle and one 15° angle, as shown. Joyce needs to find the measure of $\angle 1$ before she can make this piece of the quilt. What is the measure of $\angle 1$?

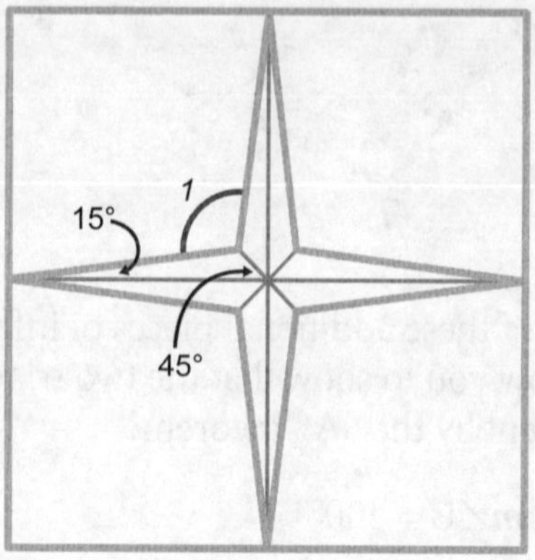

   $m\angle 1 =$ ______________°

5. The diagram shows $\triangle DOG$ and $\triangle CAT$. For these two triangles, it is known that $\overline{GO} \cong \overline{TA}$.

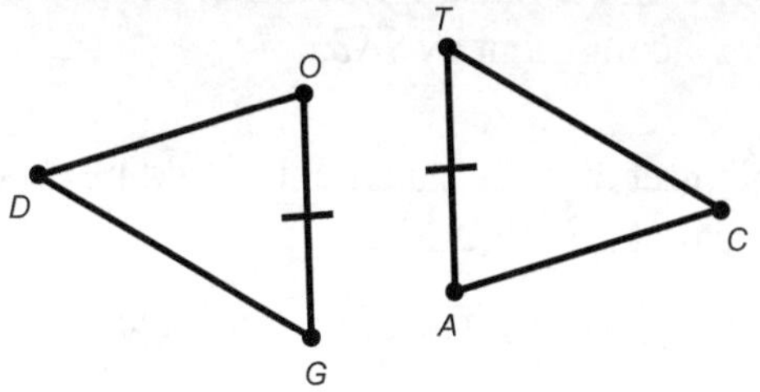

   What additional information do you need to prove that $\triangle DOG \cong \triangle CAT$ by the ASA triangle congruence theorem?

   Select all that apply.

   A. $\overline{DO} \cong \overline{CA}$

   B. $\angle ODG \cong \angle ACT$

   C. $\angle DOG \cong \angle CAT$

   D. $\overline{DG} \cong \overline{CT}$

   E. $\angle DGO \cong \angle CTA$

Practice Exercises *(continued)*

6. In an isosceles triangle, an angle opposite one of the two congruent sides has a measure of 63°.

The other two angles measure ______° and ______°.

7. Triangles *RUS* and *RTS* are shown in the diagram.

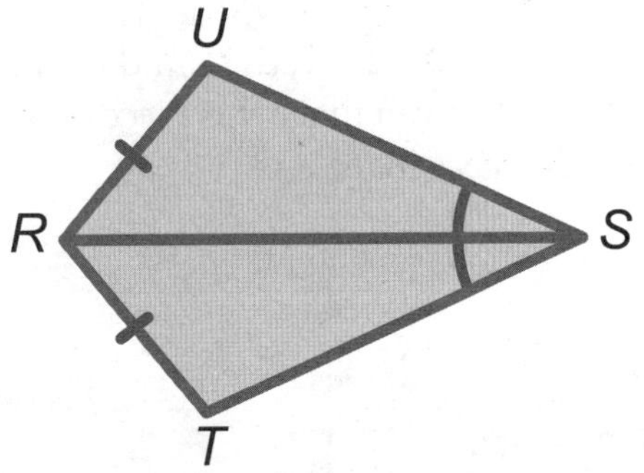

Which of these additional pieces of information will allow you to show that the two triangles are congruent by the SAS theorem?

A. $m\angle U = 100°$

B. $\overline{RU} \cong \overline{RT}$

C. $ST = 2RT$

D. $\overline{US} \cong \overline{TS}$

E. The two triangles cannot be shown congruent by SAS.

8. Which set of three statements could be used to prove that $\triangle DEF \cong \triangle RUN$?

A. $\overline{DE} \cong \overline{RU}, \overline{DF} \cong \overline{UN}, \overline{EF} \cong \overline{RN}$

B. $\angle E \cong \angle U, \overline{DE} \cong \overline{RU}, \overline{DF} \cong \overline{RN}$

C. $\angle F \cong \angle N, \angle D \cong \angle R, \overline{DF} \cong \overline{RN}$

D. $\angle F \cong \angle N, \angle E \cong \angle R, \overline{EF} \cong \overline{RN}$

9. Clara constructs two triangles such that $m\angle L = m\angle A = 90°$, $\overline{LO} \cong \overline{AS}$, and $\overline{SO} \cong \overline{FS}$.

Complete the conclusion.

$\triangle SLO \cong \triangle$ ______ by ______

10. $\overline{AB}$ and $\overline{CD}$ intersect at point *M* and are the perpendicular bisectors of each other.

Which congruence statement is not valid?

A. $\triangle AMD \cong \triangle CMB$ by SAS

B. $\triangle AMD \cong \triangle BMD$ by SSS

C. $\triangle AMD \cong \triangle AMC$ by SAS

D. $\triangle ACD \cong \triangle BCD$ by SSS

Apply

How Will Frank Lloyd Wright Inspire You?

Frank Lloyd Wright is recognized as one of the most influential architects of the 20th century. Wright designed over 900 buildings, more than 400 of which were built. His unique style, known as Prairie style, was a key element in the Arts and Crafts movement. Designers all over the world have gotten their start by imitating his work in order to study it up close and form their own unique styles.

Wright's stained glass is on display at Alright-Knox Gallery.

Research online to find one of Frank Lloyd Wright's window designs or another design you like that contains triangles, and print it out.

- Using the image you find as inspiration, create your own geometric stained glass window design that incorporates at least two congruent triangles. Your design also should include elements to illustrate at least two of the following theorems: the triangle sum theorem, the exterior angle theorem, the isosceles triangle theorem, and the equilateral triangle theorem.
- Prove congruency by using the least possible number of measurements and any of the triangle congruence theorems.
- Explain the elements of your design that illustrate your two chosen theorems.

Show what you've learned by completing the other performance tasks in the online Apply section.

Apply *(continued)*

Your answer to Apply will be assessed on the following criteria:

1. Creating a design that contains two congruent triangles and can be used to illustrate two of the triangle theorems
2. Using the design to illustrate the theorems
3. Taking the least possible number of measurements of the two triangles to prove they are congruent
4. Applying one of the triangle congruence theorems correctly to prove that two triangles are congruent

Scale / Criteria	4	3	2	1	0
	Exceeds Criteria	Meets Criteria	Progressing to Criteria	Below Expectations	No Expectation
Design	Creates a design that contains congruent triangles and can be used to illustrate two of the theorems.	Creates a design that contains congruent triangles and can be used to illustrate one of the theorems.	Creates a design that does not contain congruent triangles, but can be used to illustrate at least one of the theorems.	Creates a design without triangles.	Does not create a design.
Measurements	Lists the least possible number of measurements needed to prove congruence, and explains reasoning; takes the measurements.	Lists one extra measurement needed to prove congruence, and explains reasoning; takes the measurements.	Lists two extra measurements needed to prove congruence, and the reasoning may be unclear; takes the measurements.	Takes inadequate measurements or three or more extra measurements, and the reasoning may be unclear.	Does not list or take measurements.
Proof	Applies the triangle congruence theorems correctly to prove the triangles are congruent.	Applies the triangle congruence theorems, but has a minor error.	Applies the triangle congruence theorems, but contains two or more errors or is unclear.	Applies the triangle congruence theorems incorrectly.	Does not apply any of the triangle congruence theorems.
Explanation	Correctly explains how the design illustrates two of the theorems.	Explains how the design illustrates two of the theorems with minor errors.	Explains how the design illustrates one of the theorems.	Explains how the design illustrates the theorems with major errors.	Does not explain how the diagram illustrates the theorems.

UNIT 4: Triangle Geometry

4.2 Construct Special Triangles and Angles

photo: Getty Images

Lesson Objectives

- Construct equilateral triangles, segment bisectors, congruent angles, and angle bisectors.
- Prove that a construction is valid.

Essential Question

- What properties of geometric objects are revealed through geometric constructions?

photo: Getty Images

Investigations

The Flower of Life

Use the Dynamic Geometry Tool to create a Flower of Life design. Describe the steps for others to follow.

Constructing an Equilateral Triangle

Use the Construction Tool to create an equilateral triangle. Describe and verify the steps.

Constructing Congruent Triangles and Angles

Is that art a copy? Verify congruency using the Construction Tool.

Midpoints and Perpendicular Bisectors

An artist's tools of the trade include midpoints and perpendicular lines. Find out how.

Constructing an Angle Bisector

Continue down the construction path. Construct angles and their bisectors.

Key Vocabulary

midpoint, perpendicular bisector

Discover

As you complete Engage and the investigations, record the most important ideas you've learned.

Engage	
Investigation 1	**Investigation 2**
Investigation 3	**Investigation 4**

Name ______________________________ Date __________

Check for Understanding

Construct Special Triangles and Angles: Investigation 1

1. Circle *A* and circle *B* each have a radius of 5 cm. Circle *A* passes through the center of circle *B*, and circle *B* passes through the center of circle *A*. If the centers of the two circles and another point of intersection of the two circles are the vertices of a triangle, how many sides of the triangle must have a length of 5 cm?

A. 0 **B.** 1

C. 2 **D.** 3

2. Anne was given $\overline{CD}$ and was asked to construct an equilateral triangle. First she placed her compass on point *C* and measured the distance to point *D*. Then she swung an arc of this size without intersecting the line segment. Next, keeping the span of her compass the same, she placed the compass on point *D* and swung an arc that intersected the first arc. Finally, she drew a triangle with the point of intersection of the arcs, along with points *C* and *D*, as the vertices. *Circle the correct words in the following sentence.*

If the arcs intersect above $\overline{CD}$, the triangle [is / is not] equilateral, and if the arcs intersect below $\overline{CD}$, the triangle [is / is not] equilateral.

3. In class, Kendrick learned how to construct a regular hexagon inscribed in a circle. With a compass, he drew a circle with a center at point *O*. Then he drew a point on the circle and, with the same compass span, he placed the compass on this point and drew a small arc intersecting the circle at another point. From this new point, he repeated the process until he had six points on the circle, which were the hexagon's vertices. Next, he drew line segments connecting every other vertex of the hexagon to create $\triangle RST$ as shown.

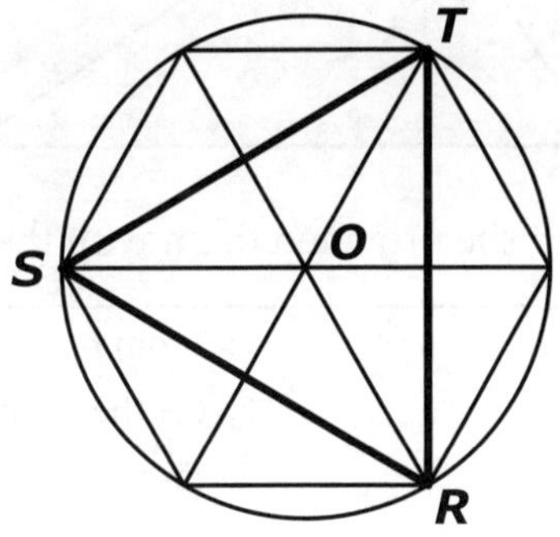

Kendrick says that $\triangle RST$ is an equilateral triangle? Is he correct? Explain your answer.

Name ______________________________ Date ______________

Check for Understanding

Construct Special Triangles and Angles: Investigation 2

1. Luke and Stacey want to construct $\angle DEF$ so that it is congruent to $\angle ABC$. To begin, they will mark a point E that will be the vertex of the angle. Then, from point E, they will draw $\overrightarrow{EF}$. Luke says that $\overrightarrow{EF}$ must be parallel to $\overrightarrow{BC}$, while Stacey says that the direction of $\overrightarrow{EF}$ must be the same as that of $\overrightarrow{BC}$. Who is correct?

   A. only Luke

   B. only Stacey

   C. both Luke and Stacey

   D. neither Luke nor Stacey

2. Suppose you constructed a triangle that is congruent to a given triangle. *Fill in the blanks in the sentence with the correct numbers.*

   The triangle you have constructed has _____ interior angles, and of these, _____ of them must be congruent to at least one of the interior angles of the given triangle.

3. Ms. Davis asked her students to construct a triangle that is congruent to $\triangle GHK$ below.

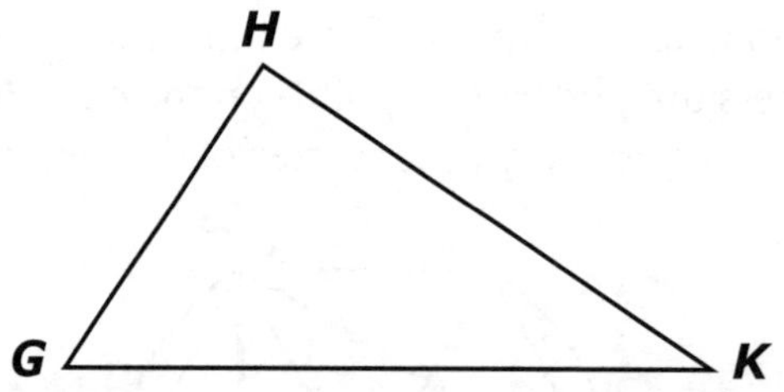

To help them with their constructions, she provided them with the following steps.

1. Mark a point L that will be a vertex.	6. From L, draw an arc where point M will be.
2. Set the compass to the length of $\overline{GK}$.	7. Set the compass to the length of $\overline{HK}$.
3. From L, draw an arc.	8. From N, draw an arc across the last one.
4. Mark a point N on this arc.	9. Mark the intersection of the arcs point M.
5. Set the compass to the length of $\overline{GH}$.	10. Draw $\overline{LM}$, $\overline{MN}$, and $\overline{LN}$.

One of Ms. Davis's students mixed up steps 5 and 7. Is the student's triangle still congruent to $\triangle GHK$? Can it be mapped onto $\triangle GHK$ with transformations? Explain.

Name ______________________________ Date ____________

Check for Understanding

Construct Special Triangles and Angles: Investigation 3

1. Frank constructed two circles that intersect each other at two points. Then he drew two line segments. The endpoints of the first segment are the centers of the circles, while the endpoints of the second segment are the points of intersection of the circles. Which of the following statements must be correct about the two line segments Frank drew?

A. One bisects the other.

B. One does not bisect the other.

C. They are perpendicular.

D. They are not perpendicular.

2. Suppose you want to construct the perpendicular bisector of $\overline{CD}$. To do so, you first place your compass on point C, and draw an arc above and an arc below $\overline{CD}$. Then you place your compass on point D, and draw another arc above and another arc below $\overline{CD}$. After that, you finish the construction. *Circle the correct words in the sentence.*

When drawing the arcs both from point C and from point D, the compass span should be [less than / equal to / greater than] half the length of $\overline{CD}$.

3. In an attempt to construct perpendicular line segments, Victoria first drew a line and marked a point O on it. Then, from O, she drew a circle with her compass so that it intersected the line at two points. She labeled these points R and T. Next, she marked point S on the circle, and drew $\overline{RS}$ and $\overline{ST}$. Her construction is shown below.

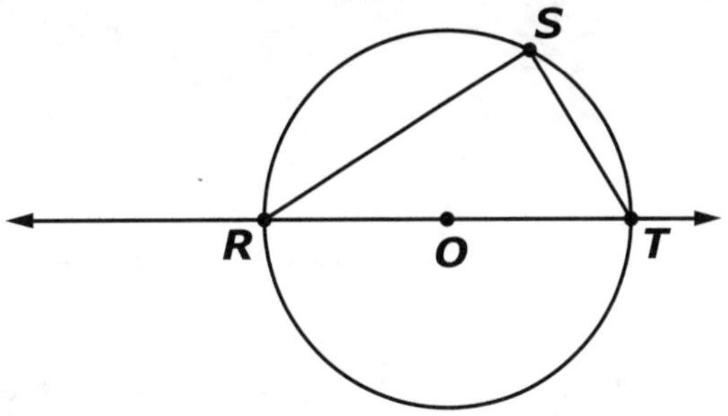

Victoria claims that $\overline{RS}$ and $\overline{ST}$ are perpendicular. Is she correct? Justify your answer.

Name ______________________________ Date ______________

Check for Understanding

Construct Special Triangles and Angles: Investigation 4

1. Suppose you constructed $\overrightarrow{BD}$, which bisects $\angle ABC$. *Fill in the blanks in the sentence.*

   If $\angle ABC$ is formed by perpendicular lines, $m\angle ABD =$ ______, and $m\angle CBD =$ ______.

2. Chris wants to construct a ray that bisects $\angle GHK$. To do so, he will place a compass on point H and follow the steps below. *Circle the correct words in the sentences that follow.*

   1. From H, draw an arc across $\overrightarrow{HG}$, one of the legs of $\angle GHK$.
   2. From H, draw an arc across $\overrightarrow{HK}$, the other leg of $\angle GHK$.
   3. From the point where the arc drawn in step 1 crosses $\overrightarrow{HG}$, draw an arc in the interior of $\angle GHK$.
   4. From the point where the arc drawn in step 2 crosses $\overrightarrow{HK}$, draw another arc in the interior of $\angle GHK$ that intersects the arc drawn in step 3.
   5. Draw a ray from H through the point of intersection of the arcs in the interior of $\angle GHK$ that were drawn in steps 3 and 4.

   It [does / does not] matter if Chris changes the span of the compass between steps 1 and 2.

   It [does / does not] matter if he changes the span of the compass between steps 2 and 3.

   It [does / does not] matter if he changes the span of the compass between steps 3 and 4.

3. Fiona drew $\overrightarrow{MN}$. Then she placed her compass on point M and drew an arc across the ray. Next, from the point where the arc crosses the ray, she drew another arc across the one she just drew without changing her compass span. Her construction is shown.

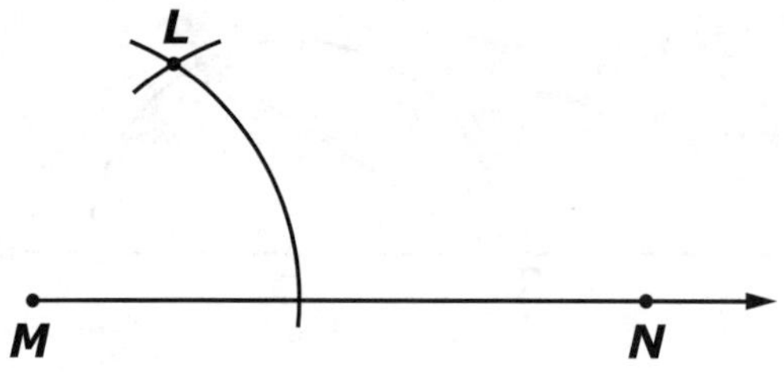

   If Fiona draws $\overrightarrow{ML}$ through the point of intersection of the arcs in the figure, what will be the measure of the angle she has constructed? Explain how you got your answer.

Summary

Before you attempt the Practice Exercises, review what you've learned.

In this lesson, you used a compass and straightedge to discover how to construct certain specific relationships among lines, segments, and angles.

Constructing the Midpoint

Midpoint or Perpendicular Bisector of a Segment

1. Construct a circle centered at one endpoint with a radius greater than half the length of the segment.

2. Construct a circle centered at the other endpoint with the same radius. The circles will intersect at two points.

3. Construct a line through the intersections. This line is the perpendicular bisector of the segment. It intersects the segment at the midpoint.

Constructing Congruent Triangles

Copy a Triangle

1. Draw a line that will contain one side of the copy of the triangle. Mark one vertex on the line.

2. Set the compass to the length of one side of the original triangle and mark the length on the line. This will be the second vertex.

3. Set the compass to the length of the second side of the triangle and construct an arc centered at the first vertex of the copy.

4. Set the compass to the length of the third side of the triangle and construct an arc centered at the second vertex of the copy that intersects the previous arc. The intersection is the third vertex of the copy.

5. Construct the triangle through the three vertices.

Constructing Equilateral Triangles

Equilateral Triangle

1. Construct a line and mark a point on the line. This is the first vertex of the triangle.

2. Construct an arc centered at the point that intersects the line. The intersection is the second vertex.

3. Using the same radius, construct one circle centered at the first vertex and another circle centered at the second vertex. Either of the intersections of the circles is the third vertex.

4. Construct segments connecting pairs of vertices to complete the triangle.

My Notes

Summary *(continued)*

Constructing Angles

Copy an Angle

1. Construct a segment connecting the two rays created by the angle. This will create a triangle.

2. Copy the triangle. The corresponding angle in the copy is congruent to the original angle.

Angle Bisector

1. Using any length as a radius, construct an arc that intersects both sides of the angle.

2. Construct two arcs with the same radius centered at the intersections of the arc with the sides of the angle. The arcs should intersect in the interior of the angle.

3. Construct the ray from the vertex of the angle through the intersection point of the two arcs.

My Notes

Practice Exercises

Review what you've learned using these practice problems. For practice problems with feedback, try the Coach and Play items in the Practice section online.

1. Tyler needs to construct an angle bisector for $\angle DEF$. Help Tyler remember the procedure for finding an angle bisector. *Use the options below to complete the steps.*

   A. is at least half the distance between the two rays

   B. is at the intersection of the two arcs

   C. is the same

   D. crosses both rays of the angle

   Draw an arc centered at point E using any radius that ________.

   Using the point where the arc intersects $\overrightarrow{ED}$ as the center, draw an arc with a radius that ________.

   From the point where the first arc intersects $\overrightarrow{EF}$, draw another arc with a radius that ________.

   Finally, connect point E to the point that ________.

2. Which statements guarantees that a constructed triangle, $\triangle DEF$, is equilateral?

   A. The radius lengths used at points E and D are equal.

   B. The arcs drawn from points E and D intersect.

   C. The distances from point E to point D, point D to point F, and point F to point E are all equal.

   D. $\angle F$ is congruent to $\angle D$.

   E. The angle bisector of $\angle F$ is also the perpendicular bisector of $\overline{ED}$.

3. Kim constructs a particular pentagon $RSTUV$ that has the perpendicular bisectors of all five sides intersecting at a single point. He labels the point of intersection of the perpendicular bisectors as point Q.

   Kim then constructs a circle with center Q such that $\overline{QR}$ is a radius of the circle.

   Which statements must be true based on this given information? *Select all that apply.*

   A. The vertices of pentagon $RSTUV$ are equidistant from point Q.

   B. $\angle UTQ \cong \angle QUT$

   C. $RS = 2UV$

   D. $\overline{QT} \perp \overline{UT}$

   E. Points S, T, U, and V lie on $\odot Q$.

4. Given $\overline{LM}$, consider a construction formed by the following steps:

   - Set the compass distance to be longer than LM.
   - Put the compass point on point L and create an arc.
   - Without changing the compass setting, put the compass point on point M and create an arc that intersects the previous arc.
   - Label the intersection of the two arcs as point N.

   Based on this construction, which of the following statements are true about $\triangle LMN$? *Select all that apply.*

   A. $LM = \frac{1}{2}NM$

   B. $NL = NM$

   C. $\angle MLN \cong \angle LMN$

   D. $\overline{NL} \perp \overline{LM}$

   E. $\triangle LMN$ is an equilateral triangle.

   F. $\triangle LMN$ is an isosceles triangle.

   G. $\triangle LMN$ is a scalene triangle.

Practice Exercises *(continued)*

5. The diagram shows Andrea's construction of $\angle LKM$.

   Andrea says she will complete her construction by using a straightedge to construct $\overleftrightarrow{KM}$.

   Which theorem can Andrea use to justify her claim that $\overleftrightarrow{KM} \parallel \overleftrightarrow{AC}$?

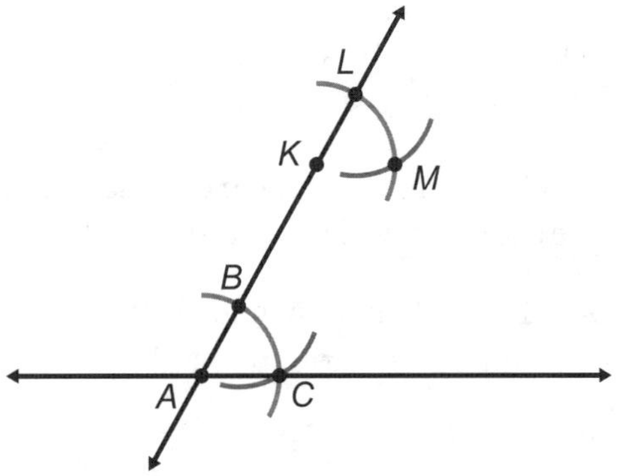

   A. If two lines are intersected by a transversal so that alternate interior angles are congruent, then the lines are parallel.

   B. If two lines are intersected by a transversal so that alternate exterior angles are congruent, then the lines are parallel.

   C. If two lines are intersected by a transversal so that consecutive interior angles are supplementary, then the lines are parallel.

   D. If two angles are intersected by a transversal so that corresponding angles are congruent, then the lines are parallel.

   E. None of the above

6. You want to find the midpoint of line segment $\overline{JK}$ by using a perpendicular bisector.

   Which of the steps below is NOT correct?

   A. Set the compass to less than halfway between *J* and *K*.

   B. Draw an arc centered on point *J* toward point *K* using the set radius.

   C. Draw a similar arc centered on point *K* toward point *J* using the set radius.

   D. Connect the two intersecting points of the arcs to create a perpendicular bisector.

7. Which of these are correct statements about the process for constructing an angle bisector? *Select all that apply.*

   A. Any two points on the rays of the angle can be used to determine an angle bisector.

   B. The same radius must be used when finding the intersection of the two arcs drawn from a point on each of the rays of the angle.

   C. From the vertex point, the same radius must be used to determine the two points on the rays of the angle.

   D. Any set of radii can be used to construct and angle bisector.

   E. A point that appears to be halfway between the two rays of the angle can be used to draw the angle bisector.

8. Carlos knows that point *F* is the midpoint of $\overline{CD}$. He has also constructed $\overline{EF}$ from point *F* by drawing arcs from points *C* and *D* using a compass with a set radius. However, he is unsure whether $\overline{EF}$ is perpendicular to $\overline{CD}$. For $\overline{EF}$ to be perpendicular to $\overline{CD}$, which two sets of distances must be equal?

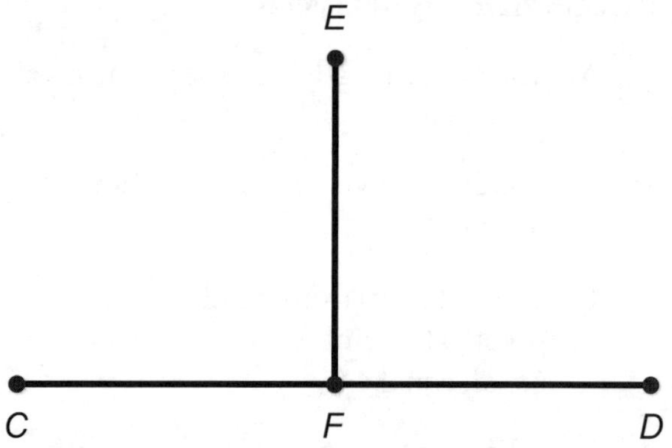

   If $\overline{EF}$ is perpendicular to $\overline{CD}$, then

   $CF =$ __________ and $CE =$ __________ .

Apply

How Can You Copy a Design with Only a Straightedge and Compass?

Your challenge is to construct an identical replica of the design below, using only a straightedge and a compass. Take photos of your work as you go along. Record your step-by-step process as you copy the design, and label your construction so you can reference its parts.

Accurately recreate this design using only a straightedge and compass!

photo: Discovery Education

Hands-On Activity: Triangle Patterns. Print the Triangle Design activity before you start your construction.

Show what you've learned by completing the other performance tasks in the online Apply section.

Apply *(continued)*

Your answer to Apply will be assessed on the following criteria:

1. Constructing an equilateral triangle using the tools
2. Constructing the "center" of the triangle and justifying the validity of this method
3. Copying the design exactly, including the correct shading
4. Describing the steps used to copy the design and documenting the process with snapshots of work in progress

Scale / Criteria	4 Exceeds Criteria	3 Meets Criteria	2 Progressing to Criteria	1 Below Expectations	0 No Expectation
Equilateral Triangle	Accurately constructs an equilateral triangle with clear evidence of using tools.	Constructs a triangle with evidence of using tools, but the resulting triangle is isosceles.	Constructs a triangle with evidence of using tools, but the resulting triangle is scalene.	Draws an equilateral triangle without evidence of using tools.	Does not construct a triangle.
Triangle "Center"	Accurately constructs line segments that meet at the center with clear evidence of using tools, and correctly justifies the method used.	Constructs line segments that meet at the center with evidence of using tools, but makes an error in reasoning in the justification.	Constructs line segments with evidence of using tools, but the segments do not meet at the center, and makes errors in reasoning in the justification.	Draws segments that appear to meet at the center without evidence of using tools, and provides little or no justification.	Does not attempt to locate the center of the triangle.
Design	Accurately copies the design, including size and shading.	Copies the design, but the shading is not correct.	Copies the design, but the copy is smaller or larger than the original.	Does not accurately copy the design.	Does not copy the design.
Description of Steps	Accurately writes a description of the steps and includes references to the labeled diagram.	Accurately writes a description of steps, but does not include references to the labeled diagram.	Writes a description that has gaps in how the design was copied, or steps are described inaccurately.	Writes a description that has inaccuracies or is missing multiple steps.	Does not describe the steps.

photo: Discovery Education

UNIT 5: Similarity

5.1 Explore Similarity and Dilation

Lesson Objectives

- Define, identify, and draw dilations informally and with coordinates.
- Define similarity in terms of dilations, and explain why dilations do not preserve congruence unless the scale factor is 1.
- Explain the relationship between similarity and proportionality.

Essential Question

- What are similarity transformations, and how do they affect the measurements of figures?

photo: Getty Images

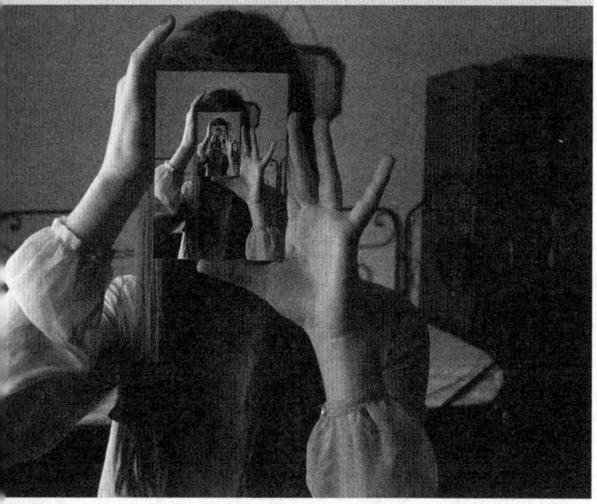

Investigations

Photo Editing

Arranging photos to fit on a page sometimes requires manipulations. How do the photos change?

Dilations

Explore dilations. What role does the center of dilation play? What properties do all dilations have?

Digital Photos

Use the precision of the coordinate plane to investigate and describe dilations with function rules.

Photo Collage

What are similar figures? Which transformations result in similar figures?

Key Vocabulary

collinear dilation, parallel, rigid motion, rotation, scale factor, similar, transformation

Discover

As you complete Engage and the investigations, record the most important ideas you've learned.

Engage	**Investigation 1**
Investigation 2	**Investigation 3**

Name______________________________ Date__________

Check for Understanding

Explore Similarity and Dilation: Investigation 1

1. If you apply a dilation to a polygon, which of these scale factors will result in an image of the polygon that is smaller than the polygon's pre-image? *Select all that apply.*

A. $\frac{7}{8}$	B. $\frac{3}{2}$	C. $\frac{10}{3}$	D. $\frac{5}{4}$
E. $\frac{9}{7}$	F. $\frac{6}{5}$	G. $\frac{8}{9}$	H. $\frac{4}{15}$

2. Jen dilated a line segment with a length of 12 units and a slope of $\frac{1}{3}$. *Fill in the blanks.*

Given that the scale factor of the dilation was 6, the length of the line segment's image is ______ units, and the slope of the line segment's image is ______.

3. Yousef applied a composition of two dilations to rectangle *WXYZ*. The first dilation had a center at point *P* and produced the rectangle *W'X'Y'Z'*. The second dilation had a center at vertex *Z'* of rectangle *W'X'Y'Z'* and produced the rectangle *W''X''Y''Z''*. In the figure below, point *P*, rectangle *WXYZ*, and rectangle *W''X''Y''Z''* are shown.

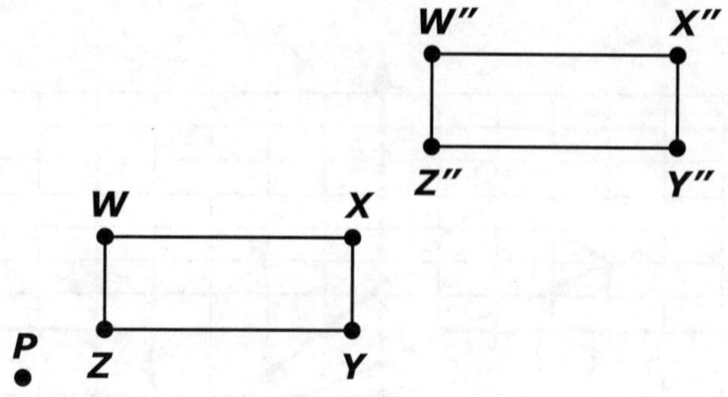

Explain what you can tell about the scale factors of the two dilations from the figure, and based on your findings, give examples of what the scale factors could be.

Name ______________________________ Date ______________

Check for Understanding

Explore Similarity and Dilation: Investigation 2

1. The coordinates of the vertices of an octagon are as follows: (−3, 1), (−1, 3), (1, 3), (3, 1), (3, −1), (1, −3), (−1, −3), and (−3, −1). If the octagon were dilated by a scale factor of 3 and with a center of dilation at the origin, which of these would be the coordinates of one of the image's vertices? *Select all that apply.*

   A. (3, −6) **B.** (−9, 1) **C.** (6, 3) **D.** (9, −3)

   E. (9, −1) **F.** (−3, −9) **G.** (−9, 3) **H.** (−3, −6)

2. Hallie applied a dilation given by the rule $D_{0.8}(x, y) = (0.8x, 0.8y)$ and centered at the origin to a quadrilateral in a coordinate plane. The quadrilateral's image has vertices with coordinates of $A'(-8, -12)$, $B'(4, -8)$, $C'(12, -4)$, and $D'(16, 8)$. *Fill in the blanks.*

   The quadrilateral's pre-image has vertices with coordinates of *A*(______, ______), *B*(______, ______), *C*(______, ______), and *D*(______, ______).

3. Bo wants to dilate △*XYZ* below by a scale factor of 0.5 and with a center of dilation at *P*.

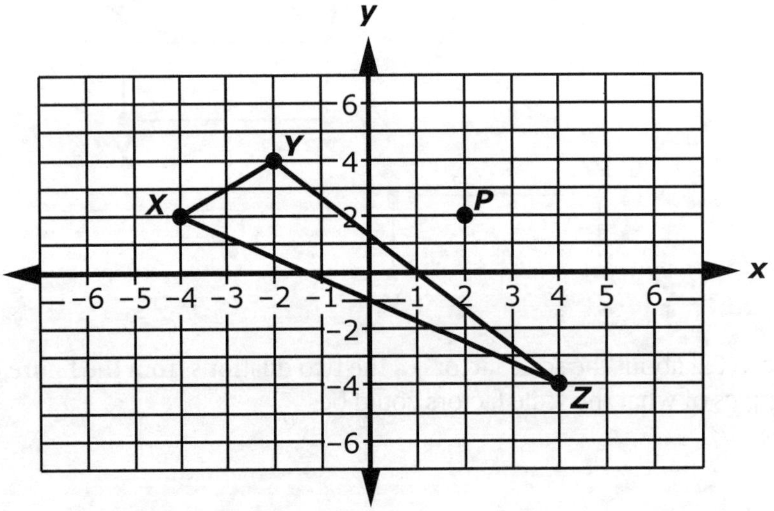

   Explain how he can use the rule for a dilation by a scale factor of *k* centered at the origin to help perform the dilation. What would be the coordinates of the vertices of △*X′Y′Z′*?

Name ______________________________ Date ____________

Check for Understanding

Explore Similarity and Dilation: Investigation 3

1. Chelsea showed that rectangles *ABCD* and *FGHK* are similar by dilating rectangle *FGHK* and then translating its image, *F'G'H'K'*, so that it coincided with rectangle *ABCD*. Suppose that the scale factor of the dilation was 4. *Fill in the blanks in the sentence.*

If the length of rectangle *ABCD* is 20 in., then the length of rectangle *FGHK* is ______ in., and if the width of rectangle *ABCD* is ______ in., then the width of rectangle *FGHK* is 4 in.

2. A composition of transformations consisting of a dilation, a translation, a rotation, and a reflection was applied to $\triangle RST$. This resulted in its image, $\triangle R'S'T'$, coinciding with $\triangle UVW$, showing that $\triangle RST \sim \triangle UVW$. Which of these statements must be correct?

A. No side of $\triangle R'S'T'$ is congruent to a side of $\triangle UVW$.

B. Each side of $\triangle RST$ is congruent to a side of $\triangle UVW$.

C. No interior angle of $\triangle RST$ is congruent to an interior angle of $\triangle UVW$.

D. Each interior angle of $\triangle R'S'T'$ is congruent to an interior angle of $\triangle UVW$.

3. To show that $\triangle LMN$ and $\triangle XYZ$ below are similar, Jorge mapped $\triangle LMN$ onto $\triangle XYZ$ by first dilating $\triangle LMN$ by a scale factor of 1.5 and with a center of dilation at *L*. He then reflected the image across $\overleftrightarrow{PQ}$. The combined transformation was $R_{\overleftrightarrow{PQ}}(D_{1.5,\,L}(\triangle LMN))$.

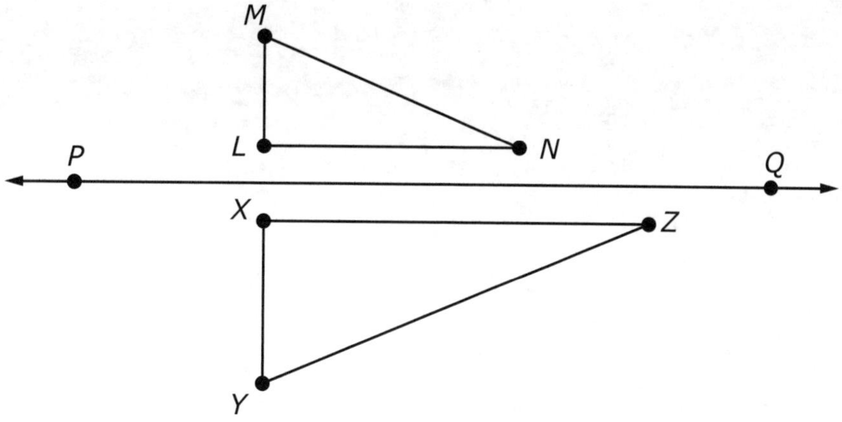

Could Jorge have switched the order of the dilation and the reflection in his composition of transformations and still have mapped $\triangle LMN$ onto $\triangle XYZ$? Explain your answer.

Summary

Before you attempt the Practice Exercises, review what you've learned.

Dilations

A **dilation** is a transformation in which a figure is enlarged or reduced by a given scale factor around a given point, called the center of dilation.

The dilated image of a figure is similar to the pre-image. Dilations have the following properties:

- A dilation with a scale factor of 1 is a congruence.
- When a segment is dilated with a scale factor of k, the length of the image is k times the length of the pre-image.
- When a figure is dilated, the sides of the image are proportional to the corresponding sides of the pre-image, and the corresponding angles are congruent.
- The dilation of a line produces a parallel line, unless the center of dilation is on the line, in which case the line is unchanged.

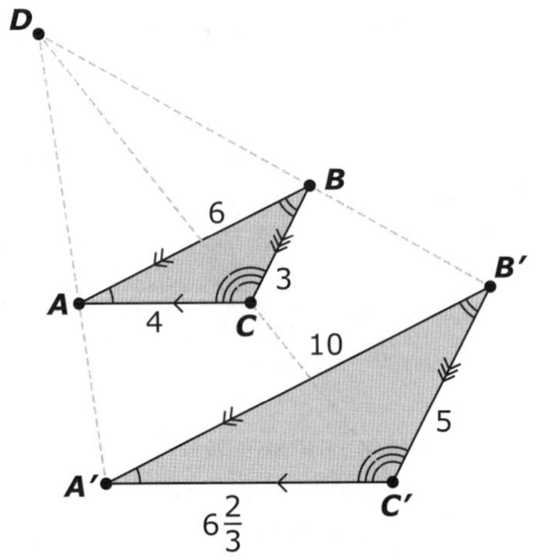

$\triangle ABC$ is similar to $\triangle A'B'C'$.

Dilations in the Coordinate Plane

A dilation in the coordinate plane with a scale factor of k centered at the origin can be written as $D_k(x, y) = (kx, ky)$.

My Notes

Summary *(continued)*

EXAMPLE: Dilation in the Coordinate Plane

Dilate $\triangle ABC$ with vertices A (−3, 4), B (2, 5), and C (4, −6) by a scale factor of $\frac{3}{2}$.

SOLUTION:

The coordinates of the image can be found using the rule $D_{\frac{3}{2}}(x, y) = (\frac{3}{2}x, \frac{3}{2}y)$.

Point A': $D_{\frac{3}{2}}(-3, 4) = (-4.5, 6)$

Point B': $D_{\frac{3}{2}}(2, 5) = (3, 7.5)$

Point C': $D_{\frac{3}{2}}(4, -6) = (6, -9)$

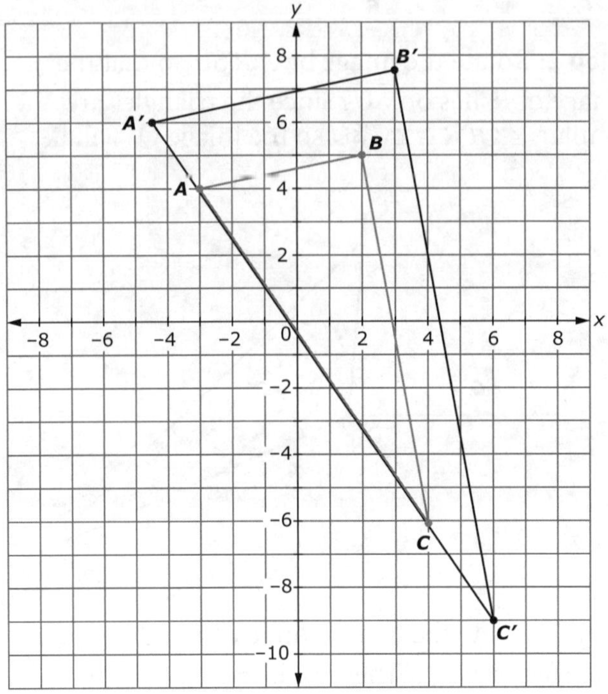

Dilation of 3 halves

A **similarity transformation** is a dilation or a combined transformation made up of a dilation and a rigid transformation.

If two figures are similar, then there is a combined transformation of rigid transformations and a dilation that maps one to the other.

My Notes

Summary *(continued)*

EXAMPLE: Mapping Similar Figures

In the figure, $\triangle PQR \sim \triangle STU$. Show a combined transformation that maps $\triangle PQR$ onto $\triangle STU$.

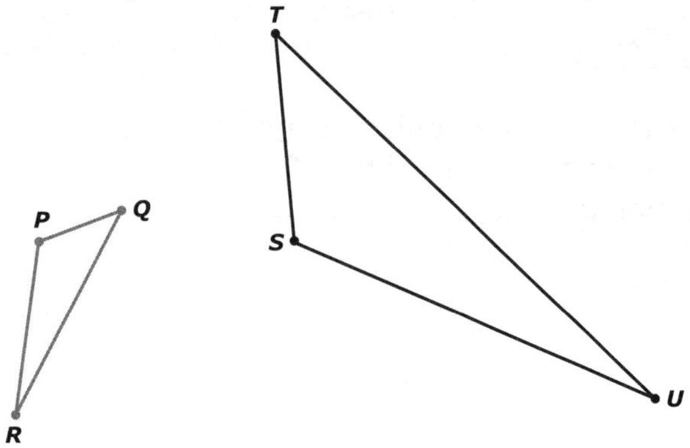

SOLUTION:

One possible mapping is shown.

Step 1: Translate $\triangle PQR$ along $\overline{PS}$ so that the image of P coincides with S.

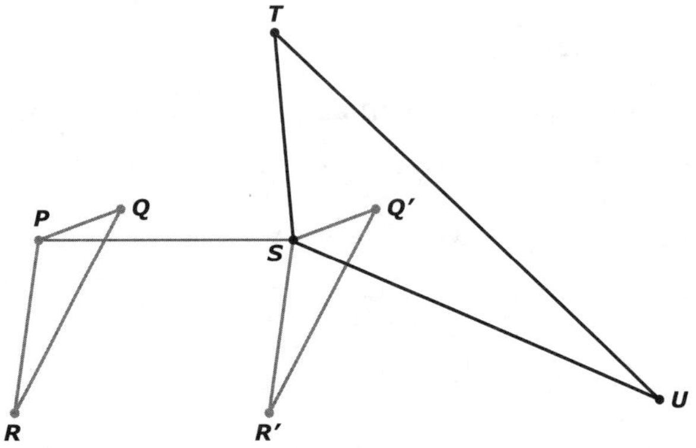

Step 2: Rotate the image by $\angle R'SU$ so that the image of R' lies on $\overline{SU}$. Since the triangles are similar, $\angle QPR \cong \angle TSU$ so the image Q'' will lie on $\overline{ST}$.

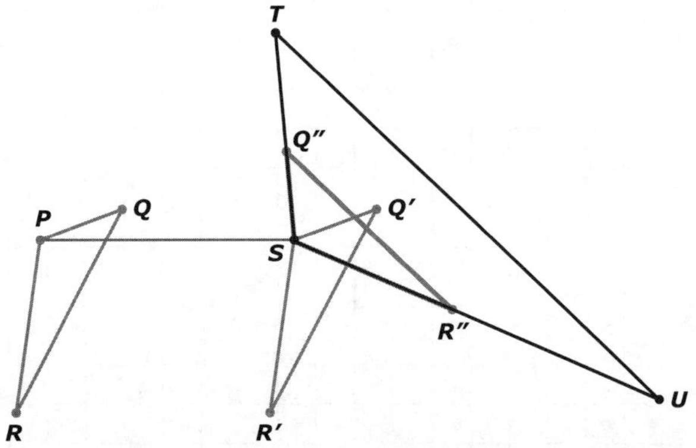

Step 3: Dilate the image with a center of S and a scale factor of $\frac{SU}{PR}$. The image coincides with $\triangle STU$.

My Notes

Practice Exercises

Review what you've learned using these practice problems. For practice problems with feedback, try the Coach and Play items in the Practice section online.

1. A triangle has vertices at coordinates (1, 5), (2, 3), and (−3, 4). Two vertices of a second similar triangle, with sides parallel to the sides of the first, are located at the coordinates (3, 4) and (6, −2). In the spaces below, enter the coordinates of the third vertex of the second triangle.

   The coordinates of the third vertex are (______, ______).

2. Thomas draws a segment with a length of 20 centimeters. The segment is dilated such that the length of the new segment is 8 centimeters.

   To find the scale factor, the lengths are written in the ratio ______.

   When simplified, the scale factor is ______.

3. A triangle has vertices at coordinates (4, 5), (−1, 3), and (5, 2). Two vertices of a second similar triangle, with sides parallel to the sides of the first, are located at coordinates (2, 2) and (−8, −2).

   The coordinates of the third vertex are ______.

4. Let a dilation be defined by a scale factor *s* and a center of dilation at the point (*x*, *y*). Each of the following represents a dilation of a polygon corresponding to the given values of *s*, *x*, and *y*.

   Which of the following represents a dilation that will produce a polygon congruent to the original? *Select all that apply.*

   A. $s = 0, x = 0, y = 0$

   B. $s = 0, x = 0, y = 1$

   C. $s = 0, x = 1, y = 0$

   D. $s = 1, x = 0, y = 0$

   E. $s = 1, x = 1, y = 2$

   F. $s = 1, x = 2, y = 1$

   G. $s = 2, x = 1, y = 1$

5. On the grid below, plot the figure that corresponds to a dilation of this five-sided polygon, with a scale factor of 0.75 and a center of dilation at (−4, −8).

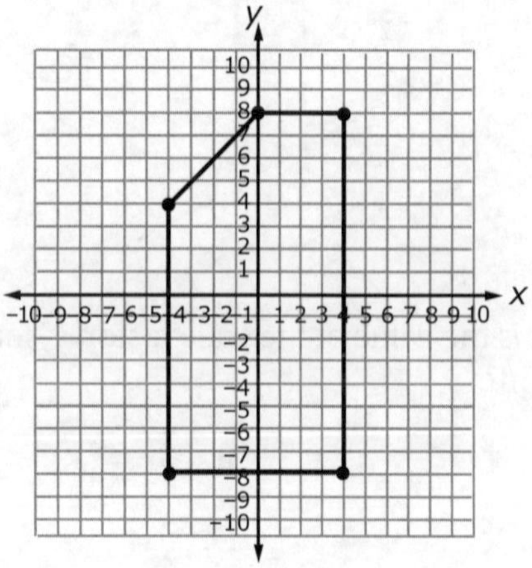

6. A polygon undergoes a series of five dilations, each with a scale factor of 1.3. A sixth dilation produces a polygon that is congruent to the original.

   Determine the scale factor for the sixth dilation.

   Round the scale factor to the nearest thousandth.

   Scale factor: __________

7. Yoshi is experimenting with an image-editing program on his computer. He rotates his picture clockwise by 60°, reflects it across the *y*-axis, moves it horizontally to the right by 153 pixels, increases the picture's magnification to 110%, and then moves it vertically downward by 82 pixels.

   Is the final image congruent to Yoshi's original? If not, why? *Choose the best response.*

   A. No; it underwent a reflection.

   B. No; it underwent a translation.

   C. No; it underwent a rotation.

   D. No; it underwent a dilation.

   E. Yes; the two are congruent.

Practice Exercises *(continued)*

8. The diagram shows $\overline{X'Y'} = D_{k,\,P}(\overline{XY})$.

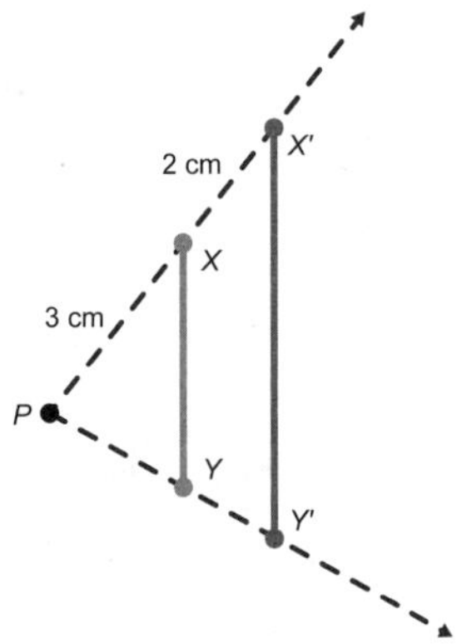

What is the value of the scale factor, k, for the dilation?

A. $\frac{2}{5}$ **B.** $\frac{3}{5}$

C. $\frac{2}{3}$ **D.** $\frac{3}{2}$

E. $\frac{5}{3}$ **F.** $\frac{5}{2}$

9. These two polygons are similar.

Determine the values of a, b, $m\angle C$, and $m\angle D$.

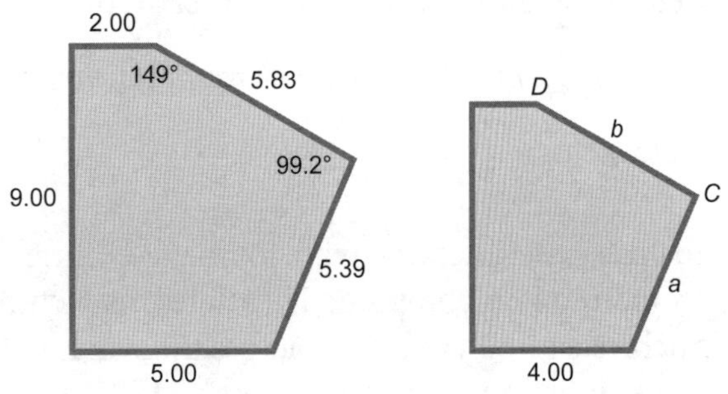

Round answers to two decimal places.

$a =$ ______ units

$b =$ ______ units

$m\angle C =$ ______ degrees

$m\angle D =$ ______ degrees

10. The graph shows a four-sided polygon on the coordinate plane.

The polygon experiences a dilation with a scale factor of 1.5 and with point A as the center of dilation.

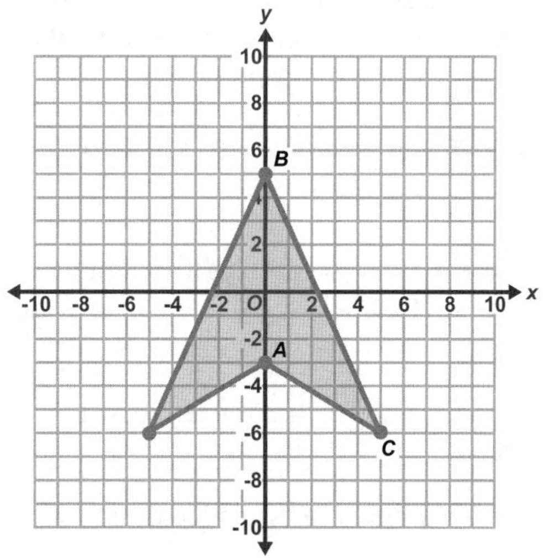

Determine the coordinates of vertices B' and C' for the dilated polygon.

Point B': ____________

Point C': ____________

Apply

Where Is the Third Vertex?

A triangle has vertices with coordinates $A(-2, 4)$, $B(-5, 0)$, and $C(3, -6)$. After a combined transformation, the image of the shortest side of $\triangle ABC$ has endpoints at $(4, -4)$ and $(10, 4)$.

How many possible images of $\triangle ABC$ are there in which the shortest side has endpoints at $(4, -4)$ and $(10, 4)$?

photo: Getty Images

How might an artist use coordinates to dilate a triangle in a mural?

- List the third vertex for each possible image.
- Describe a series of transformations that maps $\triangle ABC$ onto each possible image.
- Include the scale factor, and explain why the image is similar to the pre-image.
- Include a graph that shows all possible images of $\triangle ABC$.

Show what you've learned by completing the other performance tasks in the online Apply section.

Apply *(continued)*

Your answer to Apply will be assessed on the following criteria:

1. Determining the number of possible images of $\triangle ABC$ in which the shortest side has endpoints at (4, −4) and (10, 4), and listing the third vertex for each image
2. Describing a series of transformations that maps $\triangle ABC$ onto each possible image
3. Giving the scale factor dilation, and explaining why each image is similar to the pre-image
4. Graphing all possible images of $\triangle ABC$ in which the shortest side has endpoints at (4, −4) and (10, 4)

Scale / Criteria	4 Exceeds Criteria	3 Meets Criteria	2 Progressing to Criteria	1 Below Expectations	0 No Expectation
Image and Vertex	Correctly determines the number of possible images, and lists the coordinates of the third vertex for each image.	Determines the number of possible images, and lists the coordinates of the third vertex for each image with minor errors.	Determines the number of possible images, but only lists some of the coordinates of the third vertex for each image.	Determines the number of possible images, but only lists one possible set of coordinates of the third vertex for an image.	Does not determine the number of images or list possible coordinates for the third vertex.
Combined Transformations	Correctly describes a combined transformation that maps the triangle to each image.	Describes a combined transformation that maps the triangle to each image with minor errors.	Describes some of the possible combined transformations that map the triangle to each image.	Describes one possible combined transformation that maps the triangle to an image.	Does not describe a series of transformations for each image.
Scale Factor	Gives the correct scale factor, and explains why each image is similar to the pre-image.	Gives the correct scale factor, but explains why each image is similar to the pre-image with minor errors.	Gives the incorrect scale factor, or explains why each image is similar to the pre-image with significant logical errors.	Gives the incorrect scale factor, or does not explain why the images are similar to the pre-images.	Does not include scale factor or explain why any image is similar to the pre-image.
Graph	Correctly graphs all possible images of the given triangle.	Graphs all possible images of the given triangle with minor errors.	Graphs some of the possible images of the given triangle.	Graphs one of the possible images of the given triangle.	Does not graph images.

UNIT 5: Similarity

5.2 Prove Similarity Theorems

photo: Getty Images

Lesson Objectives

- Prove figures are similar using similarity transformations.
- Prove theorems about similar figures.
- Explain the criteria necessary to show that two triangles are similar.
- Prove and apply the triangle midsegment theorem.
- Prove the Pythagorean theorem.
- Solve mathematical problems by applying congruence and similarity criteria.

Essential Question

- How can similarity be verified and used to prove properties of geometric figures?

photo: Getty Images

Investigations

Modeling without a Blueprint

What measurements do you need to make a scale model? How can you prove a model is to scale?

Similarity Criteria

What's the least you have to know to determine if two triangles are congruent?

Putting Similarity to Use

How are side lengths of similar triangles related?

Pythagoras's New Clothes

It's as simple as building a kite! Prove the Pythagorean theorem with similar triangles.

Measuring the Unreachable

Use indirect measurement techniques involving similar triangles to measure the unreachable.

Key Vocabulary

scale factor

Discover

As you complete Engage and the investigations, record the most important ideas you've learned.

Engage	
Investigation 1	Investigation 2
Investigation 3	Investigation 4

Name ______________________________ Date __________

Check for Understanding

Prove Similarity Theorems: Investigation 1

1. In the figure below, $F \circ D_k(\triangle ABC) = \triangle FGH$, where F is a composition of rigid motion transformations and D is a dilation by a scale factor of k. What is the value of k?

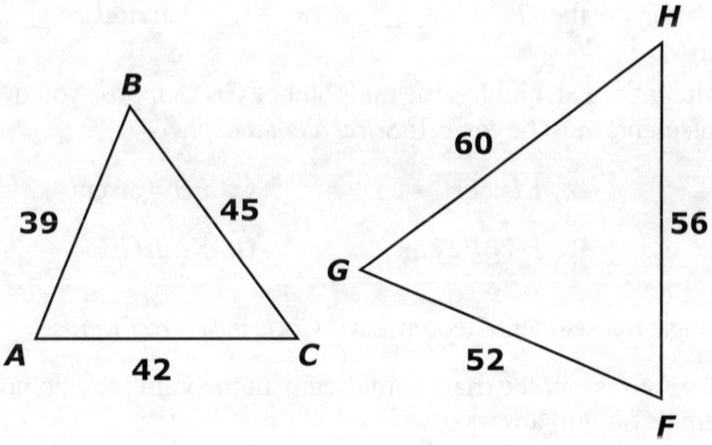

A. $\frac{7}{10}$ B. $\frac{3}{4}$ C. $\frac{4}{3}$ D. $\frac{10}{7}$

2. In class, Eileen learned about the AA similarity criteria, the SAS similarity criteria, and the SSS similarity criteria for triangles. She is wondering why she did not learn about the ASA similarity criteria and the AAS similarity criteria for triangles. *Circle the correct criteria below.*

   It was not necessary for Eileen to learn about the ASA similarity criteria and the AAS similarity criteria for triangles, because the [AA / SAS / SSS] similarity criteria already show that triangles meeting these two other sets of criteria are similar.

3. Juan believes that it is possible to show that two right triangles are similar by the HL similarity criteria. That is, if one pair of corresponding legs and the hypotenuses of two right triangles are in proportion, the triangles must be similar. To demonstrate, he drew two right triangles, $\triangle RST$ and $\triangle XYZ$, that have corresponding legs $\overline{RS}$ and $\overline{XY}$ and hypotenuses $\overline{RT}$ and $\overline{XZ}$. In his figure, $\frac{RS}{XY} = \frac{RT}{XZ}$. Is Juan correct in his belief? Explain.

Name ______________________________ Date ____________

Check for Understanding

Prove Similarity Theorems: Investigation 2

1. Samuel drew $\triangle ABC$ and then drew $\overleftrightarrow{DE}$, which intersects $\overline{AB}$ and $\overline{AC}$ at points D and E, respectively, and is parallel to $\overline{BC}$. *Fill in the blanks in the following sentence.*

   If $AD = 8$, $DB = 20$, and $AE = 6$, then $EC =$ ______, $AB =$ ______, and $AC =$ ______.

2. In $\triangle FGH$, K is the midpoint of $\overline{FH}$, and L is the midpoint of $\overline{GH}$. Suppose you drew $\overline{KL}$ and that it is 36 cm long. Which statements must be correct? *Select all that apply.*

   A. $GL = 18$ cm
   B. $\overline{KL} \parallel \overline{FH}$
   C. $FH = 36$ cm
   D. $\overline{FG} \parallel \overline{KL}$
   E. $\overline{FH} \parallel \overline{GH}$
   F. $FG = 72$ cm
   G. $\overline{KL} \perp \overline{GH}$
   H. $HK = 18$ cm

3. Consider what the triangle midsegment theorem says. *Circle the correct words.*

   Counterexamples to the converse of the triangle midsegment theorem [do / do not] exist, because the converse of the theorem [is / is not] always true.

4. As shown in the figure below, Valerie used the midpoints of $\triangle NPQ$ to divide the triangle into the following four smaller triangles: $\triangle NRT$, $\triangle RPS$, $\triangle STR$, and $\triangle TSQ$.

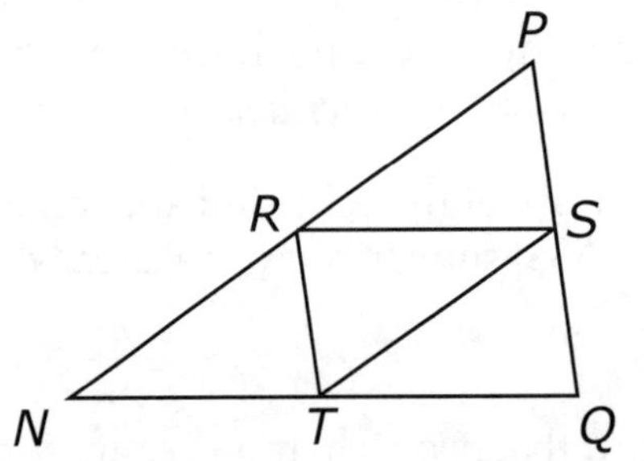

   Valerie claims that the four smaller triangles are all congruent to each other, and also that they are all similar to $\triangle NPQ$. Is she correct? Explain your answer.

Name ______________________________ Date ____________

Check for Understanding

Prove Similarity Theorems: Investigation 3

1. Harry, Rosa, and Finn are students working together in a group to prove the Pythagorean theorem. Each dilated a right triangle with a shorter leg of length a, a longer leg of length b, and a hypotenuse of length c. Harry used a scale factor of a, Rosa used a scale factor of b, and Finn used a scale factor of c. *Fill in the table with the lengths of the sides of the images of Harry's triangle, Rosa's triangle, and Finn's triangle.*

	Length of Shorter Leg	Length of Longer Leg	Length of Hypotenuse
Harry's Image			
Rosa's Image			
Finn's Image			

2. Two of the students in question 1 put the images of their triangles together so that a side of one of the images coincided with a side of the other image. The figure below shows how the images of the two triangles were put together.

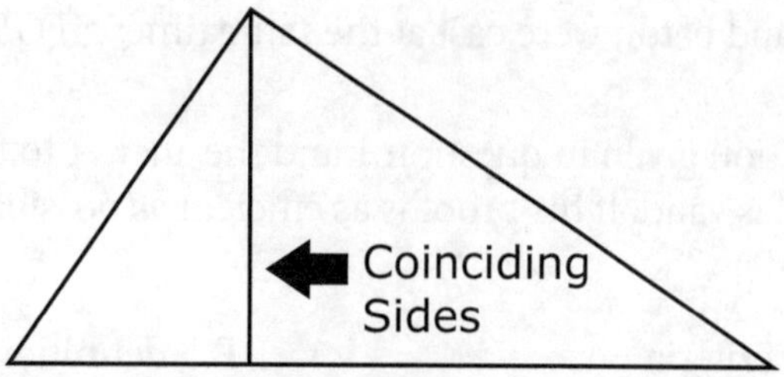

Which of these are the coinciding sides in the figure?

A. the shorter leg of Finn's image and the longer leg of Rosa's image

B. the shorter leg of Finn's image and the longer leg of Harry's image

C. the shorter leg of Harry's image and the longer leg of Rosa's image

D. the shorter leg of Rosa's image and the longer leg of Harry's image

3. Explain why the figure in question 2 proves the Pythagorean theorem.

Name ________________________________ Date ______________

Check for Understanding

Prove Similarity Theorems: Investigation 4

1. Norbert is attempting to determine the height of a tent in the shape of a square pyramid by using a straight baton perpendicular to the ground and the shadow method. In the figure below, $\triangle ABC$ is the tent, $\overline{FG}$ is the baton, $\overline{CE}$ is the tent's shadow, $\overline{FH}$ is the baton's shadow, and $\overline{BD} \perp \overline{AE}$. *Fill in the blank in the sentence that follows.*

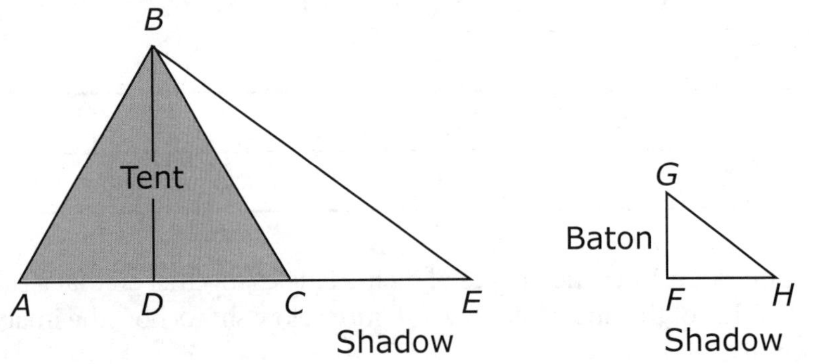

If the shadows of the tent and baton were cast at the same time, $\angle FGH \cong$ ____________.

2. Suppose that with the information given in question 1 and the answer to the question, Norbert wants to prove that the shadow method is valid. If his proof is as efficient as possible, which of the following would not be a reason for one of the steps?

A. AA triangle similarity criteria

B. definition of perpendicular lines

C. right angle congruence theorem

D. transitive property of congruence

3. At a particular time during the day, Norbert from question 1 found the baton's shadow to measure 8 in. and the tent's shadow to measure 64 in. However, he was interrupted before he could calculate the tent's height, so he came back later and found the baton's shadow to measure 4 in. *Circle the correct word(s) in the sentence below.*

When Norbert found the baton's shadow to measure 4 in., the tent's shadow measured [less than / exactly / more than] 32 in.

4. Given the information from question 3 and that $CD = 56$ in., what is the tent's height if the baton is 14 in. long? Explain how you got your answer.

Summary

Before you attempt the Practice Exercises, review what you've learned.

Triangle Similarity Criteria

In order to prove that two triangles are similar, you can use the **AA triangle similarity criteria**.

Angle-angle (AA) triangle similarity criteria: If two angles of one triangle are congruent to two angles of another triangle, then the triangles are similar.

Side-side-side (SSS) triangle similarity criteria: If the lengths of the corresponding sides of two triangles are proportional, then the triangles are similar.

Side-angle-side (SAS) triangle similarity criteria: If an angle of one triangle is congruent to an angle of another triangle and the corresponding sides forming those angles are proportional, then the triangles are similar.

We can use a dilation to show that the AA triangle similarity criteria are sufficient to prove that triangles are similar.

In the diagram, $\angle A \cong \angle D$ and $\angle B \cong \angle E$. We can show that $\triangle ABC \sim \triangle DEF$ based on this information.

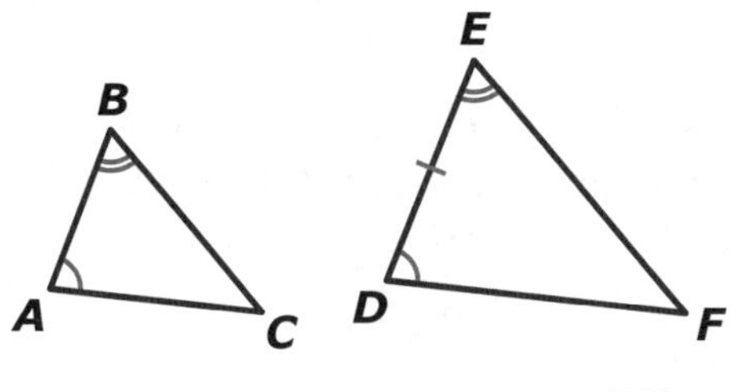

Let's dilate $\triangle ABC$ so that the image of $\overline{AB}$ is congruent to $\overline{DE}$.

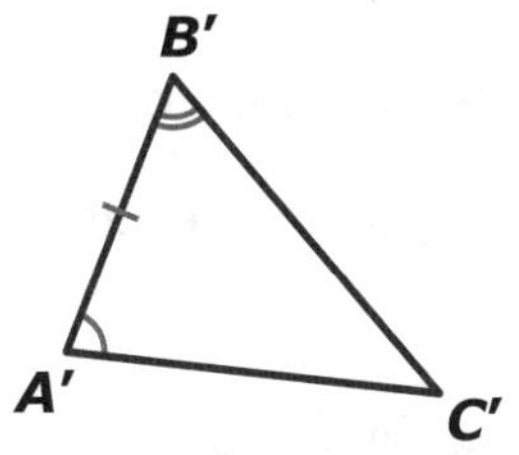

Because $\triangle A'B'C'$ is a dilation of $\triangle ABC$, $\angle A \cong \angle A'$ and $\angle B \cong \angle B'$. By the transitive property of congruence, we know that $\angle A' \cong \angle D$ and $\angle B' \cong \angle E$. We also know that $\overline{A'B'} \cong \overline{DE}$. So, by the ASA triangle congruence theorem, $\triangle A'B'C' \cong \triangle DEF$.

Triangle $A'B'C'$ is a dilation of $\triangle ABC$, so $\triangle A'B'C' \sim \triangle ABC$. We showed that $\triangle A'B'C' \cong \triangle DEF$, which means that $\triangle A'B'C'$ is also similar to $\triangle DEF$. This means there is a composition of transformations so that $F \circ D(\triangle ABC) = \triangle DEF$. Therefore, $\triangle ABC \sim \triangle DEF$.

Using Triangle Similarity to Prove Other Theorems

Triangle proportionality theorem: If a line parallel to one side of a triangle intersects the other two sides, then the line divides those sides proportionally.

My Notes

Summary *(continued)*

Converse of the triangle proportionality theorem: If a line divides two sides of a triangle proportionally, then the line is parallel to the third side of the triangle.

A **midsegment** of a triangle is a segment that connects the midpoints of two sides of the triangle.

Triangle midsegment theorem: A segment joining the midpoints of two sides of a triangle is parallel to the third side and half its length.

You can also prove the Pythagorean theorem by using triangle similarity.

Using Similar Triangles to Make Indirect Measurements

You can use the properties of similar triangles to help you make indirect measurements.

My Notes

EXAMPLE: Indirect Measurement with Similar Triangles

Use the diagram to find the distance d in meters across the river.

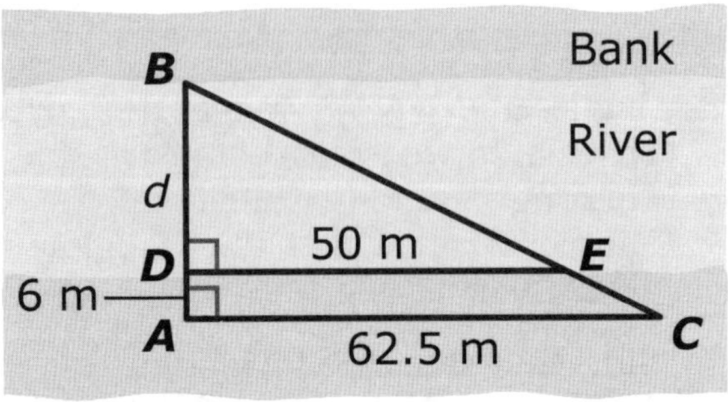

SOLUTION:

Step 1: First, we need to show that the triangles are similar.

Angle $\angle BAC$ and $\angle BDE$ are marked as right angles, so $\angle BAC \cong \angle BDE$ by the right angle congruence theorem. We also know that $\angle B \cong \angle B$ by the reflexive property of congruence.

So, we can conclude that $\triangle ABC \sim \triangle DBE$ by the AA triangle similarity criteria because we have shown that two angles of $\triangle ABC$ are congruent to two angles of $\triangle DBE$.

Step 2: Now we can use the properties of similar triangles to find d.

Corresponding side lengths of similar triangles are proportional.

$$\frac{AB}{DB} = \frac{AC}{DE}$$

Substitute d for DB, 62.5 for AC, and 50 for DE. By the segment addition postulate, $AB = 6 + d$, so substitute $6 + d$ for AB.

$$\frac{6+d}{d} = \frac{62.5}{50}$$

$\frac{6+d}{d} = \frac{62.5}{50}$	Simplify the right side of the equation.
$\frac{6+d}{d} = 1.25$	Multiply both sides by d.
$6 + d = 1.25d$	Subtract d from each side.
$6 = 0.25d$	Divide both sides by 0.25.
$24 = d$	

The distance across the river is 24 meters.

Practice Exercises

Review what you've learned using these practice problems. For practice problems with feedback, try the Coach and Play items in the Practice section online.

1. Select all of the statements that are both sufficient to prove a pair of triangles are similar and necessarily true if the triangles are given to be similar. *Select all that apply.*

A. Three pairs of corresponding sides are congruent.

B. Three pairs of corresponding sides are proportional.

C. Two pairs of corresponding sides are proportional, and the included angles are congruent.

D. One pair of corresponding sides and one pair of corresponding adjacent angles are congruent.

E. One pair of corresponding sides and the pair of opposite angles are congruent.

F. Two pairs of corresponding angles are congruent.

G. Two pairs of corresponding angles are proportional.

H. Three pairs of corresponding angles are congruent.

2. Consider ΔQRS with right $\angle S$ and point T located at the midpoint of $\overline{SQ}$. Suppose $SQ = 23.1$ and $QR = 39.47$. Points A, B, C, D, and E located on segment $\overline{RS}$. Additionally, suppose we have $SA = 16.0$; $SB = 16.3$; $SC = 19.8$; $SD = 23.9$; and $SE = 31.9$.

Based on these segment lengths, which segment is parallel to $\overline{QR}$?

A. $\overline{TA}$

B. $\overline{TB}$

C. $\overline{TC}$

D. $\overline{TD}$

E. $\overline{TE}$

3. One sunny day, Margaret raises a flag on a flagpole. She finds that her shadow is exactly 10 feet long, which is twice her height. She then discovers that the shadow of the flagpole is 30 feet long.

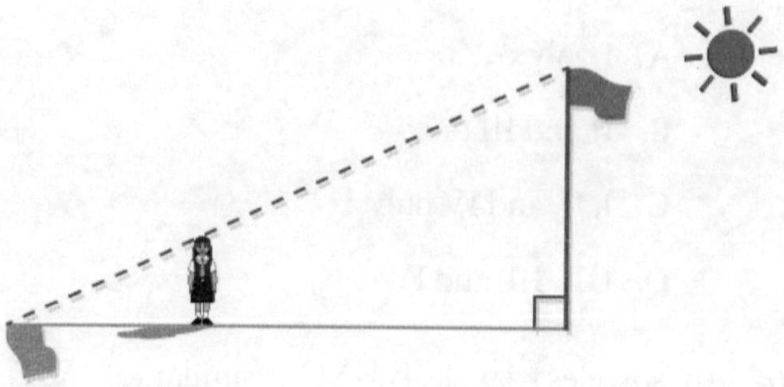

How tall is the flagpole?

A. 5 ft

B. 10 ft

C. 15 ft

D. 30 ft

E. 60 ft

4. The following figure shows triangle ΔABC with side lengths $AB = 10$, $BC = 8$, and $CA = 5$. $\overline{DE}$ is constructed to be parallel to $\overline{AB}$ and to originate at point D, the midpoint of $\overline{AC}$.

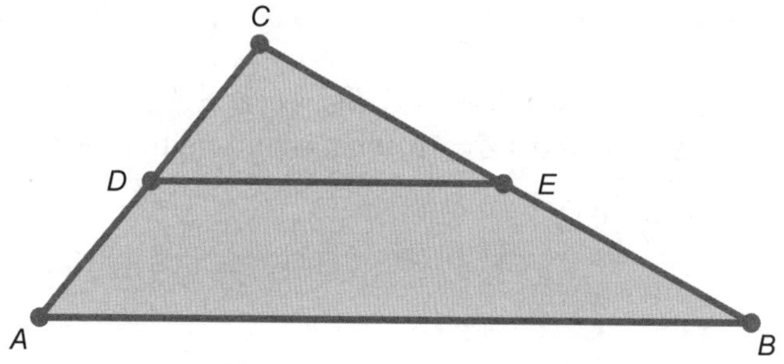

Write the lengths for the segments listed below.

$BE =$ ___________

$ED =$ ___________

$DA =$ ___________

Practice Exercises *(continued)*

5. Which of the following are triangle similarity theorems?

   I. SSS (side-side-side)

   II. SAS (side-angle-side)

   III. SS (side-side)

   IV. AA (angle-angle)

   A. I only

   B. II and III only

   C. I, II, and IV only

   D. I, II, III, and IV

6. An isosceles triangle, $\triangle MNO$, is similar to $\triangle RST$. $m\angle N = 20°$. Which of the following could be the measure of $\angle T$?

   I. 70°

   II. 80°

   III. 130°

   IV. 140°

   V. 150°

   A. I only

   B. IV only

   C. II or IV only

   D. III or V only

7. $\triangle ABC \sim \triangle XYZ$. $AB = 3$, $BC = 6$, and $AC = 4$. If $XY = 8$ and $YZ = 16$, what is the length of XZ?

   A. 9.67

   B. 10.67

   C. 11.67

   D. 12.67

8. The new school gym is under construction, and the space is filled with wheelbarrows and ladders.

   A 10-foot tall ladder makes an angle of approximately 75° with the floor. It rests against a wall, touching it at a point 9.7 feet up from the floor.

   A second ladder is twice as long and rests against the wall, touching it at a point 19.4 feet up from the floor. Which of the following angles does the second ladder make with the floor?

   A. 15°

   B. 37.5°

   C. 75°

   D. 90°

   E. 150°

9. Given $\triangle ABC$, $\overleftrightarrow{JK}$ is parallel to $\overline{BC}$ and intersects sides $\overline{AB}$ and $\overline{AC}$ at points J and K, respectively. If $AJ = 4$ ft, $KC = 9$ ft, and $JB = AK = x$ ft, determine the value of x. *Round your answer to the nearest tenth.*

   $x =$ ____________ ft

10. Sela's bridge-construction firm needs to measure the distance across a canyon. She looks at a large tree directly across the canyon. Then, she walks 70 feet in a straight line along the edge, plants a stake in the ground, and walks another 70 feet in a straight line along the edge. Then, Sela walks away at a right angle to the canyon edge until she can see that the tree lines up with her marked spot. She then measures the distance from her current location to the stake she placed in the ground. It is 250 feet.

    Sela estimates that it is about _______ feet across the canyon at the point of the large tree. Give your answer to the nearest 10 feet.

Apply

How Tall Are Those Towering Objects?

At a museum of natural history, there might be "Do not touch" signs everywhere to protect the unique artifacts. The information plaque for something like dinosaur fossils may give an average height, but measuring the actual size of a large item is difficult even when you can reach it. Think back to when you worked as part of a group to measure a tall object at your school.

Some objects are more out of reach than others.

The World Is Your Museum: View this video segment to see how to measure objects located in the Natural History Museum.

Choose something that is outside and taller than you can reach. How tall is it? Instead of using a mirror, recruit the help of the sun to find its height. Create a detailed diagram to show how you determined your measurements. Does the time of day affect the calculations? Explain why or why not.

Show what you've learned by completing the other performance tasks in the online Apply section.

Apply *(continued)*

Your answer to Apply will be assessed on the following criteria:

1. Creating a diagram that represents the problem
2. Writing the correct equation and solving for the height
3. Explaining how triangle similarity can be used to solve the problem
4. Explaining if the time of day affects the calculations and supporting your answer mathematically

Scale / Criteria	4	3	2	1	0
	Exceeds Criteria	Meets Criteria	Progressing to Criteria	Below Expectations	No Expectation
Diagram	Correctly creates a diagram to represent the problem and labels all known and unknown information.	Creates a diagram to represent the problem and labels all known and unknown information, but labels contain a minor error.	Creates a diagram to represent the problem, but incorrectly labels the diagram.	Creates a diagram, but it is missing information and does not include labels.	Does not submit a diagram.
Equation	Reports a plausible personal height, correctly sets up an equation, and finds a solution that is reasonable.	Reports a personal height, sets up an equation, and finds a solution, but the solution may be slightly off due to a minor error.	Reports a plausible personal height, but sets up an incorrect equation, so the solution is not reasonable.	Sets up an incorrect equation without using personal height, and the solution is not reasonable.	Does not write an equation to solve the problem.
Explanation	Correctly explains how triangle similarity can be used to solve the problem.	Explains how triangle similarity can be used to solve the problem with a minor error.	Explains how triangle similarity can be used to solve the problem, but with significant errors.	Attempts to explain how triangle similarity can be used to solve the problem, but the explanation is incorrect.	Does not submit an explanation.
Work and Reasoning	States whether the time of day affects the outcome of the calculations and supports the answer mathematically.	States whether the time of day affects the outcome of the calculations and supports the answer mathematically, with one error.	States whether the time of day affects the outcome of the calculations and supports the answer mathematically, with two or more errors.	States whether the time of day affects the outcome of the calculations, but does not support answer mathematically.	Does not state whether time of day affects the outcome.

UNIT 5: Similarity

5.3 Apply Similarity Theorems

photo: Getty Images

Lesson Objectives

- Understand and apply the effect of dilation on perimeter, area, and volume measurements.
- Solve real-world problems by applying congruence and similarity criteria.

Essential Question

- What relationships exist between different measurements for similar solids?

photo: Getty Images

Investigations

Scale Models

Explore similarity in 3-D. What effect does similarity have on weight?

3-D Similarity

Examine the measurements and make conjectures about the similarity of pyramids and other 3-D figures.

Container Gardening

Scale your garden to size. How will dilations affect the perimeter, area, and volume?

Key Vocabulary

scale model

Discover

As you complete Engage and the investigations, record the most important ideas you've learned.

Engage	
Investigation 1	Investigation 2

Name ______________________ Date __________

Check for Understanding

Apply Similarity Theorems: Investigation 1

1. Calvin says that all cubes are similar to each other, while Isaac says that all square pyramids are similar to each other. Who is correct?

   A. only Calvin **B.** only Isaac

   C. both Calvin and Isaac **D.** neither Calvin nor Isaac

2. Consider what it means for two 3-D objects to be similar. *Circle the correct words.*

   If two 3-D objects are not congruent, they [can / cannot] be similar, and if they are not similar, they [can / cannot] be congruent.

3. The figure shows the nets of two similar right rectangular prisms. *Fill in the blanks.*

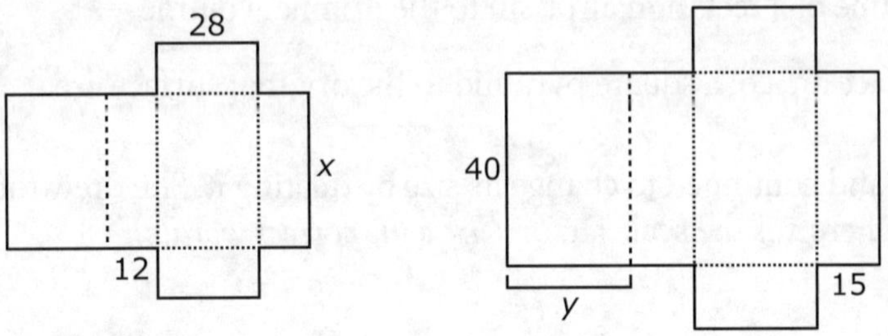

   $x =$ ______________ $y =$ ______________

4. One of the interior angles of the base of a triangular pyramid measures 120°. Which statements are correct about the base of a similar pyramid? *Select all that apply.*

   A. It may have an interior angle measuring 40°.

   B. It may have an interior angle measuring 80°.

   C. It must have an interior angle measuring 20°.

   D. It must have an interior angle measuring 60°.

   E. It must have an interior angle measuring 120°.

5. Two right prisms with the same height each have bases on their tops and bottoms that are equilateral triangles. Do the prisms have to be similar to each other? Explain.

Name ____ Date ______________

Check for Understanding

Apply Similarity Theorems: Investigation 2

1. Suppose a regular pentagon with a perimeter of 34 inches was dilated by a scale factor of 0.25. *Fill in the blanks in the following sentences. Do not round your answers.*

   The pentagon's side length before the dilation was ______ in. The new side length after the dilation was ______ in., and its perimeter after the dilation was ______ in.

2. Which ratios are equal to 64 when the scale factor is 8? *Select all that apply.*

   A. the scaled area of a triangle to the original area

   B. the scaled volume of a cube to the original volume

   C. the scaled perimeter of an octagon to the original perimeter

   D. the scaled volume of a rectangular prism to the original volume

   E. the scaled surface area of a square pyramid to the original surface area

3. Iris began with a solid and continued to change its size by dilating it. She drew the graph shown (without units for y), where x is the scale factor. *Circle the correct word(s).*

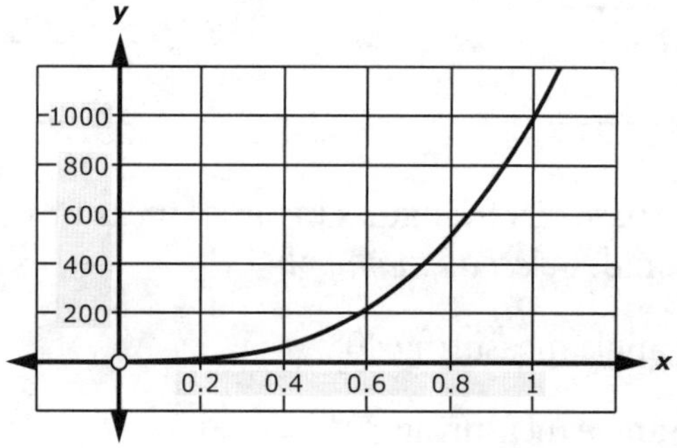

   In Iris's graph, y could be the solid's [area / perimeter / surface area / volume].

4. When a 3-D shape with a volume of 336 cm^3 was dilated by using a certain scale factor, its volume increased by 2,352 cm^3. If the shape's surface area was 292 cm^2 before the dilation, by how many square centimeters did its surface area increase? Explain.

Summary

Before you attempt the Practice Exercises, review what you've learned.

The chart lists some of the properties of similar three-dimensional figures.

Properties of Similar Figures in Three Dimensions
• The figures have the same number and type of faces.
• Corresponding faces are similar to each other.
• Ratios of corresponding edge lengths are proportional.
• Corresponding angles are congruent.

EXAMPLE: Identifying Similar Figures in Three Dimensions

Determine whether each pair of three-dimensional figures is similar.

Part A: Both figures are right rectangular prisms.

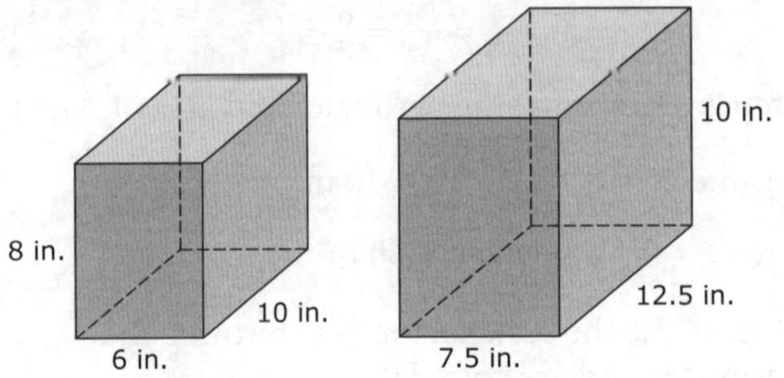

Prism 1 Prism 2

SOLUTION:

All faces are rectangles, which means that all corresponding angles are congruent because all corresponding angles are right angles.

Compare the ratios of the side lengths of corresponding faces.

Faces	Ratio of Length to Width Prism 1	Ratio of Length to Width Prism 2
Front and back	$\frac{8}{6} = \frac{4}{3}$	$\frac{10}{7.5} = \frac{20}{15} = \frac{4}{3}$
Left and right	$\frac{10}{8} = \frac{5}{4}$	$\frac{12.5}{10} = \frac{25}{20} = \frac{5}{4}$
Top and bottom	$\frac{10}{6} = \frac{5}{3}$	$\frac{12.5}{7.5} = \frac{25}{15} = \frac{5}{3}$

For each pair of corresponding faces, the corresponding side lengths are proportional.

So, the faces of Prism 1 are similar to the corresponding faces of Prism 2, which means that the prisms themselves are similar.

Part B: Both figures are triangular prisms.

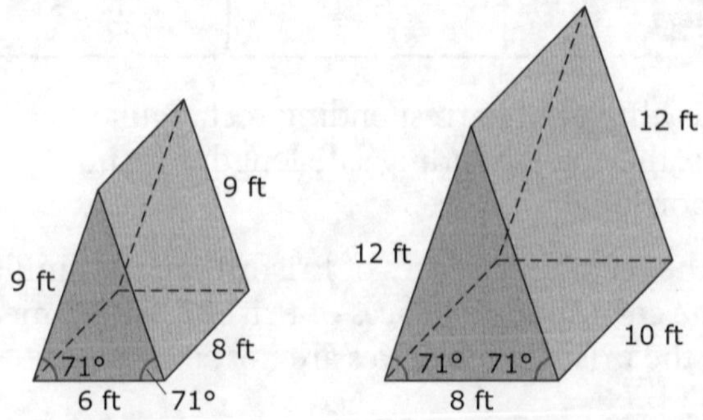

Prism 1 Prism 2

SOLUTION:

The triangular faces each have two angles that measure 71°, so the triangular faces are similar based on the AA triangle similarity theorem.

The corresponding angles of the rectangular faces are congruent because all corresponding angles are right angles.

My Notes

Summary *(continued)*

Compare the ratios of the side lengths of corresponding rectangular faces.

Faces	Ratio of Length to Width Prism 1	Ratio of Length to Width Prism 2
Left and right	$\frac{9}{8}$	$\frac{12}{10} = \frac{6}{5}$
Bottom	$\frac{8}{6} = \frac{4}{3}$	$\frac{10}{8} = \frac{5}{4}$

For each pair of corresponding rectangular faces, the corresponding side lengths are not proportional.

So, the rectangular faces of Prism 1 are not similar to the corresponding faces of Prism 2, which means that the prisms themselves are not similar.

Perimeter, Area, Surface Area, and Volume

Let k represent the scale factor that relates two similar figures. The chart shows how the perimeters, areas, surface areas, and volumes of the figures are related to the scale factor.

Measurements and Similar Figures
• The ratio of the perimeters is equal to the scale factor, k. • The ratio of the areas or surface areas is equal to the square of the scale factor, k^2. • The ratio of the volumes is equal to the cube of the scale factor, k^3.

My Notes

EXAMPLE: Applying the Scale Factor in Two Dimensions

$\triangle ABC$ is similar to $\triangle DEF$. Use the scale factor to determine the perimeter and area of $\triangle DEF$.

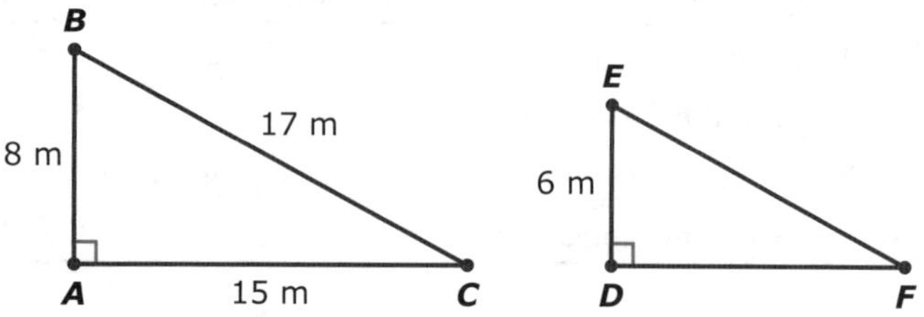

SOLUTION:

Step 1: Determine the scale factor for the similar figures.

The scale factor is equal to the ratio of corresponding side lengths.

$$k = \frac{DE}{AB} = \frac{6}{8} = \frac{3}{4}$$

Step 2: Determine the perimeter and area of $\triangle ABC$.

Perimeter = 8 + 17 + 15 = 40 m

Area $= \frac{1}{2}bh = \frac{1}{2} \cdot 15 \cdot 8 = 60 \text{ m}^2$

Step 3: Use the scale factor to determine the perimeter and area of $\triangle DEF$.

$$\text{Perimeter } \triangle DEF = k \cdot \text{Perimeter } \triangle ABC$$
$$= \frac{3}{4} \cdot 40 = 30 \text{ m}$$

$$\text{Area } \triangle DEF = k^2 \cdot \text{Area } \triangle ABC$$
$$= \left(\frac{3}{4}\right)^2 \cdot 60 = \frac{9}{16} \cdot 60 = 33.75 \text{ m}^2$$

Step 4: Check your answer.

Use the scale factor to find EF and DF.

$$\frac{EF}{BC} = k$$
$$\frac{EF}{17} = \frac{3}{4}$$
$$EF = 12.75 \text{ m}$$

$$\frac{DF}{AC} = k$$
$$\frac{DF}{15} = \frac{3}{4}$$
$$DF = 11.25 \text{ m}$$

Summary *(continued)*

Now, use *DE*, *EF*, and *DF* to find the perimeter and area of $\triangle DEF$.

Perimeter = 6 + 12.75 + 11.25 = 30 m

Area $= \frac{1}{2}bh = \frac{1}{2} \cdot 11.25 \cdot 6 = 33.75 \text{ m}^2$

So, $\triangle DEF$ has a perimeter of 30 meters and an area of 33.75 square meters.

EXAMPLE: Applying the Scale Factor in Three Dimensions

The right rectangular prisms shown are similar. Use the scale factor to determine the surface area and volume of Prism 2.

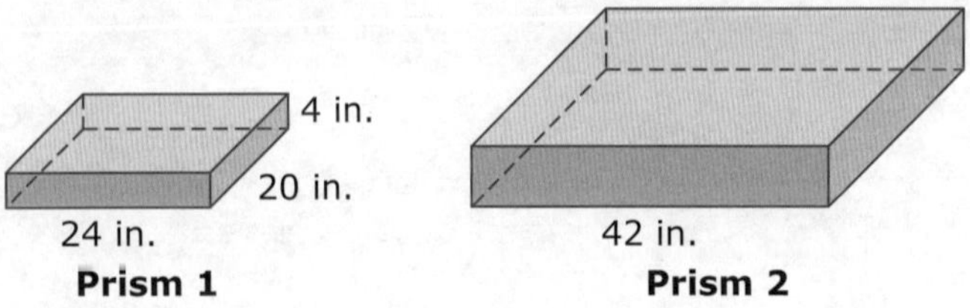

SOLUTION:

Step 1: Determine the scale factor for the similar figures.

The scale factor is equal to the ratio of corresponding edge lengths.

$$k = \frac{\text{length of Prism 2}}{\text{length of Prism 1}} = \frac{42}{24} = \frac{7}{4}$$

Step 2: Determine the surface area and volume of Prism 1.

$$\begin{aligned}\text{Surface area} &= 2lw + 2wh + 2lh \\ &= 2(24 \cdot 20) + 2(20 \cdot 4) + 2(24 \cdot 4) \\ &= 1{,}312 \text{ in.}^2\end{aligned}$$

$$\begin{aligned}\text{Volume} &= lwh \\ &= 24 \cdot 20 \cdot 4 = 1{,}920 \text{ in.}^3\end{aligned}$$

Step 3: Use the scale factor to determine the surface area and volume of Prism 2.

$$\begin{aligned}\text{Surface area Prism 2} &= k^2 \cdot \text{Surface area Prism 1} \\ &= \left(\frac{7}{4}\right)^2 \cdot 1{,}312 = \frac{49}{16} \cdot 1{,}312 \\ &= 4{,}018 \text{ in.}^2\end{aligned}$$

$$\begin{aligned}\text{Volume Prism 2} &= k^3 \cdot \text{Volume Prism 1} \\ &= \left(\frac{7}{4}\right)^3 \cdot 1{,}920 = \frac{343}{64} \cdot 1{,}920 \\ &= 10{,}290 \text{ in.}^3\end{aligned}$$

Step 4: Check your answer.

Use the scale factor to find the width and height of Prism 2.

$$\frac{\text{width Prism 2}}{\text{width Prism 1}} = k \qquad \frac{\text{height Prism 2}}{\text{height Prism 1}} = k$$

width Prism 2 = 35 in. height Prism 2 = 7 in.

Now, use the length, width and height of Prism 2 to find its surface area and volume.

$$\begin{aligned}\text{Surface area} &= 2lw + 2wh + 2lh \\ &= 2(42 \cdot 35) + 2(35 \cdot 7) + 2(42 \cdot 7) \\ &= 4{,}018 \text{ in.}^2\end{aligned}$$

$$\begin{aligned}\text{Volume} &= lwh \\ &= 42 \cdot 35 \cdot 7 = 10{,}290 \text{ in.}^3\end{aligned}$$

So, Prism 2 has a surface area of 4,018 square inches and a volume of 10,290 cubic inches.

A scale model is similar to the figure it represents.

My Notes

Summary *(continued)*

EXAMPLE: Solving Problems Involving Similar Figures

The swept area of a wind turbine is the area of the circle traced by the tips of its blades. An engineer makes a scale model of a wind turbine using a scale of 1 : 200. The swept area of the model is 0.31 m^2. What will be the swept area of the actual turbine once it is constructed?

Scale Model of a Wind Turbine

photo: Getty Images

The three blades of a wind turbine are attached to a central hub. The tips of the blades move in a circle about the hub as they rotate.

Step 1: Use the scale of the model to find the scale factor.

k = length in model corresponding length in actual turbine = 1200

Step 2: Use the scale factor to determine the swept area A of the actual turbine.

$$0.31 = \left(\frac{1}{200}\right)^2 \cdot A$$

$$0.31 = \frac{1}{40{,}000} \cdot A$$

$$12{,}400 = A$$

The actual turbine will have a swept area of 12,400 square meters (which is more than 2 football fields).

My Notes

Practice Exercises

Review what you've learned using these practice problems. For practice problems with feedback, try the Coach and Play items in the Practice section online.

1. The diagram shows two similar pyramids. The larger one has a volume of 2,197 cubic centimeters. The smaller one has a volume of 729 cubic centimeters.

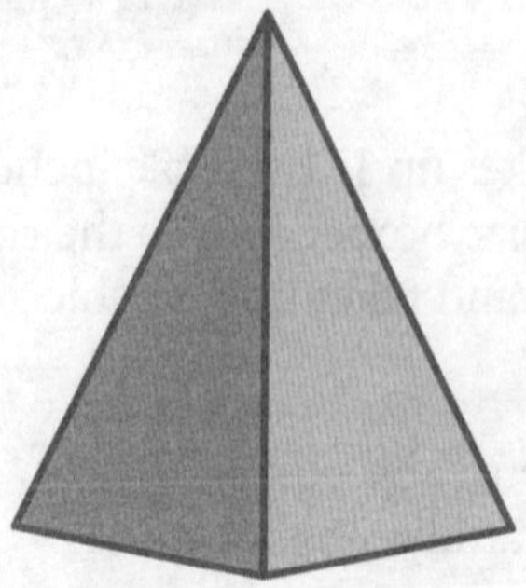

What is the ratio of their heights?

A. 1.5 : 1

B. 1 : 3

C. 13 : 9

D. 169 : 81

E. 2,197 : 729

2. Laurie is attending a French pastry school. She generally uses a large cone-shaped bag to squeeze icing onto her cakes.

Today, another student was using the larger bag, so Laurie is using a similar cone-shaped bag that is half as big in each dimension.

How many times will Laurie need to fill the smaller bag to decorate the same size cake that the larger bag could do with one filling?

Laurie will need to fill the smaller bag ________ times.

3. Marty is making props for the school play that are right cylinders. He originally designs the cylinders to be 5 feet tall, but the cost of the fabric to wrap around the sides of the cylinders will be too expensive. The play's producer asks Marty to reduce the dimensions proportionally so that the cylinders are 2 feet tall.

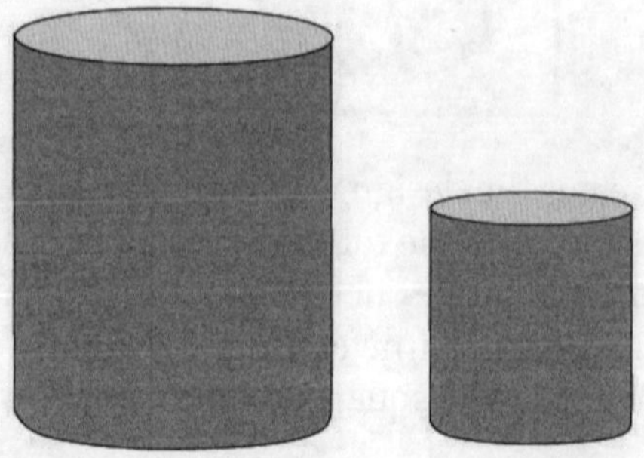

By what percentage should Marty multiply his estimate for the amount of the fabric in order to give the producer an accurate idea of the new cost?

The smaller cylinder's surface area is ________ % of the surface area of the larger cylinder.

4. A sculpture is made of two similar cones with parallel bases, joined at their tips. The sculpture is 15 inches tall, and its base is 12 inches in diameter.

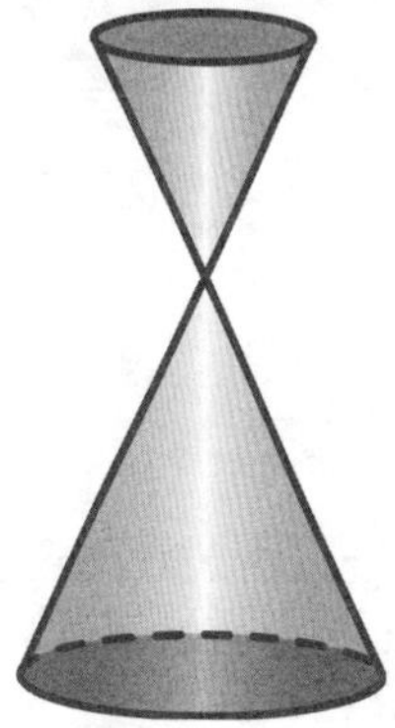

If the ratio of the volumes of the two cones is 8 : 27, how many inches from the base of the sculpture is the meeting point of the cones?

The point is ________ inches above the base of the sculpture.

Practice Exercises *(continued)*

5. Jamie is designing a model for art class. The model is a cube-shaped box with a circular image on one face of the box.

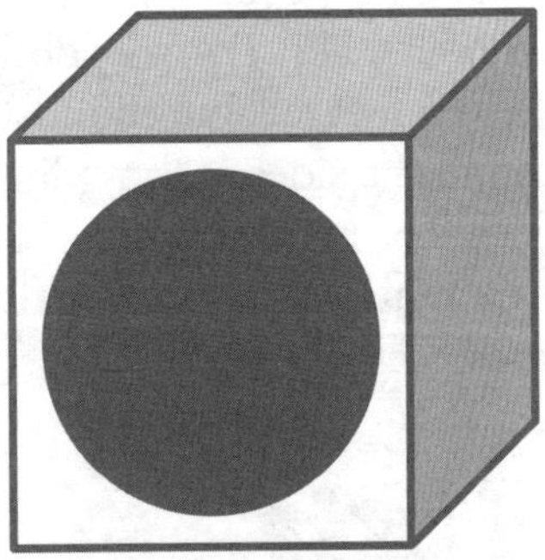

She creates the design with a smaller scale model before building the full-size piece. The smaller scale model of the cube has edge lengths each measuring 4 centimeters, and the circular image has an area of 9π square centimeters.

How large will the cube need to be in the full-size art piece so that the circular image will have an area of 36π square centimeters?

Each edge of the cube should be __________ centimeters.

6. $\triangle ABC \sim \triangle DEF$. The scale factor between $\triangle ABC$ and $\triangle DEF$ is 1 : 3. If the area of $\triangle ABC$ is x, then find the area of $\triangle DEF$.

   A. $\frac{1}{3}x$

   B. $3x$

   C. $\frac{1}{2}(9x)$

   D. $9x$

7. $\triangle MNO \sim \triangle RST$

   $\triangle MNO$ has an area of 750 cm^2. $\triangle RST$ has an area of 30 cm^2. What is the ratio of similitude between $\triangle MNO$ and $\triangle RST$?

   A. 5 to 1

   B. 25 to 1

   C. 50 to 1

   D. 250 to 1

8. A company that bakes sweet treats is introducing a family-size box of cookies.

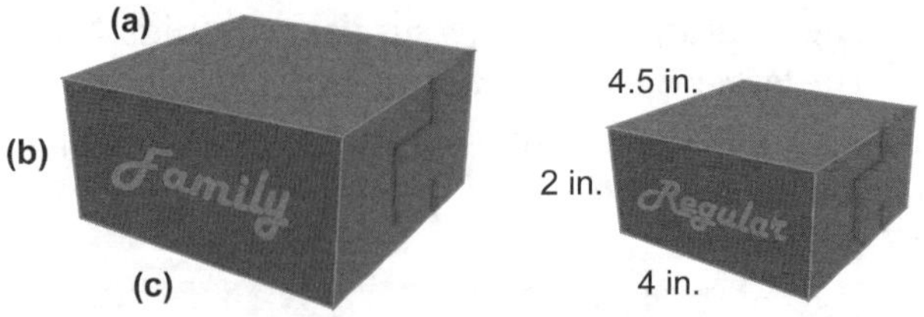

The family-size box takes up 121.5 cubic inches of space. The regular-size box, shown to the right, holds a dozen cookies and takes up 36 cubic inches.

If the boxes are geometrically similar, the measurements of the family-size box are

(a) ______ in. by **(b)** ______ in. by **(c)** ______ in.

9. Alex wants to triple the volume of a rectangular prism. By what scale factor (approximately) should she multiply each side length of the original prism to determine the side lengths of the new larger prism?

   A. 0.33

   B. 0.69

   C. 1.44

   D. 1.50

   E. 1.73

   F. 3.00

10. Ethan is working on a project in his architecture class. He has created a room (in the shape of a rectangular prism) with a volume of 1,400 cubic feet. His professor asks him to increase the size of the room proportionally so that it has a volume of 2,420 cubic feet.

    By what factor should Ethan scale the length, width, and height of the room? *Express your answer to the nearest tenth.*

    He should scale each dimension by __________.

Apply

How Much Bigger Is That Box?

Shipping boxes come in many sizes and shapes.

Corrugated cardboard boxes come in an incredibly wide variety of sizes. In fact, one online retailer states that they have more than 1,300 box sizes in stock! What would have to be true for two boxes to be similar? If they were similar, how would their surface areas and volumes be related? How would you advise the company to price similar boxes?

Research shipping boxes online and find two pairs of similar boxes with different scale factors (they cannot be cubes). Compute and compare the ratios of their dimensions, surface areas, and volumes and generalize the patterns you see. Compare the prices and discuss the best measure for determining pricing, including mathematical evidence.

Similar Boxes

Explain your answer below.

Show what you've learned by completing the other performance tasks in the online Apply section.

Apply *(continued)*

Your answer to Apply will be assessed on the following criteria:

1. Researching to find similar boxes and demonstrating similarity
2. Calculating surface areas and comparing the ratios
3. Calculating volumes and comparing the ratios
4. Comparing prices and stating which measure is best for pricing

Scale / Criteria	4	3	2	1	0
	Exceeds Criteria	**Meets Criteria**	**Progressing to Criteria**	**Below Expectations**	**No Expectation**
Demonstration of Similarity	Correctly identifies two pairs of non-cubic similar boxes and proves their similarity.	Identifies two pairs of non-cubic similar boxes and proves their similarity with minor errors.	Correctly identifies one pair of non-cubic similar boxes and proves their similarity.	Identifies pairs that are cubic or are not similar.	Does not attempt.
Surface Area Ratios	Correctly calculates the surface area of each box and compares the ratio of each pair.	Calculates the surface area of each box and compares the ratio of each pair with minor errors.	Calculates the surface area of each box, but does not compare ratios or makes significant calculation errors.	Calculates the surface area and ratios incompletely or incorrectly.	Does not attempt.
Volume Ratios	Correctly calculates the volume of each box and compares the ratio of each pair.	Calculates the volume of each box and compares the ratio of each pair with minor errors.	Calculates the volume of each box, but does not compare ratios or makes significant calculation errors.	Calculates the volume and ratios incompletely or incorrectly.	Does not attempt.
Price Comparison	Correctly compares the prices and states which measure is best for pricing.	Compares the prices and states which measure is best for pricing with minor errors in calculations or reasoning.	Compares the prices and states which measure is best for pricing with significant errors in calculations or reasoning.	Compares the prices but does not state which measure is best or states which measure is best without justification.	Does not attempt.

UNIT 6: Trigonometry

6.1 Investigate Right Triangle Trigonometry

photo: Getty Images

Lesson Objectives

- Understand that angle measure determines ratio of side length in all right triangles.
- Define sine, cosine, and tangent ratios for acute angles.
- Use trigonometry to calculate the area of triangles.
- Apply trigonometric ratios and the Pythagorean theorem to solve real-world problems.

Essential Question

- How can you solve problems that involve right triangles?

Investigations

Adventure Racing!

Plan a strategic adventure racing route over the river and through the woods.

Gaining an Advantage

Consider route options—straight line or multilegged. Either way, bring a triangle along to help you.

Off Course

How are the angles and sides of right triangles related? Use the Triangle Data Collector to find out.

Crossing the Divide

Use trigonometric relationships to strategize adventure racing routes.

Mapping Out a Course

Keep the race area contained. Use trigonometry to find the area of next year's adventure race.

photo: Getty Images

Key Vocabulary

cosine, hypotenuse, included angle, included side, Pythagorean theorem, sine, special right triangle, triangle sum theorem, trigonometric ratios

Discover

As you complete Engage and the investigations, record the most important ideas you've learned.

Engage	
Investigation 1	**Investigation 2**
Investigation 3	**Investigation 4**

Name ______________________________ Date ____________

Check for Understanding

Investigate Right Triangle Trigonometry: Investigation 1

1. Suppose that you want to find the perimeter and the area of an isosceles right triangle with a hypotenuse measuring $8\sqrt{2}$ inches. *Fill in the blanks in the sentence below.*

The legs of the triangle measure ______ in. and ______ in., so the triangle's perimeter is ________________ in., and its area is ______ in^2.

2. Gwendolyn drew $\triangle FGH$, which is a 30-60-90 right triangle, with $FG = 5$ centimeters. Which of the following statements are correct about her triangle? *Select all that apply.*

A. If $\overline{FG}$ is the shorter leg, then the length of one of the other sides is $\frac{5\sqrt{3}}{3}$ cm.

B. If $\overline{FG}$ is the hypotenuse, then the length of one of the other sides is $5\sqrt{3}$ cm.

C. If $\overline{FG}$ is the longer leg, then the length of one of the other sides is 5 cm.

D. If $\overline{FG}$ is the shorter leg, then the length of one of the other sides is 10 cm.

E. If $\overline{FG}$ is the hypotenuse, then the length of one of the other sides is $\frac{5\sqrt{3}}{2}$ cm.

3. A park's trail system is shown. Waldron Trail is 4 kilometers long, and all the trails are straight.

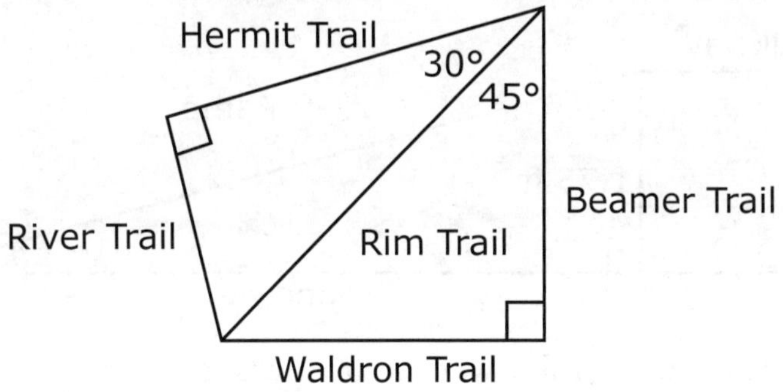

What are the lengths of Beamer Trail, Hermit Trail, Rim Trail, and River Trail? Explain.

Name ______________________________ Date ____________

Check for Understanding

Investigate Right Triangle Trigonometry: Investigation 2

1. Match each of the values on the left with one on the right that has an equal value.

A. cos 22°	**I.** sin 53°
B. sin 71°	**II.** cos 31°
C. cos 37°	**III.** sin 42°
D. sin 64°	**IV.** cos 19°
E. cos 48°	**V.** sin 68°
F. sin 59°	**VI.** cos 26°

2. The tangent of one of the acute angles in a right triangle is $\frac{3}{10}$. What is the value of the tangent of the other acute angle in the triangle?

A. $\frac{3}{10}$ **B.** $\frac{7}{10}$ **C.** $\frac{10}{7}$ **D.** $\frac{10}{3}$

3. Larry constructed a 6.75-foot ramp to a walkway 1.75 feet above the ground as shown in the figure. *Fill in the blank in the sentence. Round your answer to two decimal places.*

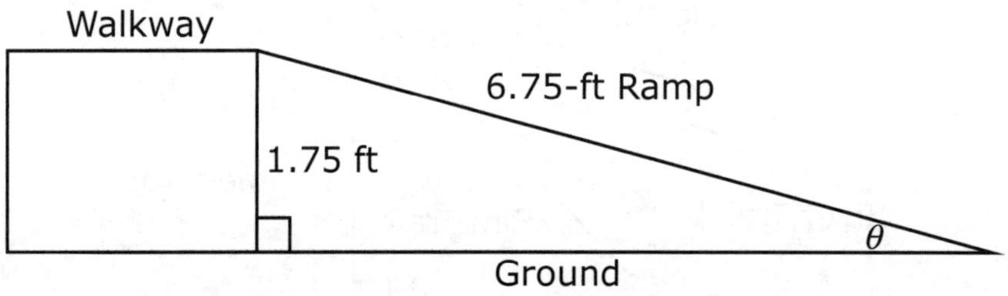

The sine of angle θ that the ramp makes with the ground is approximately ______.

4. If the sine of an acute angle in a right triangle is equal to $\frac{35}{37}$, what is the value of the cosine of the angle? Explain by using the definitions of sine and cosine in your answer.

Name ______________________________ Date ____________

Check for Understanding

Investigate Right Triangle Trigonometry: Investigation 3

1. A different right triangle was presented to each of three students in a class.
   - Florence was given the measures of the two acute angles of her triangle.
   - Regina was given the measure of one of the acute angles and the length of one of the legs of her triangle.
   - Nancy was given the lengths of the two legs of her triangle.

   Which of the students has enough information to completely solve her triangle?

   A. only Regina **B.** only Florence and Regina

   C. only Regina and Nancy **D.** Florence, Regina, and Nancy

2. Examine the figure below. *Fill in the blanks in the sentence that follows.*

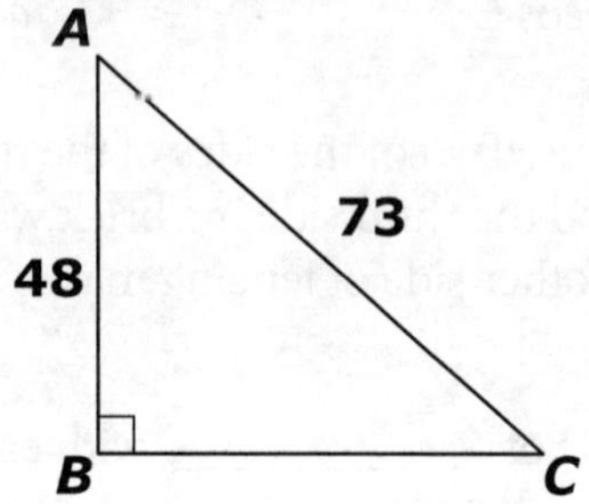

   To the nearest tenth of a degree, $m\angle BAC \approx$ ________, and $m\angle BCA \approx$ ________.

3. On a scavenger hunt, Jon and Kelly split up to look for a four-leaf clover. Starting simultaneously from the same location, Jon walked directly north at 65 meters per minute, while Kelly walked directly east at 80 meters per minute. After a certain number of minutes, Kelly spotted a clover, so she sent Jon a text message, and they stopped at exactly the same instant. Kelly then bent down to pick the clover, turned, and walked in a straight line to where Jon was waiting. Explain why you don't have to know how many minutes Jon and Kelly had been walking in order to find the angle bearing at which Kelly walked back to Jon. Then calculate the angle bearing to the nearest tenth of a degree.

Name ______________________________ Date ______________

Check for Understanding

Investigate Right Triangle Trigonometry: Investigation 4

1. The area of $\triangle ABC$ shown in the figure below is given by the formula Area $= \frac{1}{2}bh$.

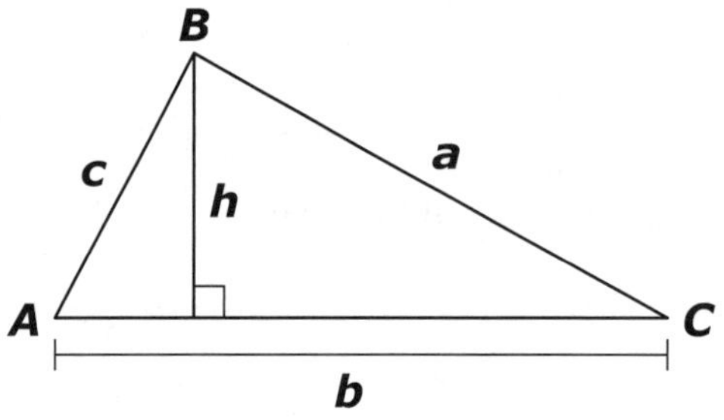

Which expressions can be substituted for h in the formula? *Select all that apply.*

A. $a \cos B$ **B.** $c \sin A$ **C.** $b \cos A$ **D.** $a \sin B$

E. $c \sin B$ **F.** $b \cos C$ **G.** $a \sin C$ **H.** $c \cos B$

2. A flower bed is in the shape of a triangle. Two of the sides of the triangle, with lengths of 4 feet and 7.5 feet, respectively, are formed by fencing, and the third side is a brick wall. One of the sides of fencing makes a 24° angle with the brick wall, and the other side of fencing makes a 49° angle with the brick wall. *Fill in the blanks in the sentence.*

The two sides of fencing make an angle of ____________ with each other, so to two decimal places, the area of the flower bed is ____________ ft^2.

3. Simon derived the formula Area $= \frac{1}{2}ab \cos C \tan C$ for the area of a triangle, where a and b are the lengths of two sides of the triangle, and C is the measure of the included angle. His teacher correctly pointed out that the formula doesn't work for all angle classifications. Use your calculator to test Simon's formula with an acute, obtuse, and right angle. For which angle type(s) doesn't the formula work, and why? Explain.

Summary

Before you attempt the Practice Exercises, review what you've learned.

Ratios in Special Right Triangles

45-45-90 Triangle	30-60-90 Triangle
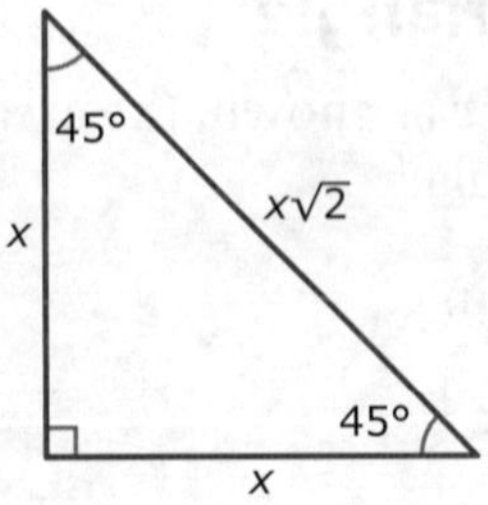	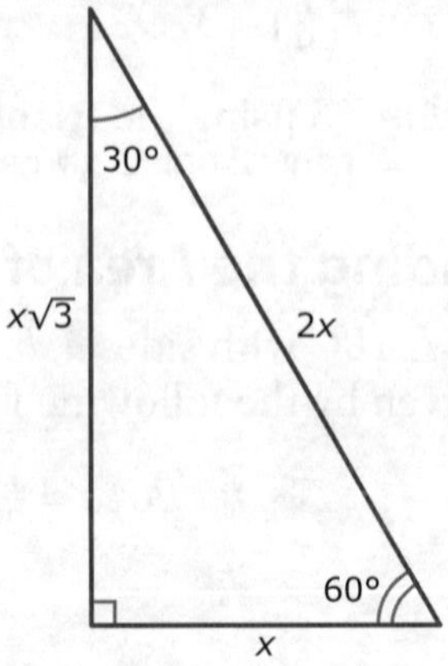
If the legs of an isosceles right triangle, or 45-45-90 triangle, have length x, then the hypotenuse has length $x\sqrt{2}$.	If the shorter leg of a 30-60-90 triangle has length x, then the longer leg has length $x\sqrt{3}$, and the hypotenuse has length $2x$.

EXAMPLE: Finding Missing Measures in Special Right Triangles

Find the lengths of the legs of a right triangle if the length of the hypotenuse is $4\sqrt{2}$ inches and one of the angles is 45°.

SOLUTION:

If the hypotenuse has length $4\sqrt{2}$, then using the special ratios of 45–45–90 right triangles, $4\sqrt{2} = x\sqrt{2}$, so $x = 4$. The legs each have a length of 4 inches.

Ratios in Any Right Triangle

When trying to find missing measures of right triangles, use the following trigonometric relationships.

Sine	$\sin A = \dfrac{\text{length of side opposite } \angle A}{\text{length of hypotenuse}}$
Cosine	$\cos A = \dfrac{\text{length of side adjacent } \angle A}{\text{length of hypotenuse}}$
Tangent	$\tan A = \dfrac{\text{length of side opposite } \angle A}{\text{length of side adjacent } A}$

In order to use any of these relationships, you must know either the lengths of two sides, or an acute angle and one of the sides.

EXAMPLE: Solving Right Triangles

Solve the given triangle.

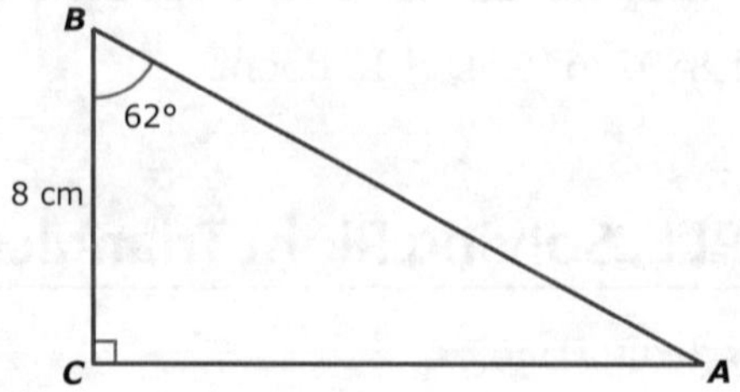

SOLUTION:

By the triangle sum theorem, $m\angle A = 180° - 90° - 62° = 28°$.

You can find the lengths of AB and AC using trigonometric ratios.

For $\angle B$, the opposite side is $\overline{AC}$, the adjacent side is $\overline{BC}$, and the hypotenuse is $\overline{AB}$.

My Notes

Summary *(continued)*

Find AB : $\cos B = \frac{\text{adjacent}}{\text{hypotenuse}}$, so $\cos 62° = \frac{8}{AB}$.

Solve the equation for AB : $AB \cos 62° = 8$.

Using a calculator, $\cos 62° \approx 0.4694715628$.

$AB \approx \frac{8}{0.4694715682} \approx 17.04$ cm.

Find AC : $\tan B = \frac{\text{opposite}}{\text{adjacent}}$, so $\tan 62° = \frac{AC}{8}$.

Solve the equation for AC : $8 \tan 62° = AC$.

Using a calculator, $\tan 62° = 1.8807264654$.

$AC \approx 8(1.8807264654) \approx 15.05$cm.

EXAMPLE: Solving Right Triangles

Solve the given triangle.

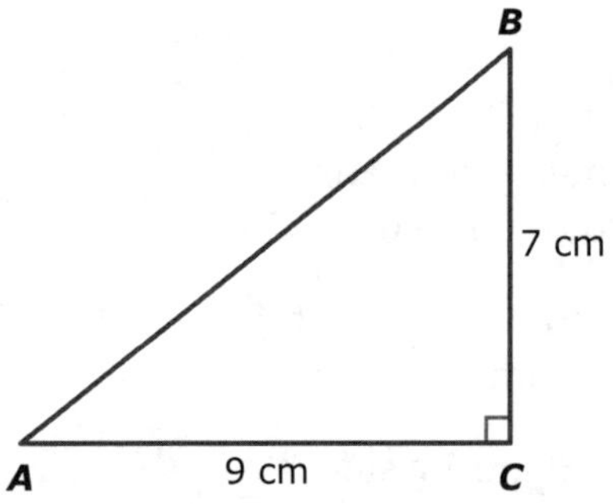

My Notes

SOLUTION:

Find AB using the Pythagorean theorem:

$AB = \sqrt{9^2 + 7^2} = \sqrt{130} \approx 11.4$ cm

Find $m\angle A$ using the inverse tangent: $\tan A = \frac{7}{9}$, so $A = \tan^{-1}\left(\frac{7}{9}\right) \approx 37.9°$

Find $m\angle B$ using the triangle sum theorem: $m\angle B \approx 180° - 90° - 37.9° \approx 52.1°$

Finding the Area of a Triangle

For $\triangle ABC$ with sides a, b, and c as shown, the area is given by the following formula:

$$Area = \frac{1}{2}ab \sin C$$

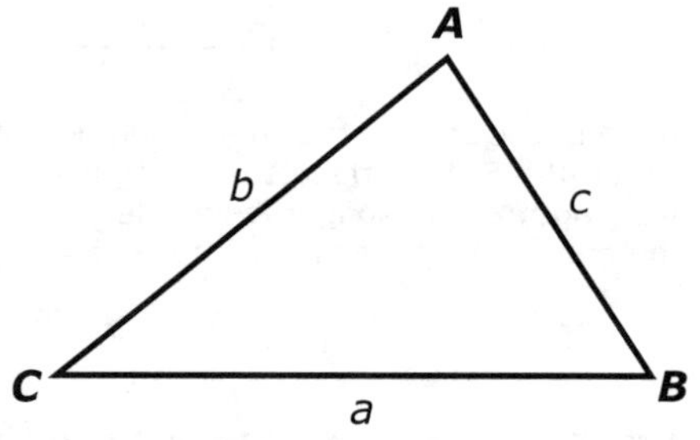

EXAMPLE: Finding the Area of a Triangle

Find the area of the given triangle.

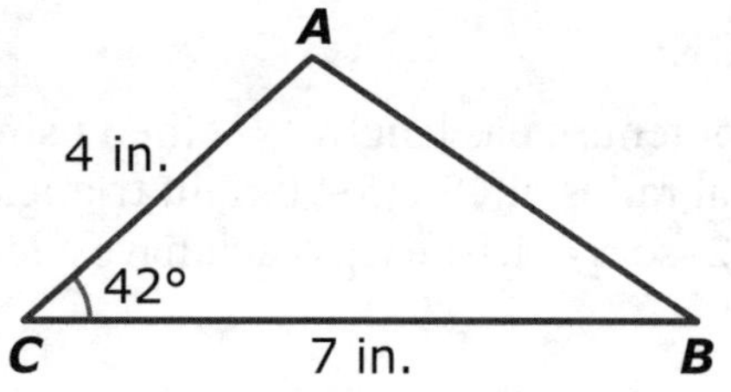

SOLUTION:

$$\begin{aligned} \text{Area} &= \frac{1}{2}ab \sin C \\ &= \frac{1}{2}(7)(4)\sin 42° \\ &\approx 14(0.66913) \\ &\approx 9.37 \text{ in.}^2 \end{aligned}$$

Practice Exercises

Review what you've learned using these practice problems. For practice problems with feedback, try the Coach and Play items in the Practice section online.

1. Harold is moving a 5-foot-tall refrigerator. He tilts it onto its rear edge, forming an angle between the back of the refrigerator and the floor; the top edge of the tilted refrigerator is exactly 4 feet above the floor as shown in the diagram.

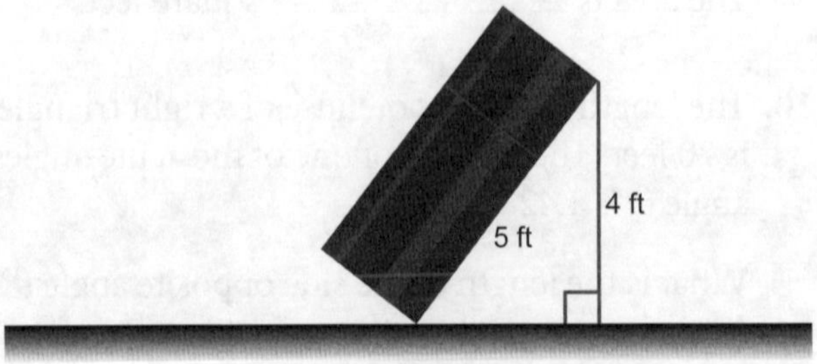

What is the cosine of the angle the refrigerator makes with the floor?

A. $\frac{4}{5}$

B. $\frac{5}{4}$

C. $\frac{4}{3}$

D. $\frac{3}{4}$

E. $\frac{3}{5}$

2. A 10-foot-tall ladder rests against a wall, touching it at a point 9 feet above the ground.

Approximately what angle does the ladder make with the floor?

Round your answer to the nearest degree.

A. 82°

B. 80°

C. 72°

D. 64°

E. There is not enough information to determine the value.

3. Two sides of a right triangle form acute angle *A*. One of the sides is twice as long as the other side.

What is the measure of $\angle A$?

$m\angle A =$ ______ degrees

4. $\triangle XYZ$, shown in the diagram, was created by constructing two adjacent triangles, $\triangle XYW$ (a 30–60–90 right triangle) and $\triangle ZYW$ (a 45–45–90 right triangle).

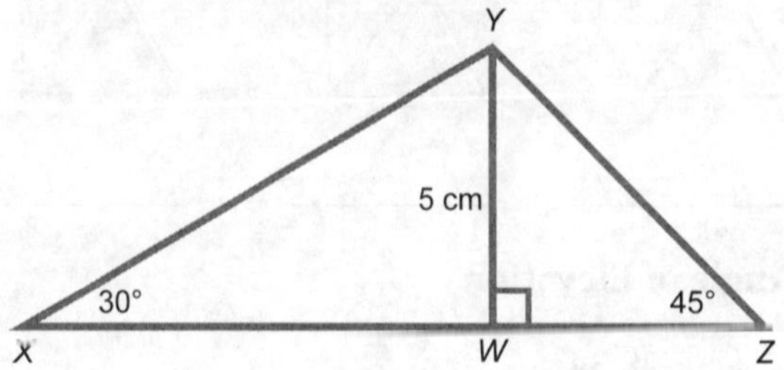

If $WY = 5$, what is XZ? *Write the exact answer.*

$XZ =$ ____________ centimeters

5. Javier built a skateboard ramp by elevating one end of an 8-foot piece of plywood so that it makes a 20° angle with the ground.

Approximately how high did Javier elevate the raised end of the ramp?

A. 2.1 ft

B. 2.3 ft

C. 2.7 ft

D. 2.9 ft

Practice Exercises *(continued)*

6. A company wants to design a tablet cover that folds into a stand, so that it holds a screen at a comfortable angle for viewing.

Use the dimensions of the tablet cover designs to determine the angle of elevation that the tablet makes with the base of the stand. What is the angle of elevation for each tablet cover?

Tablet Cover

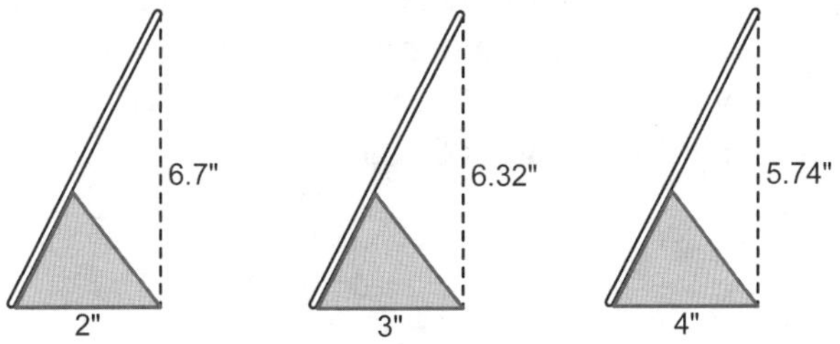

__________ __________ __________

Angle of Elevation

A. 28.34°

B. 44.18°

C. 55.13°

D. 64.61°

E. 72.63°

F. 73.38°

7. The sine and cosine of one of the acute angles in a right triangle are equal. What is the tangent of that angle?

The tangent of the angle is __________.

8. The cosine of the smallest angle in a right triangle is 0.8.

What is the sine of the other acute angle?

The sine of the other acute angle is __________.

9. Brianna made a design with chalk on the playground in the shape of a triangle.

Two sides of the triangle measuring 5 feet and 8 feet meet at a 40° angle.

What is the area of the triangle?

Round your answer to the nearest tenth.

The area is __________ square feet.

10. The length of the hypotenuse of a right triangle is 20 feet. The tangent of one of the acute angles, angle θ, is 1.42.

What is the length of the side opposite angle θ? *Round your answer to the nearest tenth of a foot.*

__________ feet

11. A plane departs the runway at an angle of elevation of 12.7°. What will the altitude of the plane be when it has covered a ground distance of 10,000 feet?

A. 1,572.9 feet

B. 2,253.6 feet

C. 2,978.1 feet

D. 3,471.8 feet

12. A telephone pole is 30 feet in height. The pole casts a shadow 46 feet long. Approximate the sun's angle of depression.

A. 26.4°

B. 33.1°

C. 40.7°

D. 49.3°

Apply

How Will You Design a Ramp?

The field of architectural design is based on geometry. Watch the video to start thinking about what considerations are important in making a facility accessible to all people.

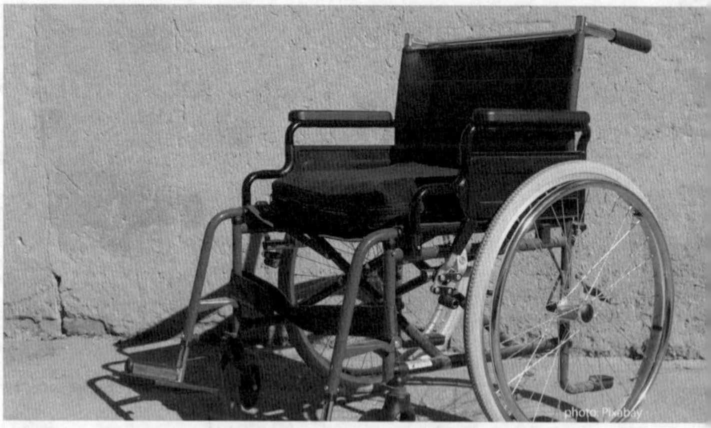

What needs to be considered when designing a ramp?

The Americans with Disabilities Act (ADA) includes guidelines for building ramps with an angle of inclination that is safe as well as comfortable to use. Research online to find the recommended guidelines.

Overcoming Barriers: View this video segment about access ramps.

How can you design a ramp for a location in your neighborhood according to the ADA specifications? Sketch the dimensions of the potential ramp.

Consider the following:

- the total height that the ramp must rise
- the number of turns needed
- Is it possible to build a ramp within the amount of space available, given the number of turns needed?

Show what you've learned by completing the other performance tasks in the online Apply section.

Apply *(continued)*

Your answer to Apply will be assessed on the following criteria:

1. Detailing the requirements for building an ADA-approved ramp
2. Outlining a plan for the ramp using location measurements and ADA-recommended guidelines
3. Drawing a sketch of the ramp labeled with its proposed dimensions
4. Calculating the required measurements using trigonometry

Scale / Criteria	4	3	2	1	0
	Exceeds Criteria	**Meets Criteria**	**Progressing to Criteria**	**Below Expectations**	**No Expectation**
Requirements	Correctly details all of the basic measurements that should be part of the ramp design.	Details the basic measurements that should be part of the ramp design with minor errors.	Details some of the basic measurements that should be part of the ramp design.	Attempts to detail guidelines for building a wheelchair ramp, but does not use relevant information.	Does not provide basic required measurements.
Plan Outline	Writes a plan that accommodates the total rise and run of the ramp, and includes all of the recommendations.	Writes a plan that accommodates the total rise and run of the ramp, and includes all but one of the recommendations.	Writes a plan that accommodates the total rise and run of the ramp, and includes only some of the recommendations.	Attempts to write a plan for building the ramp, but does not include the recommendations.	Does not submit a plan.
Sketch	Draws a sketch that accurately represents the ramp and the building, and meets the recommendations.	Draws a sketch that accurately represents the ramp and building, and meets all but one of the recommendations.	Draws a sketch that represents the ramp and building, but does not apply some of the recommendations.	Attempts to draw a sketch of the ramp, but does not apply the recommendations.	Does not include a sketch.
Ramp Dimensions	Accurately calculates the dimensions of the ramp that meet all ADA guidelines, and shows that it fits within the chosen location.	Calculates the dimensions of the ramp, and shows that it fits within the chosen location, but makes a minor calculation error.	Calculates the dimensions of the ramp, and shows that it fits within the chosen location, but makes several calculation errors.	Attempts to calculate the dimensions of the ramp, but does so inaccurately or using incorrect formulas.	Does not submit ramp dimensions.

UNIT 6: Trigonometry

6.2 Investigate Laws of Sines and Cosines

photo: Getty Images

Lesson Objectives

- Develop and prove the law of sines and the law of cosines.
- Use the law of sines and law of cosines to solve problems involving side length and angle measure of triangles.

Essential Question

- How can trigonometric ratios be used to solve non-right triangles?

photo: Getty Images

Investigations

The Last Leg of the Journey

Do you have the angle to find the distance? Use trigonometry to find the distance between checkpoints.

A Shortcut or a Wrong Turn?

Follow the sines to determine the length of the last leg.

Choose Your Own Adventure

If sine has its own law, cosine must too. Ask Pythagoras to point the way.

Law of Sines—The Ambiguous Case

How do you know you have enough information to use the law of sines?

Solving the Ambiguous Case

When a triangle is ambiguous, test the possible cases.

A Triangular Adventure

Help the director plan a race route. Use the laws to help you find the angles.

Key Vocabulary

cosine, hypotenuse, included angle, included side, law of cosines, law of sines, Pythagorean theorem, sine, tangent (trigonometry)

Discover

As you complete Engage and the investigations, record the most important ideas you've learned.

Engage

Investigation 1

Investigation 2

Investigation 3

Investigation 4

Investigation 5

Name ____________________ Date __________

Check for Understanding

Investigate Laws of Sines and Cosines: Investigation 1

1. Gene knows two of the side lengths and one of the angle measures of a scalene triangle. Can he use the law of sines to find the measure of another angle in the triangle?

A. He can use it only if the angle is formed by the two sides.

B. He can use it only if the angle is not formed by the two sides.

C. Yes, he can use it whether the angle is formed by the two sides or not.

D. No, he cannot use it whether the angle is formed by the two sides or not.

2. Examine the triangle below and then fill in the blanks in the sentence that follows. *Round side lengths to the nearest tenth of a centimeter, when necessary.*

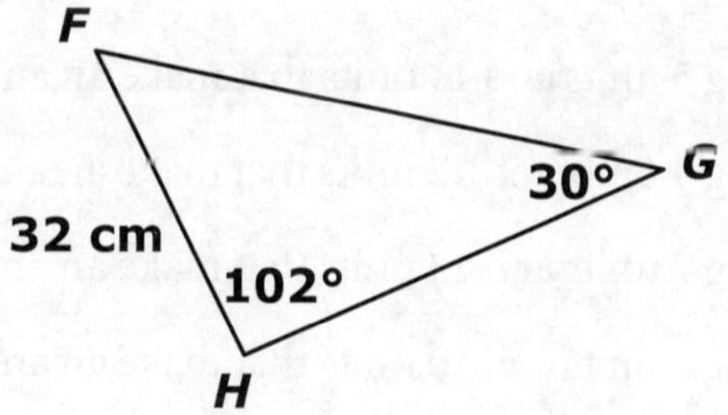

In the triangle, $m\angle F =$ ______, $GH =$ ______ cm, and $FG =$ ______ cm.

3. Tracey used the law of sines with $\triangle XYZ$ to find that $m\angle X = \sin^{-1}\left(\frac{4\sin 26°}{3}\right)$. If $m\angle Y = 26°$, which pairs of side lengths could be correct? *Select all that apply.*

A. $XZ = 21$ and $YZ = 28$

B. $XZ = 16$ and $YZ = 12$

C. $XZ = 24$ and $YZ = 16$

D. $XZ = 6$ and $YZ = 8$

E. $XZ = 18$ and $YZ = 27$

F. $XZ = 20$ and $YZ = 15$

4. Explain how you can use the formula for the area of a triangle to prove the law of sines.

Name ______________________________ Date ______________

Check for Understanding

Investigate Laws of Sines and Cosines: Investigation 2

1. So far, Jeremiah has determined the lengths, a and b, of two sides of a triangle. Can he use the equation $c^2 = a^2 + b^2 + 2ab \cos C$ to solve for an unknown in the triangle?

A. no

B. only if he can determine the length, c, of the third side

C. yes

D. only if he can determine the measure of the included angle, C

2. Which triangles have a side that is more than 15 units long? *Select all that apply.*

A. one with sides measuring 6 units and 11 units that make an angle of 105°

B. one with sides measuring 5 units and 14 units that make an angle of 83°

C. one with sides measuring 4 units and 12 units that make an angle of 117°

D. one with sides measuring 7 units and 13 units that make an angle of 99°

E. one with sides measuring 8 units and 10 units that make an angle of 121°

3. A triangular sandbox has sides measuring 20 ft, 25 ft, and 30 ft. *Fill in the blank.*

To the nearest tenth of a degree, the side of the sandbox measuring 20 ft makes an angle of approximately ____________ with the side measuring 25 ft.

4. Hannah's teacher told her that the square of the cosine of an angle plus the square of the sine of the same angle equals 1. Explain how Hannah can use this fact, along with the figure shown below, to prove the law of cosines for an acute triangle.

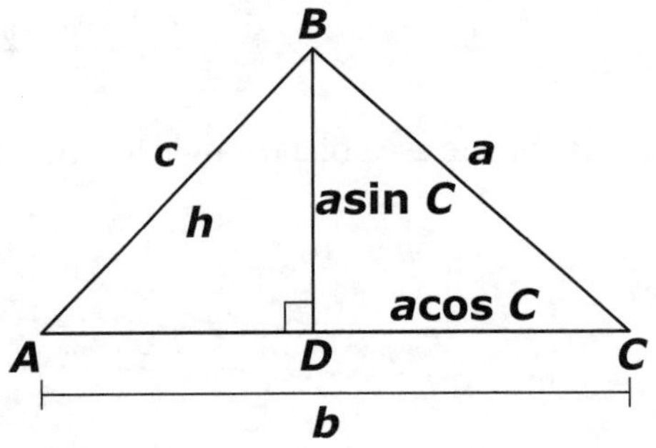

Name ______________________________ Date __________

Check for Understanding

Investigate Laws of Sines and Cosines: Investigation 3

1. Consider a triangle with a side of length a opposite $\angle A$, a side of length b opposite $\angle B$, and a side of length c opposite $\angle C$. One such triangle is shown in the figure below.

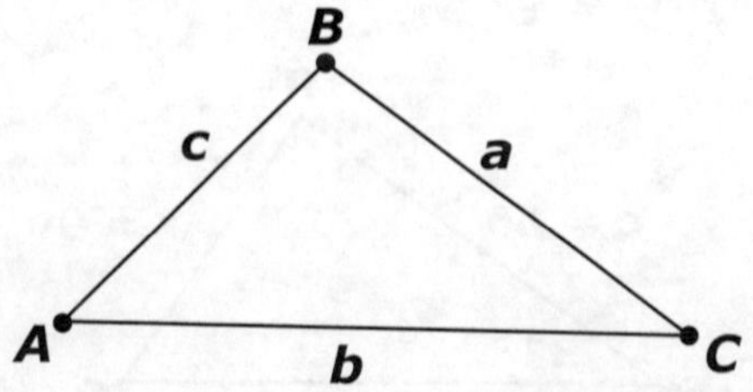

Suppose you were given $m\angle A$ and the values of a and c, and you wanted to completely solve the triangle by using the law of sines. For which measures of $\angle A$ might there be an ambiguous case, depending on the values of a and c? *Select all that apply.*

A. 99° **B.** 155° **C.** 10° **D.** 124°

E. 173° **F.** 90° **G.** 102° **H.** 66°

2. In the table below, $m\angle A$ and the values of a and c are given for different triangles of the type described in question 1. If you wanted to completely solve each triangle by using the law of sines, which would be an ambiguous case, and which would be a non-ambiguous case? Put a checkmark in the appropriate column for each triangle.

$m\angle A$	a	c	Ambiguous Case	Non-Ambiguous Case
26°	29	19		
117°	21	33		
41°	19	25		
140°	17	42		
38°	36	23		
15°	27	39		

Based on your answers to questions 1 and 2, what must be true about $m\angle A$ and a for there to be an ambiguous case when solving a triangle with the law of sines? Explain.

Name ________________________________ Date ______________

Check for Understanding

Investigate Laws of Sines and Cosines: Investigation 4

1. Joe and Mary each drew a triangle with vertices of A, B, and C, where $BC = a$, $AC = b$, and $AB = c$. Joe's triangle, shown below, is acute, while Mary's is obtuse. *Fill in the blanks in the sentence below with angle measures rounded to a tenth of a degree.*

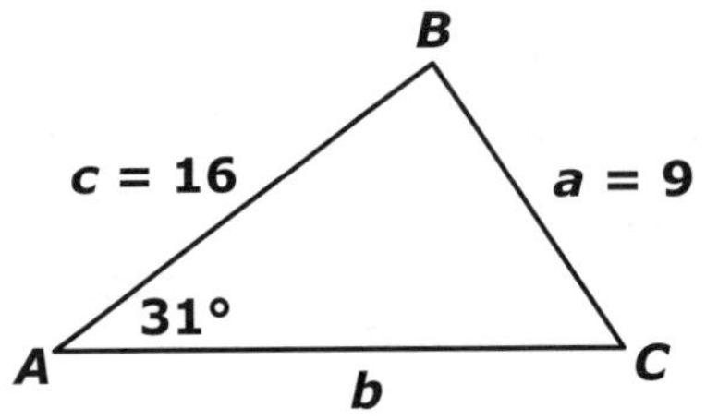

In Joe's triangle, $m\angle B =$ ______ and $m\angle C =$ ______, and if Mary's triangle has the same values for $m\angle A$, a, and c, then in it, $m\angle B =$ ______ and $m\angle C =$ ______.

2. In question 1, how much longer to a tenth of a unit is b in Joe's triangle than in Mary's?

A. 2.9 **B.** 7.2 **C.** 10.1 **D.** 17.3

3. On a test, Dirk was told that in $\triangle XYZ$, the measure of $\angle X$ is 35°, the length, x, of $\overline{YZ}$ is 25 units, and the length, z, of $\overline{XY}$ is 23 units. He was asked to find the measure of $\angle Z$, so he began by assuming that the angle is acute and used the law of sines as shown.

$$\frac{\sin 35^\circ}{25} = \frac{\sin(m\angle Z)}{23}$$

$$23\sin 35^\circ = 25\sin(m\angle Z)$$

$$\frac{23\sin 35^\circ}{25} = \sin(m\angle Z)$$

$$\sin^{-1}\left(\frac{23\sin 35^\circ}{25}\right) = m\angle Z$$

$$m\angle Z \approx 31.8^\circ$$

Next, Dirk assumed that $\angle Z$ is obtuse and subtracted 31.8° from 180° to get 180° – 31.8° = 148.2° for the approximate measure of the angle. Therefore, he concluded that the measure of $\angle Z$ could be about 31.8° or about 148.2°. However, he didn't get full credit for his answer on the test. What did he do wrong? Explain.

Name ______________________________ Date ____________

Check for Understanding

Investigate Laws of Sines and Cosines: Investigation 5

1. Thad wants to solve $\triangle ABC$ shown below. Which of the following statements is correct?

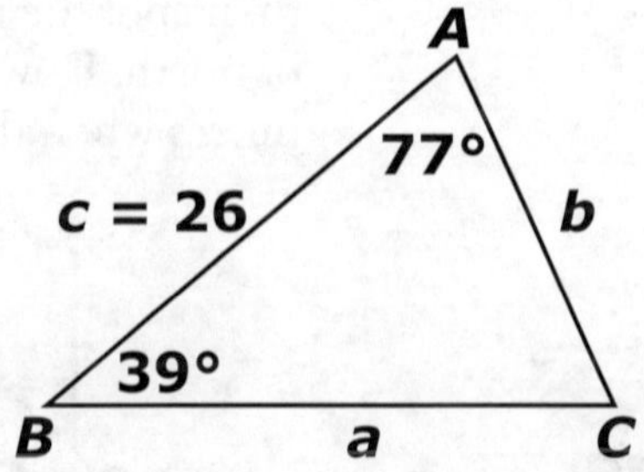

A. He must determine the value of a before he determines $m\angle C$.

B. He must determine $m\angle C$ before he determines the value of a.

C. He must determine the value of a before he determines the value of b.

D. He must determine the value of b before he determines the value of a.

2. Sienna was given $\triangle FGH$, with $GH = f$, $FH = g$, and $FG = h$. She knows that $m\angle F = 38°$, $m\angle G = 96°$, and $h = 15$ in. She intends to use both the law of sines and the law of cosines in the process of solving for the remaining unknowns, which are f, g, and $m\angle H$. Which of these could not be possible at some point? *Select all that apply.*

A. using the law of sines to solve for $m\angle H$

B. using the law of cosines to solve for $m\angle H$

C. using the law of sines to solve for the value of f

D. using the law of sines to solve for the value of g

E. using the law of cosines to solve for the value of f

F. using the law of cosines to solve for the value of g

3. Suppose the length of the base of an isosceles triangle is 12 cm and the measure of one of the base angles is 35°. Solve the triangle, and explain the steps that you used.

Summary

Before you attempt the Practice Exercises, review what you've learned.

The Law of Sines

The law of sines relates the sines of the angles of a triangle to the lengths of its sides. It states that for any triangle $\triangle ABC$ as shown in the diagram:

$$\frac{\sin A}{a} = \frac{\sin B}{b} = \frac{\sin C}{c}$$

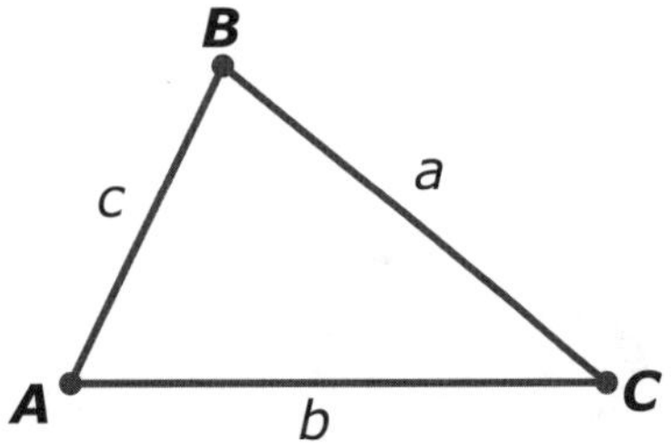

The length of the side opposite $\angle A$ is a, the length of the side opposite $\angle B$ is b, and the length of the side opposite $\angle C$ is c.

My Notes

EXAMPLE: Applying the Law of Sines

A group of hikers at point A spot a rock formation at point C at an angle of 42° east of north. The hikers travel 500 meters due north to point B and then spot the rock formation at an angle of 72° east of north. How far from the rock formation are the hikers when they reach point B?

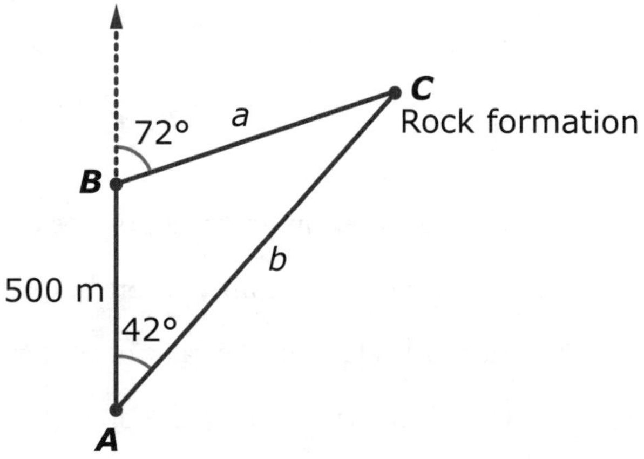

SOLUTION:

To solve the problem, we need to find a, the distance of the hikers from the rock formation when they are at point B.

Step 1: Determine the other angle measures of $\triangle ABC$.

$\angle ABC$ forms a linear pair with the 72° angle, so these angles are supplementary.

$$m\angle ABC + 72° = 180°$$
$$m\angle ABC = 108°$$

Next, use the triangle sum theorem to find the measure of $\angle C$.

$$m\angle A + m\angle ABC + m\angle C = 180°$$
$$42° + 108° + m\angle C = 180°$$
$$m\angle C = 30°$$

Summary *(continued)*

Step 2: Use the law of sines to find a.

$$\frac{\sin A}{a} = \frac{\sin B}{b}$$

$$\frac{\sin 42°}{a} = \frac{\sin 30°}{500}$$ Substitute the known values.

$$\sin 42° = \left(\frac{\sin 30°}{500}\right)a$$ Multiply both sides by a.

$$\left(\frac{500}{\sin 30°}\right)\sin 42° = a$$ Multiply both sides by $\frac{500}{\sin 30°}$.

$$669 \approx a$$ Use a calculator to solve for a.

When the hikers reach point B, they are about 669 meters from the rock formation.

The Law of Cosines

The law of cosines relates the cosine of an angle of a triangle to the lengths of its sides. It states that for any triangle $\triangle ABC$ as shown in the diagram:

$$c^2 = a^2 + b^2 - 2ab \cos C$$

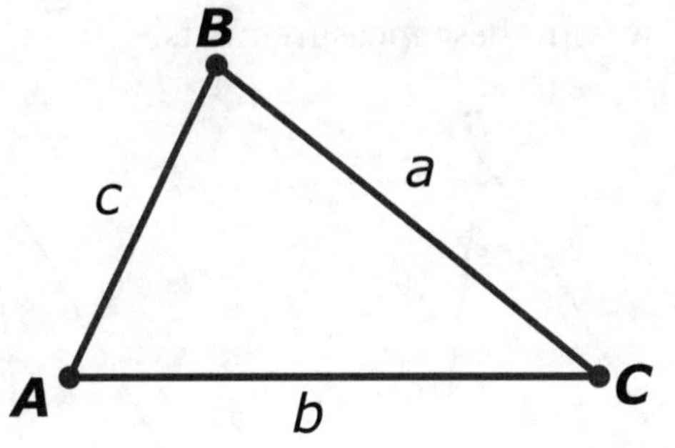

The length of the side opposite $\angle A$ is a, the length of the side opposite $\angle B$ is b, and the length of the side opposite $\angle C$ is c.

The law of cosines also can be written in these forms: $a^2 = b^2 + c^2 - 2bc \cos A$ and $b^2 = a^2 + c^2 - 2ac \cos B$.

As with the law of sines, we can use right triangles to help prove the law of cosines.

EXAMPLE: Applying the Law of Cosines

A group of hikers at point A travels 2.4 kilometers due north to point B. Then, they turn and head 118° east of north for 3.2 kilometers, stopping at point C. What distance b in kilometers will the hikers need to travel to return directly to their starting point?

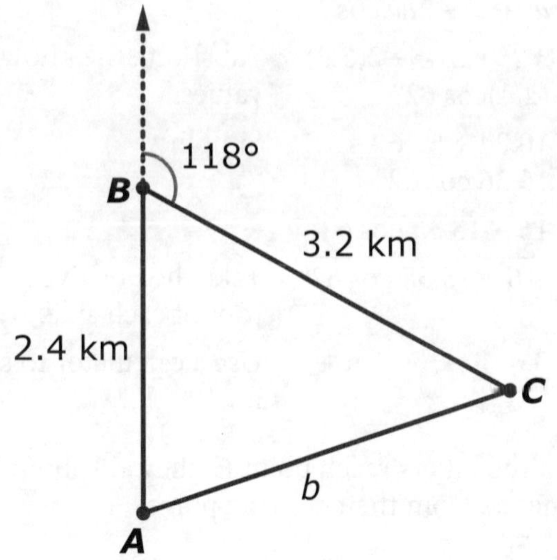

My Notes

Summary *(continued)*

SOLUTION:

Step 1: Determine the measure of $\angle ABC$.

$\angle ABC$ forms a linear pair with the 118° angle, so these angles are supplementary.

$$m\angle ABC + 118° = 180°$$
$$m\angle ABC = 62°$$

Step 2: Use the law of cosines to find b.

$b^2 = a^2 + c^2 - 2ac \cos B$	
$b^2 = 3.2^2 + 2.4^2 - 2(3.2)(2.4)\cos 62°$	Substitute the known values.
$b^2 = 10.24 + 5.76 - 15.36 \cos 62°$	Simplify.
$b^2 = 16 - 15.35 \cos 62°$	
$b = \sqrt{16 - 15.35 \cos 62°}$	Take the positive square root of each side.
$b \approx 3.0$	Use a calculator to solve for b.

When the hikers reach point C, they are about 3.0 kilometers from their starting point.

My Notes

The Ambiguous Case

You can use the laws of sines and cosines to solve triangles when you are given the following information:

- All three side lengths (SSS), or
- Two side lengths and the measure of the included angle (SAS), or
- Two angle measures and the length of the included side (ASA), or
- Two angle measures and the length of a non-included side (AAS)

However, when you are given two side lengths and the measure of a non-included angle (SSA), there may be more than one triangle that can have those measurements. For this reason, SSA is known as the ambiguous case.

For example, suppose you know that for $\triangle ABC$, $a = 6$ centimeters, $c = 7$ centimeters, and $m\angle A = 58°$. There are two possible triangles you can draw with these measurements.

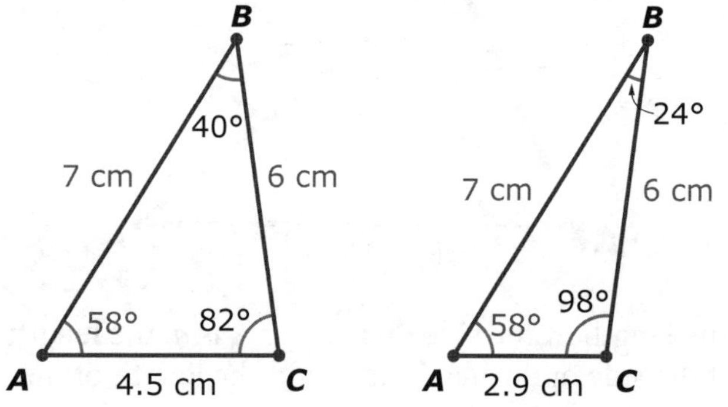

Triangle 1 **Triangle 2**

When you are given SSA information about a triangle, you may not be able to find a unique solution when you attempt to solve the triangle.

Practice Exercises

Review what you've learned using these practice problems. For practice problems with feedback, try the Coach and Play items in the Practice section online.

1. When solving ΔABC, you will find that there are two solutions.

What is the measure of side b for each of the possible solutions?

Round your answers to the nearest tenth.

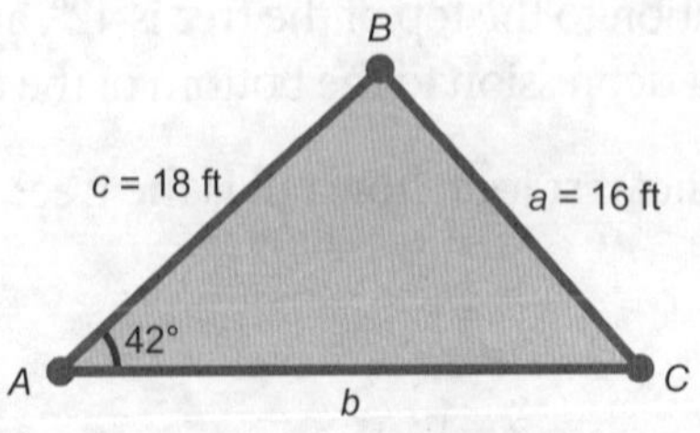

Side b could measure ____________ feet or ____________ feet.

2. Consider the given measurements for ΔABC in this diagram.

Which of the following statements about solving for the unknown measurements in this diagram are true? *Select all that apply.*

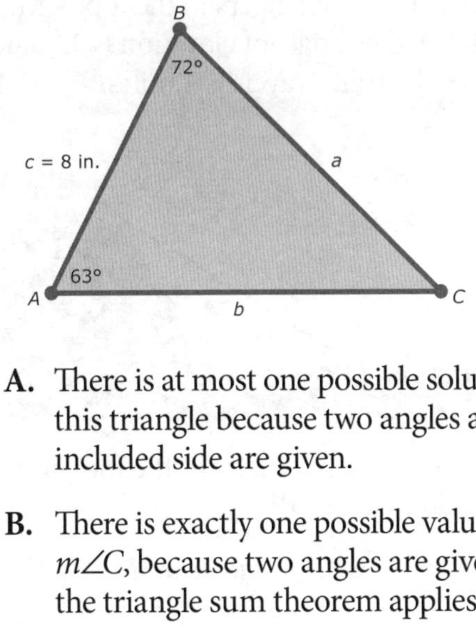

A. There is at most one possible solution for this triangle because two angles and an included side are given.

B. There is exactly one possible value for $m\angle C$, because two angles are given and the triangle sum theorem applies.

C. There are two possible values for $m\angle C$ based on the given information.

D. A triangle with these given values cannot exist.

3. Tam and Anna are playing a geometry game in which Anna tries to draw a triangle that matches Tam's triangle based on clues that Tam provides. In one such game, Tam tells Anna that she has drawn ΔABC where side $\overline{AC}$ is 6.5 cm long, side $\overline{AB}$ is 7 cm long, and $\angle B$ measures 66°.

Anna is unsure if she has enough information to replicate Tam's exact triangle. Based on this information, which of these may be the length of side $\overline{BC}$? *Select all that apply.*

A. 1.7 cm

B. 2.6 cm

C. 4.0 cm

D. 6.4 cm

E. 7.4 cm

F. 9.6 cm

G. No such triangle can exist, so there is no solution for the length of side $\overline{BC}$.

4. Campground C is 12 miles east of boat dock D. A boat leaves the campground and sails 9 miles at a bearing of 72° west of north. At this time, the boat is sighted from an observer at the dock at an angle of 28° east of north.

How far is the boat from the dock, to the nearest hundredth of a mile?

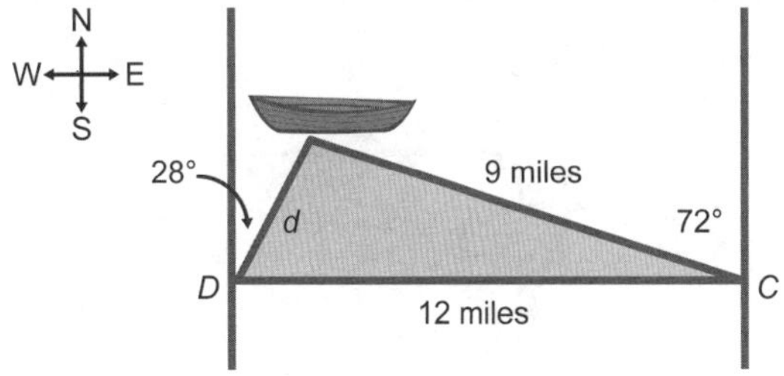

The boat is ____________ miles from the dock.

Practice Exercises *(continued)*

5. Jonathan had to construct a triangle. He wants to construct triangle $\triangle ABC$ with $m\angle A = 72°$, side $a = 28$, and side $c = 32$.

Give $m\angle C$ below. If it is not possible for Jonathan to construct this triangle, write "DNE".

$m\angle C =$ ____________

6. Three subdivisions border a lake as shown in the diagram below. New Carl is 50 miles due north of Barborsville. Smithville and Barborsville are 60 miles apart and New Carl and Smithville are 48 miles apart.

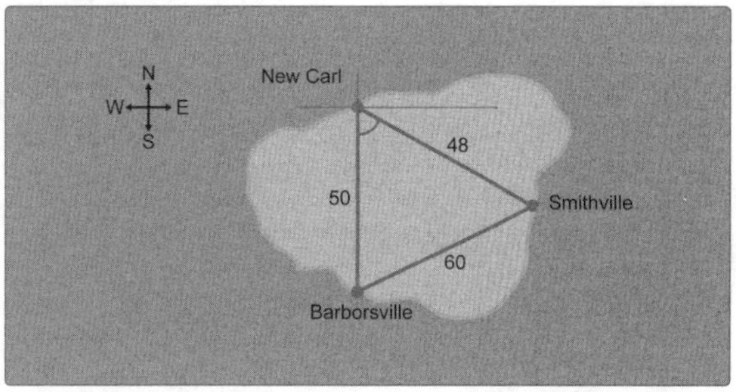

If a boat travels from New Carl to Smithville, what should its bearing be?

Round your answer to the nearest tenth of a degree.

A boat from New Carl to Smithville should travel at a bearing of ____________ degrees east of south.

7. A surveyor needs to find the distance between two points, A and C, on opposite sides of a river. She knows that the distance from A to B is 400 yards and that the angle made with $\overline{AB}$ and $\overline{AC}$ is 68° and the angle made with $\overline{AB}$ and $\overline{BC}$ is 47°, as shown in the diagram.

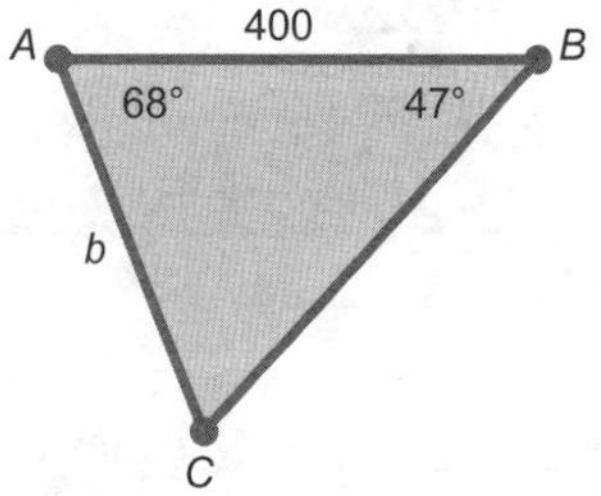

What is the distance from C to A to the nearest tenth of a yard?

The distance from C to A is ____________ yards.

8. When using the law of cosines to find a missing side length, what theorem do you end up with when the angle across from that missing side length is a right angle?

The ____________ theorem

9. A tree is growing on a hill. The angle the tree makes with the hill is 72°. From a point on the ground 70 feet up the hill from the base of the tree, the angle of elevation to the top of the tree is 42°, and the angle of depression to the bottom of the tree is 20°.

To the nearest foot, how tall is the tree?

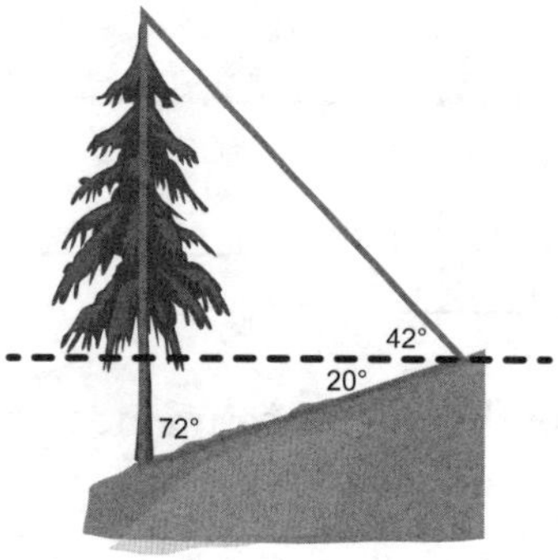

The tree is ____________ feet tall.

10. A tram carries passengers to the top of Mt. Goodview. The angle of elevation is 32° and the distance the tram travels is 4 miles.

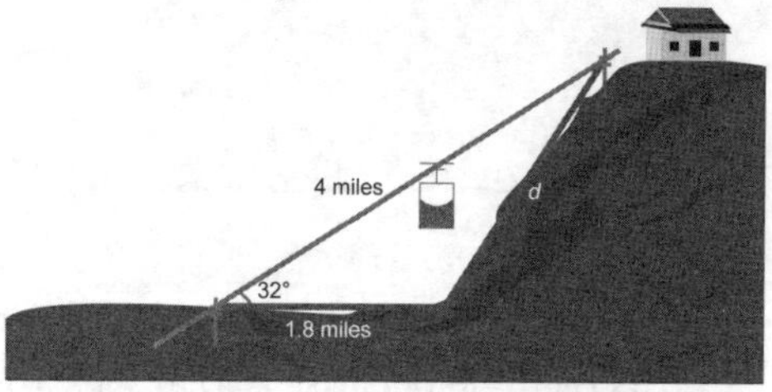

What is the distance, d, from the foot of the mountain to the top?

Round your answer to the nearest tenth of a mile.

The distance from the foot of the mountain to the top is ____________ miles.

Apply

How Far Apart Are the Stars?

How far apart are the stars in Orion's belt from each other? Using a combination of Techbook tools of your choice and research, find all six measurements of the stars in Orion's belt.

To get you started on your first astronomical analysis, here are two approximations about Orion's belt.

- From Earth, the star on the left and the star on the right appear to be about 3° apart.
- If an angle was formed by the two outer stars and Earth, with Earth as the vertex, the middle star would lie on the angle bisector.

Zeus placed Orion in the sky after Orion was either killed by a giant scorpion or by Artemis, the daughter of Zeus.

The Triangle Solution

Explain your findings below, along with any additions you made to the original drawing that helped you find your measurements.

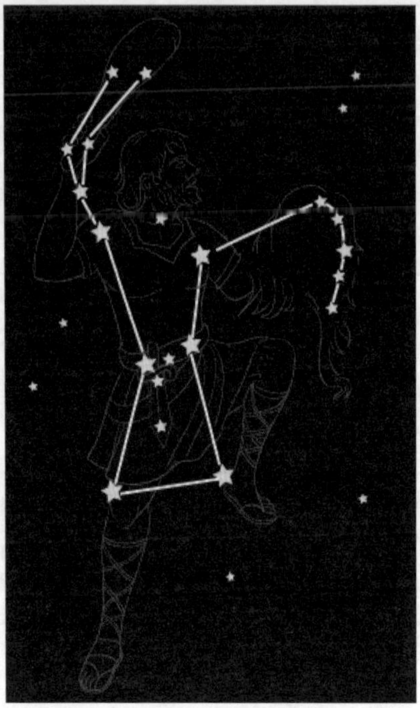

The three stars in Orion's belt may look linear to us, but that is only because of our perspective and the vastness of space.

Show what you've learned by completing the other performance tasks in the online Apply section.

Apply *(continued)*

Your answer to Apply will be assessed on the following criteria:

1. Properly labeling the drawing based on research
2. Choosing a valid method for solving for the measurements
3. Finding appropriate measurements for each angle between stars
4. Finding appropriate measurements for each distance between stars

Scale / Criteria	4	3	2	1	0
	Exceeds Criteria	**Meets Criteria**	**Progressing to Criteria**	**Below Expectations**	**No Expectation**
Labels	Accurately draws and labels a diagram.	Draws and labels a diagram with minor errors.	Draws and labels a diagram with significant errors.	Draws a diagram that does not reflect the problem.	Does not attempt draw or label a diagram.
Choice of Method	Chooses a valid method to determine each of the six measurements.	Chooses a valid method to determine only four or five of the measurements.	Chooses a valid method to determine only two or three of the measurements.	Chooses a valid method to determine only one or none of the measurements.	Does not choose any methods.
Angles	Finds all three angles correctly.	Finds two of the angles correctly.	Finds one of the angles correctly.	Attempts to find angle measurements, but none are correct.	Does not attempt to find the angles.
Distances	Finds all three distances correctly.	Finds two of the distances correctly.	Finds one of the distances correctly.	Attempts to find distances, but none are correct.	Does not attempt to find the distances.

UNIT 7: The Geometry of Circles

7.1 Investigate Circles and Parts of Circles

photo: Getty Images

Lesson Objectives

- Define and describe circles.
- Prove that all circles are similar.
- Classify lines and segments as chords, secants, tangents, radii, and diameters.
- Discover and apply relationships between inscribed, circumscribed, and central angles, radii, and chords, and show that they are true for any circle.

Essential Question

- How are the parts of a circle related to the circle and to each other?

photo: Getty Images

Investigations

What Do You See?

Circles have their own language. Who knew there were so many parts?

Hubcap Designs

Could straight lines and arcs be related? Take a look at angle and arc measurements to find out.

What's Your Angle?

How are the measures of angles and the arcs they intercept related?

The Longest Path

Lines that cross circles have laws of their own. Discover how they are related.

Return to Center

Find your way to the center of the circle. There are many routes. Choose your path.

Circles in the Sky

Are all circles similar? Prove it!

Key Vocabulary

arc, center, central angle, chord, circle, diameter, inscribed angle, major arc, minor arc, radius, sector of a circle

Discover

As you complete Engage and the investigations, record the most important ideas you've learned.

Engage	**Investigation 1**
Investigation 2	**Investigation 3**
Investigation 4	**Investigation 5**

Name____ Date__________

Check for Understanding

Investigate Circles and Parts of Circles: Investigation 1

1. Each combination in the table is for the same circle. Decide if an intersection could occur inside, on, or outside the circle. Then put a checkmark in the appropriate column(s).

Combination	Inside Circle	On Circle	Outside Circle
chord and chord			
secant and secant			
chord and secant			
tangent and tangent			
chord and tangent			
secant and tangent			

2. Examine the figure shown. Which of the statements that follow is not correct?

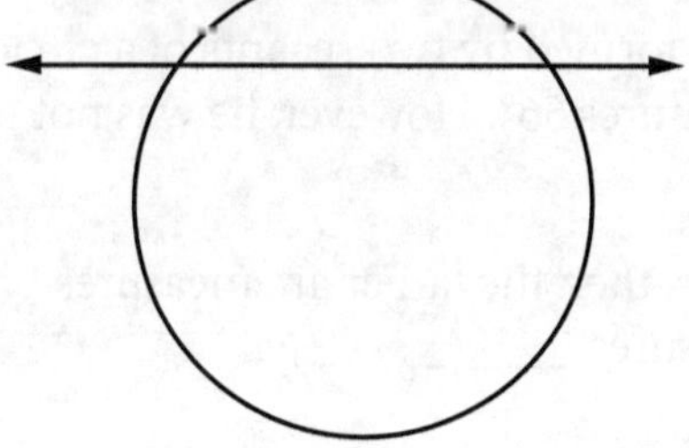

A. It is possible to draw two different chords of the circle that intersect the secant and also intersect each other.

B. It is possible to draw two different tangents of the circle that intersect the secant and also intersect each other.

C. It is possible to draw two different chords of the circle that intersect the secant but that do not intersect each other.

D. It is possible to draw two different tangents of the circle that intersect the secant but that do not intersect each other.

3. Martin and Keiko will work together to draw an intersecting chord and tangent of the same circle. Martin will draw the chord, and Keiko will draw the tangent. If Martin draws the chord first, at how many possible points could the chord and tangent intersect? If Keiko draws the tangent first, at how many possible points could they intersect? Explain.

Name ______________________________ Date ____________

Check for Understanding

Investigate Circles and Parts of Circles: Investigation 2

1. Suppose that $\angle AOC$, a central angle of a circle, and $\angle ABC$, an inscribed angle of the same circle, intercept the same arc. *Circle the correct word(s) in the sentence.*

   The measure of $\angle AOC$ is [half / equal to / double] the measure of $\angle ABC$.

2. Two secants of a circle intersect inside the circle. Which of these statements is correct about the measure of the angle formed by the two secants?

   A. If the measures of the intercepted arcs are 40° and 128°, it is 44°.

   B. If the measures of the intercepted arcs are 74° and 176°, it is 51°.

   C. If the measures of the intercepted arcs are 88° and 140°, it is 114°.

   D. If the measures of the intercepted arcs are 22° and 118°, it is 140°.

3. Garrett was told that an 18° angle is formed by two secants of a circle intersecting outside the circle, and that one of the intercepted arcs measures 56°. However, he was not told if this is the measure of the smaller or larger arc. *Fill in the blanks.*

   If it is the measure of the smaller arc, then the larger arc measures ______, and if it is the measure of the larger arc, then the smaller arc measures ______.

4. In the figure, $\overline{WY}$ and $\overline{XY}$ are chords of circle O, with $\overline{WY}$ passing through the center. Also, $\overleftrightarrow{YZ}$ is tangent to the circle. How could you find $m\angle XYZ$? What is the relationship between $m\angle XYZ$ and the measure of the 122° arc intercepted by chord $\overline{XY}$?

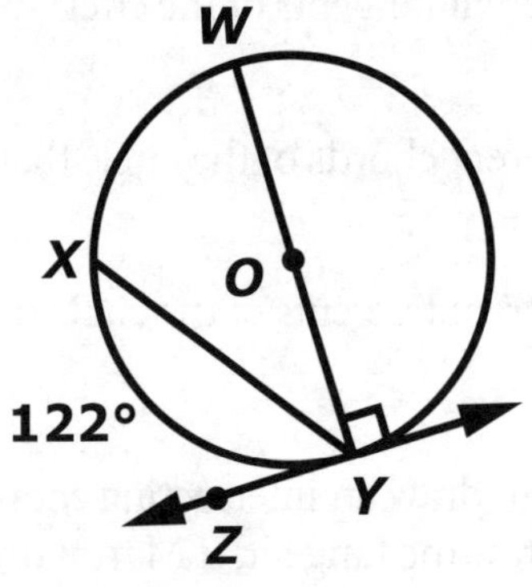

Name ______________________ Date __________

Check for Understanding

Investigate Circles and Parts of Circles: Investigation 3

1. A circle with a center at point O has chords of $\overline{AB}$ and $\overline{CD}$ that intersect arcs with the same measure. Point F lies on $\overline{AB}$, and point G lies on $\overline{CD}$. Both $\overline{OF}$ and $\overline{OG}$ have been drawn, with $\overline{OF}$ perpendicular to $\overline{AB}$, and $\overline{OG}$ perpendicular to $\overline{CD}$. Which of these pairs of line segments must have the same length? *Select all that apply.*

A. $\overline{AF}$ and $\overline{OF}$ **B.** $\overline{OF}$ and $\overline{OG}$ **C.** $\overline{AB}$ and $\overline{CD}$ **D.** $\overline{DG}$ and $\overline{OG}$

E. $\overline{CG}$ and $\overline{OG}$ **F.** $\overline{AF}$ and $\overline{BF}$ **G.** $\overline{BF}$ and $\overline{OF}$ **H.** $\overline{CG}$ and $\overline{DG}$

2. Alexandra drew $\overleftrightarrow{PQ}$, which is tangent to a circle with radius $\overline{OP}$. Which of the following statements is correct about the measure of $\angle OPQ$?

A. It may be 60°.

B. It must be 90°.

C. It may be 120°.

D. It must be 180°.

3. In the figure, $\overline{OZ}$ is a radius of the circle, and $\overline{WX}$ is a chord. The length of $\overline{OZ}$ is equal to r, the length of $\overline{WX}$ is equal to c, and the length of $\overline{YZ}$ is equal to h. Develop a formula that gives r in terms of c and h, and explain how you got your answer.

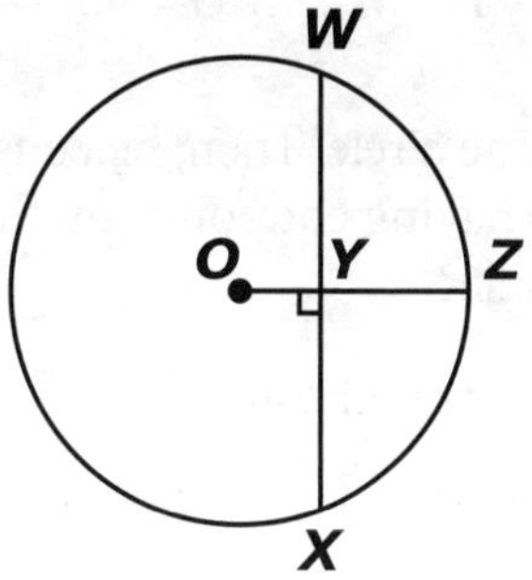

Name ______________________ Date ____________

Check for Understanding

Investigate Circles and Parts of Circles: Investigation 4

1. Two chords of a circle intersect at the circle's center. *Circle the correct words below.*

   The length of [only one / each] of the two chords has to be [less than / equal to] the length of the longest possible chord of the circle.

2. In the figure below, $\overline{AB}$, $\overline{CD}$, $\overline{FG}$ and $\overline{HK}$ are chords of the circle. The intersection of $\overline{AB}$ and $\overline{FG}$ is point M, and the intersection of $\overline{CD}$ and $\overline{HK}$ is point N. Additionally, the length of $\overset{\frown}{AF}$ is half the length of $\overset{\frown}{AB}$ and $\overset{\frown}{CH} \cong \overset{\frown}{HD}$. If point O is the center of the circle, which of the following must be true? *Select all that apply.*

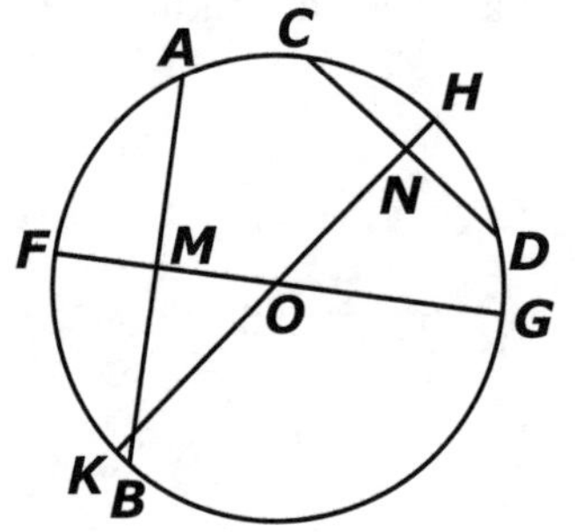

   A. $m\angle GOH = 60°$ B. $AM = BM$ C. $m\angle CNK = 90°$

   D. $2FM = GO$ E. $CN = DN$ F. $m\angle GOK = 120°$

   G. $2HN = NO$ H. $m\angle AMG = 90°$

3. Huey drew two parallel chords of the same circle. Then, he constructed the perpendicular bisector of each, with both perpendicular bisectors having endpoints on the circle. Why will he not be able to use his construction to find the center of the circle?

   A. because the perpendicular bisectors will coincide

   B. because the perpendicular bisectors will not coincide

   C. because neither perpendicular bisector will pass through the circle's center

   D. because only one perpendicular bisector will pass through the circle's center

4. Condoleezza drew two non-parallel tangent lines of the same circle. Then, for each tangent line, she drew a chord of the circle that is perpendicular to the tangent line at the point of tangency. She claims that the intersection of the chords must be the center of the circle. Is Condoleezza's claim correct? Explain your answer.

Name ______________________________ Date ____________

Check for Understanding

Investigate Circles and Parts of Circles: Investigation 5

1. Evan says that when proving two circles are similar by mapping one onto another, a reflection is never needed. Miley says that a rotation is never needed. Who is correct?

   A. only Evan
   B. only Miley
   C. both Evan and Miley
   D. neither Evan nor Miley

2. Suppose you want to prove that ⊙*A*, with a radius of 20 cm, and ⊙*B*, with a radius of 8 cm, are similar by using a combined transformation. *Fill in the blanks with decimals.*

   If ⊙*A* is transformed, the scale factor of the dilation in the combined transformation will be ____________, and if ⊙*B* is transformed, the scale factor will be ____________.

3. Paige's teacher showed her class the figure below. In it, the two circles, ⊙*X* and ⊙*Y*, touch at a single point, with ⊙*X* having a radius of r_1, and ⊙*Y* having a radius of r_2.

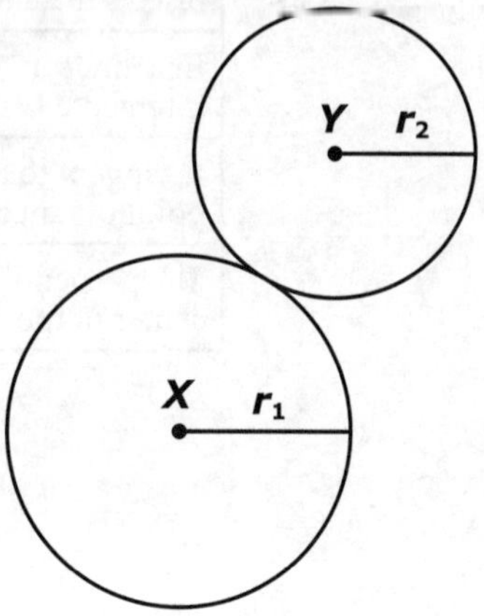

Paige proved that the circles are similar. First, she translated ⊙*X* to the right $\frac{r_1 + r_2}{2}$ units. Then, she translated the image of ⊙*X* up. Finally, she dilated the new image, with the center of dilation being at its center. How many units up did Paige translate the image of ⊙*X*? What was the scale factor when she dilated the new image? Explain.

Summary

Before you attempt the Practice Exercises, review what you've learned.

Parts of Circles

A **circle** is a set of points equidistant from a fixed point, called the **center**.

Some parts of circles are shown in the figure below.

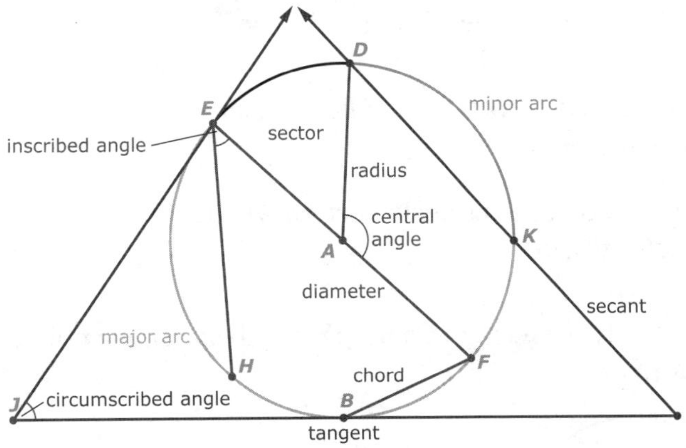

Relationships in Circles

Definition or Theorem
The **measure of an arc** is defined as the measure of its central angle.
If an angle is inscribed in a circle, then the measure of the angle is equal to $\frac{1}{2}$ the measure of its intercepted arc.
If an inscribed angle of a circle intercepts a semicircle, then the angle is a right angle.
If two secants intersect in the interior of a circle, then the measure of an angle formed is $\frac{1}{2}$ the sum of the measures of the intercepted arcs.
If two secants, a secant and a tangent, or two tangents intersect in the exterior of a circle, then the measure of the angle formed is $\frac{1}{2}$ the positive difference of the measures of the intercepted arcs.
In a circle, if a radius is perpendicular to a chord, then it bisects the chord and the arc.
In a circle, two chords are congruent if and only if their intercepted arcs are congruent.
A tangent to a circle is perpendicular to the radius at the point of tangency.
The perpendicular bisectors of two chords intersect at the center of the circle.

My Notes

Summary *(continued)*

EXAMPLE

In the figure, $m\angle CBD = 35°$. Find the following measurements:

$m\angle BDE$

$m\overparen{CD}$

$m\overparen{BE}$

$m\overparen{BCD}$

$m\overparen{BC}$

$m\overparen{DE}$

$m\angle BFE$

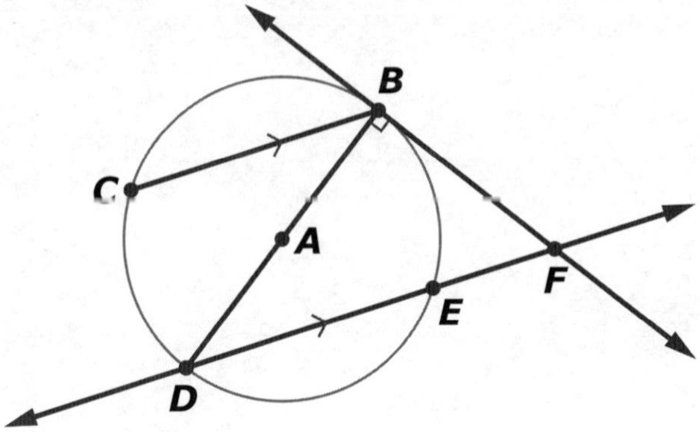

SOLUTION:

By the alternate interior angles theorem, $m\angle BDE = m\angle CBD = 35°$.

The measure of the inscribed angles is half the measure of the intercepted arc, so $m\overparen{CD} = m\overparen{BE} = 70°$.

$\overline{BD}$ is a diameter, so $\overparen{BCD}$ is a semicircle; $m\overparen{BCD} = 180°$.

$$m\overparen{BC} = m\overparen{BCD} - m\overparen{CD} = 180° - 70° = 110°$$

$$m\overparen{DE} = 360° - 70° - 70° - 110° = 110°$$

$m\angle BDE$ is half the difference of the intercepted arcs, or $\frac{1}{2}\left(m\overparen{BCD} - m\overparen{BE}\right) = \frac{1}{2}(180° - 70°) = 55°$.

My Notes

Summary *(continued)*

Circle Similarity

All circles are similar.

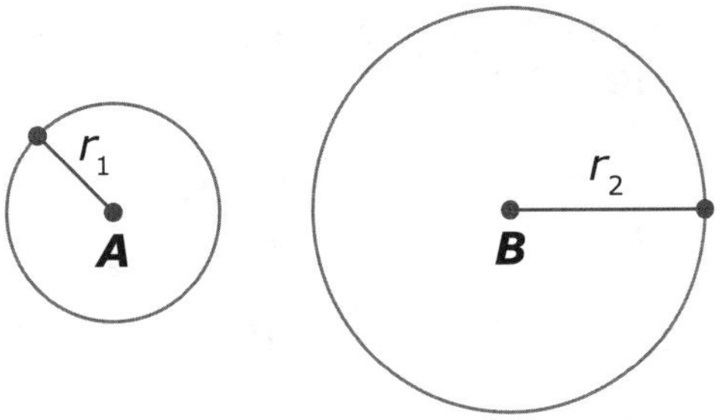

Proof:

Given two circles, $\odot A$ with radius r_1 and $\odot B$ with radius r_2, it is always possible to translate $\odot A$ so that the image is at point B. The image of the circle is the set of all points that are a distance of r_1 from point B. If this image is dilated with center B by a scale factor of $\frac{r_2}{r_1}$, then the second image is the set of points that are a distance of $\frac{r_2}{r_1}(r_1) = r_2$ from point B. By definition, this set of points is $\odot B$. So, there is a combined transformation of $\odot A$ to $\odot B$, therefore the circles are similar by definition of similarity.

My Notes

Practice Exercises

Review what you've learned using these practice problems. For practice problems with feedback, try the Coach and Play items in the Practice section online.

1. Match each term with its definition by writing its letter on the appropriate line.

A. a segment whose endpoints are on the circle

B. a line that intersects the circle at two points

C. a segment with one endpoint on the circle and the other at the circle's center

D. a line that intersects the circle at exactly one point

E. a segment passing through the circle's center whose endpoints are on the circle

Diameter: ____________

Radius: ____________

Chord: ____________

Tangent: ____________

Secant: ____________

2. What is the name for a portion of a circle intercepted by a central angle?

A. center　　**B.** arc

C. radian　　**D.** circumference

3. An angle with its vertex at the center of a circle intercepts an 80° arc of that circle.

What is the measure of the angle?

A. 20°

B. 40°

C. 80°

D. 160°

4. The diagram shows a circle with two lines (*k* and *l*), and a number of points (*A*, *B*, *C*, etc.).

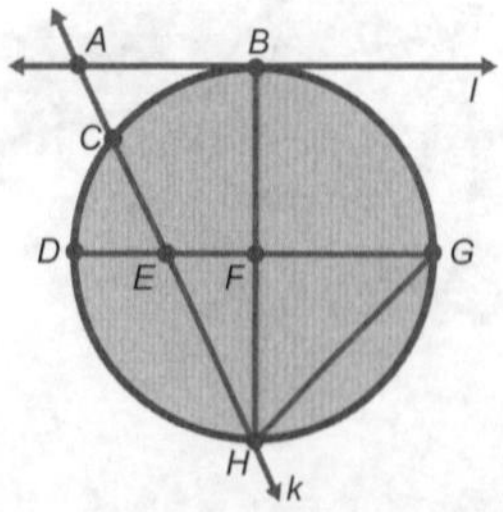

Use the diagram to match each line or line segment with the term that best describes it.

Diameter: ____________

Radius: ____________

Chord: ____________

Secant: ____________

Tangent: ____________

A. k　　**B.** l

C. $\overline{CH}$　　**D.** $\overline{DF}$

E. $\overline{DG}$　　**F.** $\overline{FB}$

G. $\overline{GH}$

5. The figure shows a circle centered at point *P*, with seven points identified around the circle.

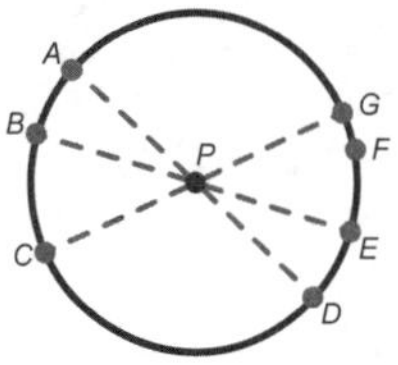

Which of the following are right angles with a vertex at point *F*? *Select all that apply.*

A. $\angle AFD$　　**B.** $\angle AFE$

C. $\angle BFD$　　**D.** $\angle BFE$

E. $\angle CFE$　　**F.** $\angle CFG$

Practice Exercises *(continued)*

6. The diagram shows two overlapping circles with centers A and B. The measures of angles and arcs created by secants are labeled 1 through 5.

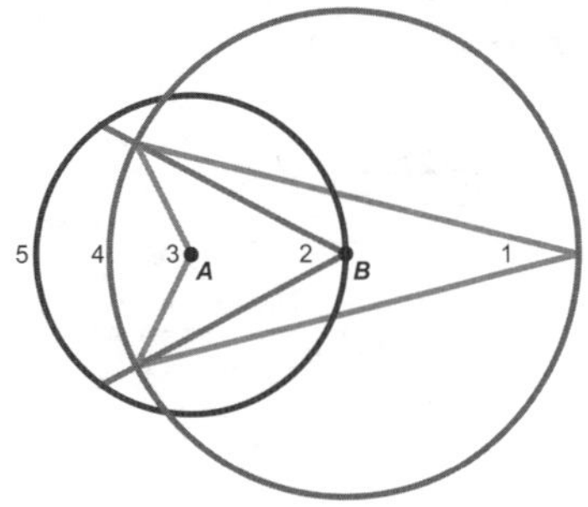

Which of the following statements are true? *Select all that apply.*

A. $m\angle 3 = m\widehat{4}$

B. $m\angle 1 = \frac{1}{2}m\angle 2$

C. $m\angle 2 = \frac{1}{2}m\angle 3$

D. $m\angle 3 = m\widehat{5}$

E. $m\angle 1 = \frac{1}{2}m\widehat{4}$

7. The figure shows a circle centered at point B with a diameter $\overline{AC}$ and chord $\overline{CE}$. The diameter of the circle is $5\sqrt{2}$ units, and the length of chord $\overline{CE}$ is 5 units. *The figure is not drawn to scale.*

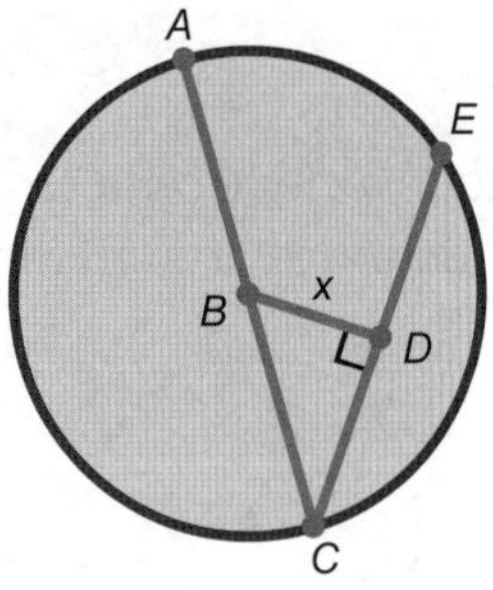

Suppose that $\overline{BD} \perp \overline{CE}$ so that x is the minimum distance from the center of the circle to the chord. Find the distance x and the measure of $\widehat{AE}$.

$x =$ ____________ units

$m\widehat{AE} =$ ____________ °

8. The figure shows a circle with center at point C and two radii forming an angle x. Two chords are drawn from point A on the circle to the endpoints of the two radii, and the two radii are extended to meet intersection points with the chords, B and D. Point A is located such that $m\angle CDA$ equals $m\angle ABC$.

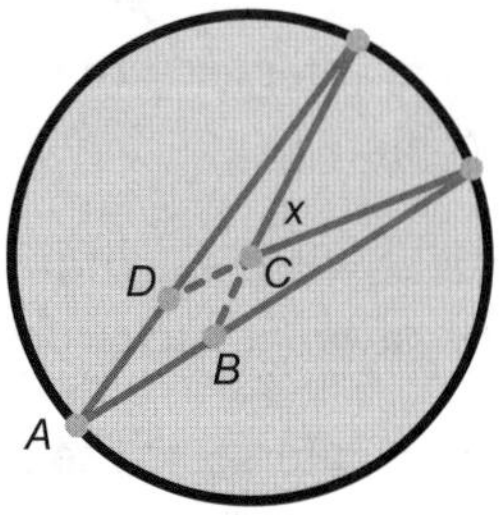

Write the measures of $\angle DAB$, $\angle ABC$, and $\angle BCD$, given that $m\angle x = 30°$.

$m\angle DAB =$ ____________ °

$m\angle ABC =$ ____________ °

$m\angle BCD =$ ____________ °

9. The figure shows a circle with several inscribed angles, with the measure of one angle given as 53°. Note that two of the line segments are perpendicular to each other.

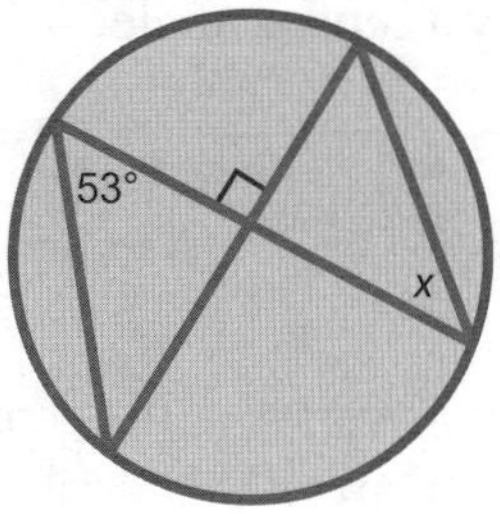

Find the value of x.

$x =$ ____________ °

Apply

How Will You Redesign a Courtyard?

A school wants to clean out its storage shed by beautifying the courtyard with leftover building materials. Watch the video to learn more.

What will make your design unique?

Courtyard Facelift!: View this video segment about the courtyard design.

Suppose the school has a square courtyard with sides that are 52 feet long. School administrators have asked students to come up with design ideas that use odds and ends they have on hand. If your design looks great and makes the most of the materials, it might be brought to life.

Rolls of Sod

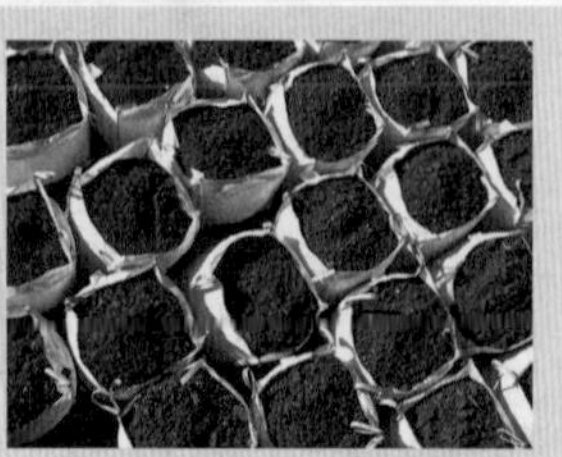

Bags of Soil

Pile of Gravel

Stack of Decking

The school has the following supplies:

- Topsoil that can fill 648 square feet of garden space
- Wooden boards that can make 389 square feet of decking for a seating area
- Sod (rolls of grass) to cover 1,215 square feet
- Gravel to make 711 square feet of walkways

The creative details are up to you, but the administration would like the design to include the following:

- A circle that is as large as possible.
- A seating area that is a circle.
- At least two walkways that cross to opposite sides of the courtyard.

Ideally, most of the materials should be used up. Label your drawing to show areas, angle measures, and chord lengths. Be prepared to describe the reasoning behind your ideas. Use any method you like to create your design, such as the Dynamic Geometry Tool or pencil and paper.

Show what you've learned by completing the other performance tasks in the online Apply section.

Apply *(continued)*

Your answer to Apply will be assessed on the following criteria:

1. Calculating the area of circles and any other shapes created
2. Identifying and calculating the measure of angles
3. Identifying and calculating the measure of chords
4. Proving that the entire courtyard is covered, and determining how much material is left over

Criteria / Scale	4 Exceeds Criteria	3 Meets Criteria	2 Progressing to Criteria	1 Below Expectations	0 No Expectation
Areas	Correctly calculates the area of the large circle, small circle, and all other sections of the courtyard.	Calculates the area of the large circle, small circle, and all other sections of the courtyard, but contains one error.	Calculates the area of the large circle, small circle, and all other sections of the courtyard, but contains two or more errors.	Attempts but does not correctly calculate any of the areas in the courtyard.	Does not calculate the areas.
Angles	Identifies and calculates the measures of all angles in the design.	Identifies and calculates the measures of all angles in the design; contains one error.	Identifies and calculates the measures of all angles in the design; contains two or more errors.	Attempts but does not correctly identify and calculate any angles in the design.	Does not identify angles or calculate their measures.
Chords	Correctly identifies all chords and correctly calculates the length of each.	Identifies all chords and calculates the length of each, but has minor calculation errors.	Identifies all chords and calculates the length of each, but has major calculation errors.	Attempts but does not correctly identify the chords.	Does not identify any chords or calculate their lengths.
Leftover Materials	Shows that the materials used cover the entire courtyard or determines the amount of leftover materials.	Shows that the materials used cover the entire courtyard or determines the amount of leftover materials, but has one error.	Shows that the materials used cover the entire courtyard or determines the amount of leftover materials, but has two or more errors.	Shows that the materials used cover the entire courtyard or determines the amount of leftover materials, but not both.	Does not determine if the materials used cover the entire courtyard or the amount of leftover materials.

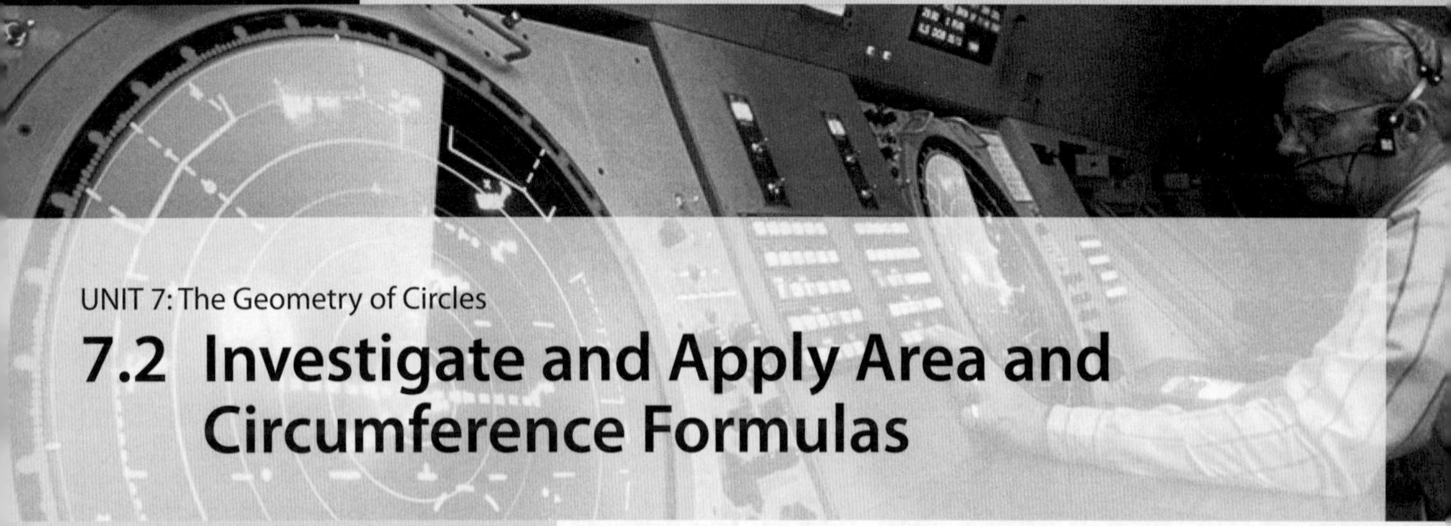

UNIT 7: The Geometry of Circles

7.2 Investigate and Apply Area and Circumference Formulas

photo: Getty Images

Lesson Objectives

- Develop and informally prove the formulas for area and circumference of a circle.
- Solve problems involving area and circumference.
- Develop and apply the formulas for arc length and area of a sector.
- Define the radian measure of an angle.

Essential Question

- How do measurements of a circle relate to one another?

photo: Getty Images

Investigations

Take to the Skies

What area of the sky can an air traffic controller monitor in one view?

A Circle in Squares

Square the circle. Use squares to approximate the area of a unit circle.

Four-Leaf Clover

Prove the formula for the area of a circle. Similarity can help you!

Sectors in Action

Circles, sectors, and lunes—they all take up their own part of the interrelated pie.

A Ride through History

Polygons inside and outside circles can help you approximate area and circumference.

Crafting with Curves

Design a stained-glass window. Find the lengths of the arcs needed for the piece.

Key Vocabulary

area, chord, circumference, image, preimage, radian, scale factor

Discover

As you complete Engage and the investigations, record the most important ideas you've learned.

Engage	Investigation 1
Investigation 2	Investigation 3
Investigation 4	Investigation 5

Name ______________________________ Date ____________

Check for Understanding

Investigate and Apply Area and Circumference Formulas: Investigation 1

1. Terry and Claire each estimated the area of a unit circle drawn on a piece of grid paper by counting squares and finding their total area. Terry's estimate for the area of a unit circle was 3.4 square units, while Claire's was 3.2 square units. Who could have included the squares that were mostly, but not entirely, within the circle when counting?

   A. only Terry
   B. only Claire
   C. both Terry and Claire
   D. neither Terry nor Claire

2. Consider an estimate of the area of a unit circle drawn on a piece of grid paper that is the total area of only the squares entirely within the circle. *Circle the correct words.*

   The [fewer / more] squares there are entirely within the circle, the closer the estimate is to the actual area of the circle, but the estimate is too [large / small].

3. Juaquin drew a unit circle on a piece of grid paper as shown in the figure below.

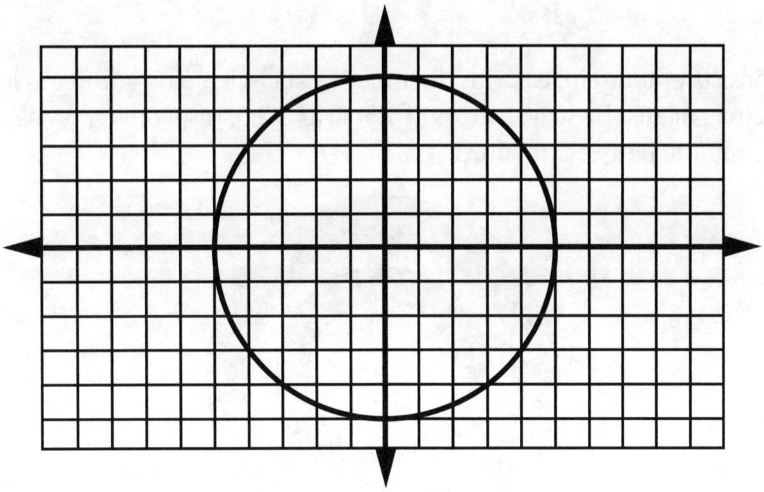

First, he counted the number of squares that appear to be entirely within the circle, and then he found the total area of all these squares. Next, he counted the number of squares that appear to be only partially within the circle. Based upon his knowledge of the area of a unit circle, he calculated the approximate percentage, on average, of each of these squares that is within the circle. Assuming that Juaquin did everything correctly, what did he get for the average percentage? Explain how you got your answer.

Name ______________________________ Date ______________

Check for Understanding

Investigate and Apply Area and Circumference Formulas: Investigation 2

1. Suppose a unit circle was dilated by each of the scale factors given in the following table. Use your calculator to help fill in the area of each of the resulting images as a decimal to the nearest hundredth square unit.

Scale Factor	Area of Resulting Image (square units)
3	
6	
8	
12	
14	
18	

2. Gwen dilated a unit circle. The unit circle's image has an area that is only 0.25% the area of the unit circle. What scale factor did Gwen use in the dilation?

A. $\frac{1}{400}$ **B.** $\frac{1}{250}$ **C.** $\frac{1}{25}$ **D.** $\frac{1}{20}$

3. The figure below consists of four circles, with the innermost circle being a unit circle. The other three circles are the unit circle dilated by scale factors of 1.5, 2, and 2.5, respectively. What is the combined area of the shaded regions in the figure? Explain.

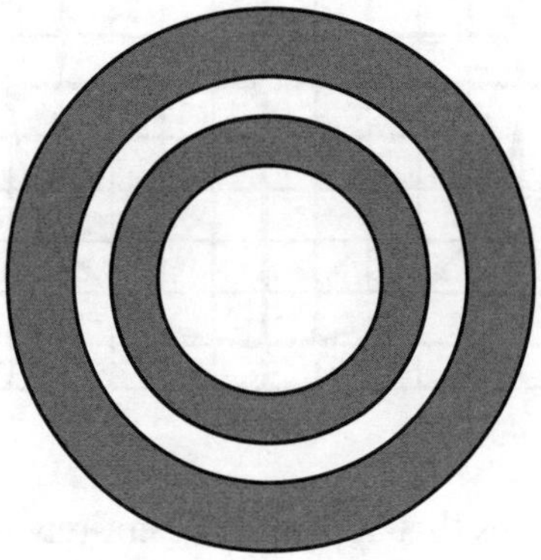

Name____ Date____________

Check for Understanding

Investigate and Apply Area and Circumference Formulas: Investigation 3

1. Cecily drew a sector with a central angle of 72° in a unit circle. *Fill in the blanks in the sentence that follows. Reduce fractions to lowest terms.*

The area of the sector is ______ that of the unit circle, so it is ______ square units.

2. Each entry in the left column below gives the radius, r, of a circle and the measure of a central angle, θ, of a sector of the circle. Match each entry on the left with the corresponding sector area on the right. The areas are rounded to two decimal places.

A. $r = 7; \theta = 36°$	**I.** 16.96 square units
B. $r = 5; \theta = 81°$	**II.** 12.72 square units
C. $r = 4; \theta = 95°$	**III.** 15.39 square units
D. $r = 9; \theta = 18°$	**IV.** 13.26 square units
E. $r = 6; \theta = 54°$	**V.** 14.52 square units
F. $r = 8; \theta = 26°$	**VI.** 17.67 square units

3. The following figure shows a sector with a central angle of $x°$ in a circle of radius r. The shaded region is a segment of the circle. Develop a formula that gives the area, A, of the shaded segment in terms of r and x, and explain how you got your answer.

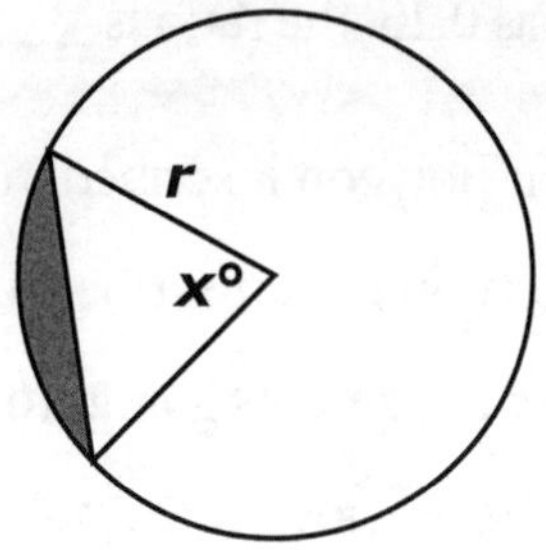

Name ______________________________ Date ______________

Check for Understanding

Investigate and Apply Area and Circumference Formulas: Investigation 4

1. Raul inscribed a regular polygon in a circle as shown to approximate the circle's area.

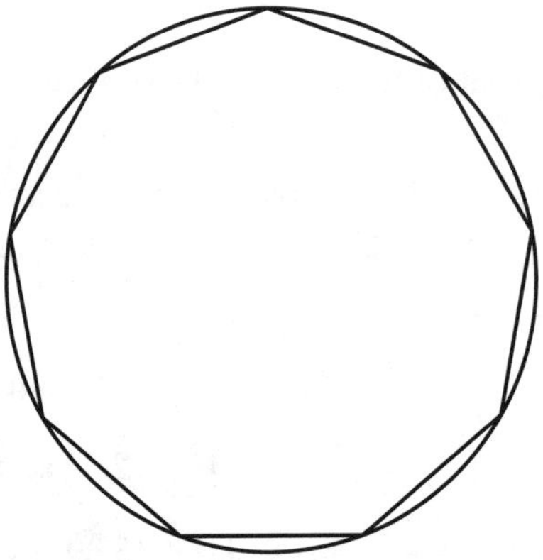

When inscribed in the circle, the area of which of these shapes would give a better approximation of the area of the circle than the area of the polygon inscribed by Raul?

A. a regular decagon
B. a regular hexagon
C. a regular octagon
D. a regular pentagon

2. Suppose you dilated a unit circle. *Fill in the blanks in the following sentence.*

If the scale factor was 4, then the ratio of the circumference of the circle's image to that of the circle is ____________, and if the scale factor was 0.25, the ratio is ____________.

3. The formula for the perimeter of a regular polygon inscribed in a unit circle is $P = 2\sin\left(\frac{\theta}{2}\right) \cdot n$, while the formula for the perimeter of a regular polygon circumscribed around a unit circle is $P = 2\tan\left(\frac{\theta}{2}\right) \cdot n$. In both formulas, θ is the angle formed by line segments going from the center of the circle to two adjacent vertices of the polygon, and n is the number of sides of the polygon. Based upon your knowledge of the circumference of a unit circle, does the circumference of a 180-sided inscribed regular polygon or the circumference of a 180-sided circumscribed regular polygon better approximate a unit circle's circumference? *Use your calculator to justify your answer.*

Name ____________________ Date __________

Check for Understanding

Investigate and Apply Area and Circumference Formulas: Investigation 5

1. In the figure shown, $\angle AOB$ is a central angle of the circle. Which of these statements are correct about the length of the arc intercepted by $\angle AOB$? *Select all that apply.*

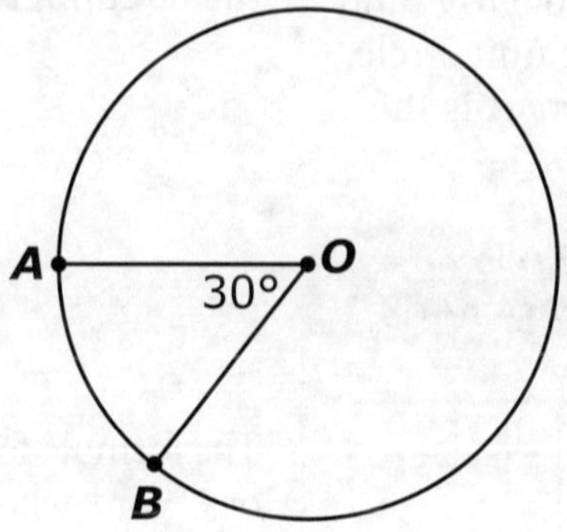

   A. It is 5π units if the length of the circle's radius is 30 units.

   B. It is 20π units if the length of the circle's radius is 60 units.

   C. It is 30π units if the length of the circle's radius is 90 units.

   D. It is 40π units if the length of the circle's radius is 240 units.

   E. It is 100π units if the length of the circle's radius is 300 units.

   F. It is 120π units if the length of the circle's radius is 360 units.

2. As a car stuck in traffic moved slightly forward, each of its tires rotated 81°, and a patch on the surface of one of the tires moved in a circular arc a total distance of 20.2 inches. Assuming the tires are all the same size, which is the approximate radius of each?

   A. 12.7 inches B. 13.2 inches C. 13.4 inches D. 14.3 inches

3. The measure of a central angle of a circle is 1 radian when the length of the arc intercepted by the angle is equal to the length of the circle's radius. Use this fact, along with the arc length formula, to find the exact number of degrees equal to 1 radian. Based on what you found, how many radians are in a circle? Explain your answer.

Summary

Before you attempt the Practice Exercises, review what you've learned.

The Circumference and Area of a Circle

The area of a unit circle is equal to π. Because all circles are similar, we can translate and dilate a unit circle by a scale factor of r to map to any other circle. The image will have a radius equal to r, and the area will be r^2 times the area of the unit circle. This leads to the formula $A = \pi r^2$, where A is the area and r is the radius of the circle.

Unit Circle	Unit Circle Dilated by a Scale Factor of 2
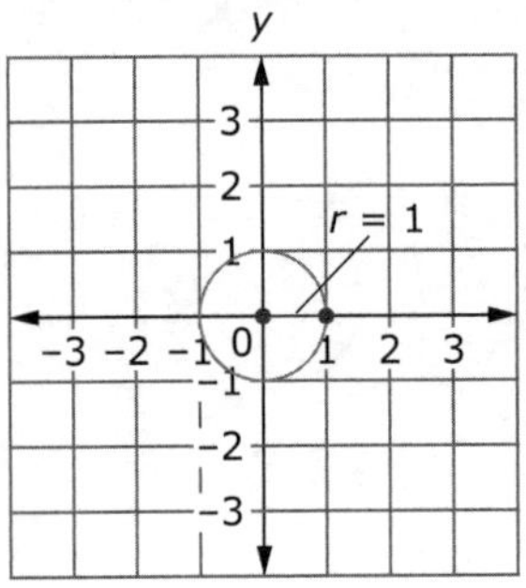	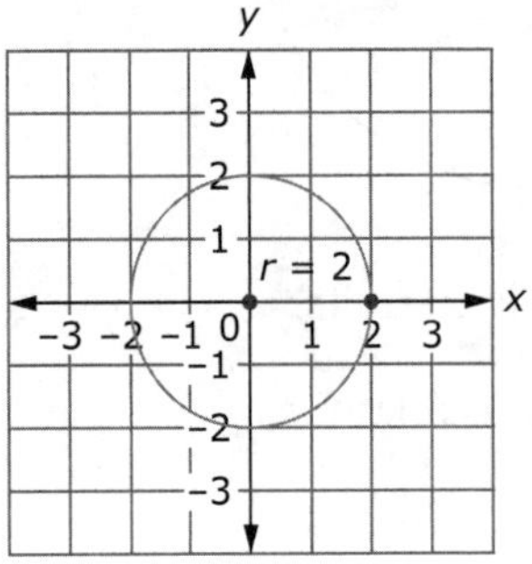

Another representation that can be derived for the area of a circle is $A = \frac{1}{2}Cr$, where C is the circumference of the circle. We can substitute $\frac{1}{2}Cr$ into the formula $A = \pi r^2$ to discover the formula for the circumference of a circle.

$$A = \frac{1}{2}Cr$$

$$\pi r^2 = \frac{1}{2}Cr$$

$$2\pi r^2 = Cr$$

$$2\pi r = C$$

The formula for the circumference of a circle is $C = 2\pi r$, or $C = \pi d$, where d is the diameter of the circle.

EXAMPLE

A circular pool is surrounded by a circular walkway. The diameter of the pool is 15.5 feet, and the width of the walkway is 6.5 feet. Find the circumference and area of the walkway.

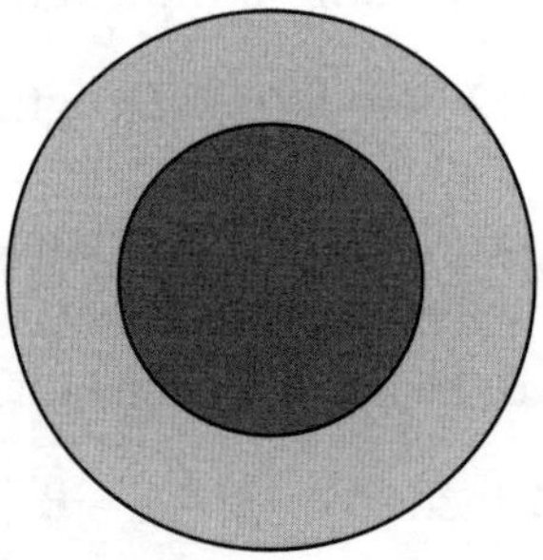

SOLUTION:

Circumference:

The width of the walkway is 6.5 feet and the radius of the pool is 7.75 feet, so the combined radius is 7.75 feet + 6.5 feet = 14.25 feet.

$$C = 2\pi r$$

$$= 2\pi r(14.25)$$

$$= 28.5\pi$$

The circumference of the walkway is 28.5π feet, or about 89.54 feet.

My Notes

Summary *(continued)*

Area:

The area of the walkway can be found by subtracting the area of the pool from the area of the entire circular region.

Find the area of the pool.

The radius of the pool is 7.75 feet.

$$A = \pi r^2$$
$$= \pi(7.75)^2$$
$$= 60.0625\pi$$

The area of the pool is 60.0625π square feet.

Find the area of the entire region.

The radius of the circle that represents the entire region is 14.25 feet.

$$A = \pi r^2$$
$$= \pi(14.25)^2$$
$$= 203.0625\pi$$

The area of the entire region is 203.0625π square feet.

Find the area of the walkway.

Subtract the areas:

$$203.0625\pi - 60.0625\pi = 143\pi$$

The area of the walkway is 143π square feet, or about 449.25 square feet.

The Area of a Sector

The area S of a sector is $S = \frac{n}{360}\pi r^2$, where $n°$ is the central angle measure of the sector, and r is the radius.

EXAMPLE

A pulsating sprinkler is set to rotate 135° with a radius of 6 feet. What is the area of the region it waters?

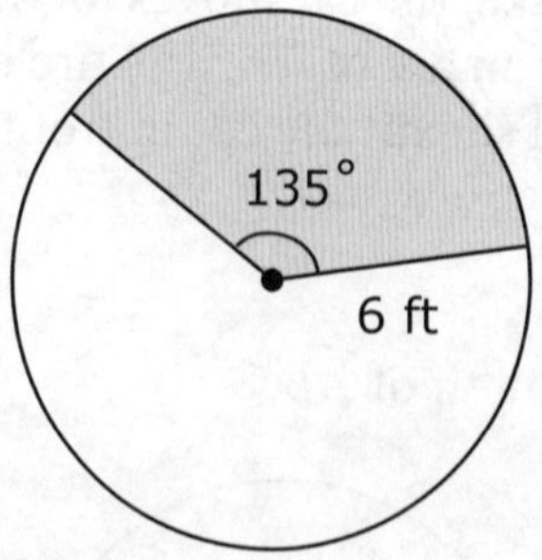

SOLUTION:

$$A = \frac{m}{360°} \cdot \pi r^2$$
$$= \frac{135°}{360°} \cdot \pi(6)2$$
$$\sim 42.41$$

The area of the region is about 42.41 square feet.

Arc Length

The formula for the distance along an arc, called arc length, is $L = \frac{m}{360°} \cdot 2\pi r$, where L is the arc length, m is the measure of the arc, and r is the radius.

My Notes

Summary *(continued)*

EXAMPLE

A driver is adjusting the position of a car in a parallel parking space. She is currently 20 inches from the car in front of her. When she takes her foot off the brake, the car moves forward so that the tires roll along an arc of 75°. The tire diameter is 25 inches. Will she hit the car in front of her?

SOLUTION:

Find the arc length of $\widehat{AB}$.

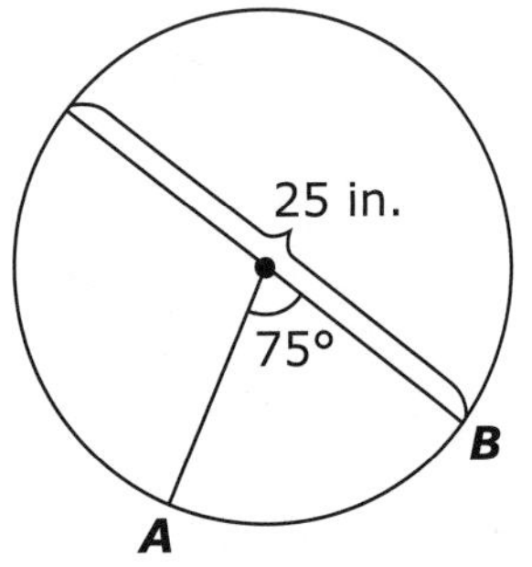

$$L = \frac{m}{360°} \cdot 2\pi r$$
$$= \frac{75°}{360°} \cdot 2\pi(12.5)$$
$$\approx 16.36$$

The car rolls about 16.36 inches. She will not hit the car in front of her.

My Notes

Radian Measure

A radian is a unit of angle measure. One radian is the angle measure that is formed when the radius and arc length are the same length.

There are 2π radians in one full circle, which is equivalent to the circumference of a circle with a radius of 1 unit.

These radian measures have corresponding degree measures in a unit circle.

Because the arc length is proportional to the circumference of the circle, the arc length formula can be rewritten as: $L = \left(\frac{m \cdot 2\pi}{360}\right) r = \left(\frac{m\pi}{180}\right) r$, where $\left(\frac{m\pi}{180}\right)$ is the constant of proportionality. This constant of proportionality is called the radian measure of the angle. It is the ratio of the length of the arc to the radius of the circle. Notice that when the measure of the arc is 360°, the length of the arc is $L = \left(\frac{360 \cdot 2\pi}{360}\right) r = 2\pi r$, which is the circumference of the circle. Because of this important relationship, some references describe the arc length as $s = r\theta$, where s is the length of the arc, r is the radius, and θ is the radian measure of the angle. θ is the Greek letter "theta."

EXAMPLE

In $\odot C$, the length of $\widehat{AB}$ is 16π centimeters, and the radius of the circle is 64 centimeters. What is the radian measure of $\angle ACB$?

SOLUTION:

Substitute the values of the length of the arc and the radius of the circle:

$$s = r\theta$$
$$16\pi = (64)\theta$$
$$\frac{\pi}{4} = \theta$$

The radian measure of $\angle ACB$ is $\frac{\pi}{4}$.

Practice Exercises

Review what you've learned using these practice problems. For practice problems with feedback, try the Coach and Play items in the Practice section online.

1. A circular racetrack has a diameter of 0.8 mile. A driver completes one lap in 50 seconds.

What is the driver's average speed during that lap?

A. 58 mph

B. 90 mph

C. 115 mph

D. 181 mph

E. 226 mph

2. Two runners are jogging around a circular quarter-mile track. The circumference of the lane one runner is using is 0.25 mile. The second runner is staying a constant 5 feet to the outside of the first runner.

How much faster, in percent, must the second person run in order to keep up with the runner in the inside lane? *Use 5,280 feet = 1 mile.*

A. 0.4% **B.** 0.8%

C. 1.2% **D.** 1.6%

E. 2.4% **F.** 4.8%

3. The diameter of Earth is about 7,920 miles. A satellite orbits 200 miles above Earth's surface.

To three significant figures, calculate the circumference of Earth and the distance the satellite travels in one orbit around Earth.

Earth circumference: ____________ mi

Satellite orbit: ____________ mi

4. A circle has a central angle of 90° and a diameter of 16 inches. What is the approximate length of the arc subtended by the central angle?

A. 3.14 inches

B. 6.28 inches

C. 8.98 inches

D. 12.57 inches

5. A circle has a central angle of 150° and a radius of 3 inches. What is the approximate length of the arc subtended by the central angle?

A. 2.61 inches

B. 3.93 inches

C. 7.85 inches

D. 11.78 inches

6. Diana would like to plant a circular flower garden with an area of 500 square feet.

What will the radius of the garden be, and how many feet of fence will be needed to enclose the garden? *Use 3.14 for pi.*

Express your answers to three significant figures.

Radius: ____________ ft

Fencing needed: ____________ ft

Practice Exercises *(continued)*

7. A pizza with a 16-inch diameter is cut into 8 equal slices.

How many equal slices should a 24-inch diameter pizza be cut into to make each slice the same area as those from the 16-inch pizza?

Number of slices for the 24-inch pizza:

____

8. Jenny installed a new windshield wiper on the back window of her car. The wiper rotates through a 120° angle as its rubber blade cleans the windshield. The wiper blade begins cleaning at a distance of 4 inches from the point of rotation. The wiper blade is 16 inches long.

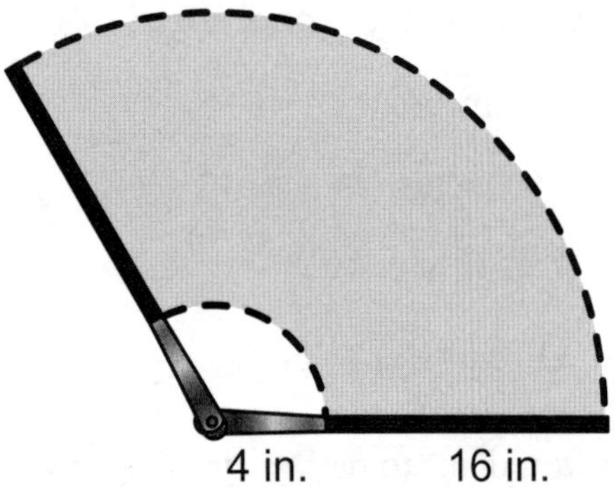

What is the area cleaned by the wiper blade?

Give your answer in terms of π.

The area cleaned by the wiper blade is __________ square inches.

9. You have been asked to create a pie chart that will display four categories. The four categories represent 16%, 18%, 32%, and 34% of the whole. What is the angle for the sector representing 34%?

A. 9 degrees

B. 34 degrees

C. 68 degrees

D. 122 degrees

10. Suppose a circle has a radius R and an area A, expressed in meters and square meters, respectively, so that the numerical values of R and A are equal.

Enter the values for the circumference C and radius R of such a circle.

Express your answers to three significant figures.

$C =$ __________ m

$R =$ __________ m

11. The diagram shows a sector of a circle defined by unknown angle x. The area of the sector is equal to the area of the square.

Find the measure of $\angle x$.

Express your answer to the nearest tenth of a degree.

$x =$ __________

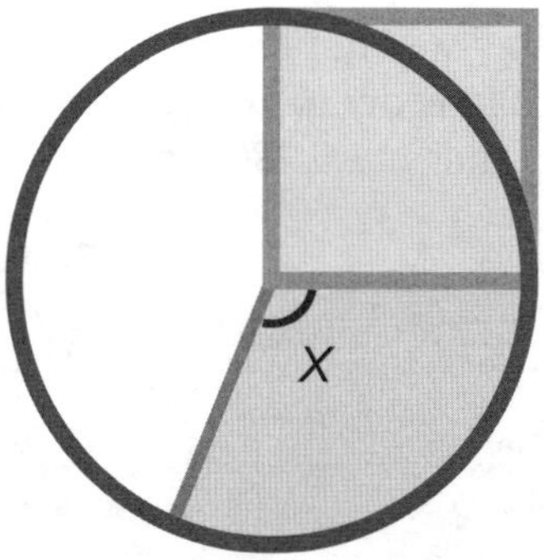

12. What is the arc length of a quarter circle with a radius of 6 inches?

A. π iches

B. 2π inches

C. 3π inches

D. 4π inches

Apply

What Is the Perfect Amount of Pizza Crust?

How much pizza crust do you like?

Everyone seems to have a different opinion on how much crust a pizza should have. What do you think is the most popular ratio of the area of plain crust to the area with toppings on a slice of pizza? Does it vary depending on the size of the pizza?

Research two sizes of pizzas from two different pizza makers. Assume that a slice is a sector equal to one-eighth of the pizza.

Calculate the area of the crust and the area of the part with toppings on one slice. Determine the ratio of crust area to topping area. Is it constant for all of the pizzas? What is your preferred ratio of crust to toppings?

Show what you've learned by completing the other performance tasks in the online Apply section.

Apply *(continued)*

Your answer to Apply will be assessed on the following criteria:

1. Researching the diameters and crust widths of four pizzas and finding the area of each slice
2. Determining the areas of the four parts with toppings
3. Calculating the areas of the four parts that are plain crust
4. Finding the ratios of crust area to topping area and making a conclusion

Criteria \ Scale	4 Exceeds Criteria	3 Meets Criteria	2 Progressing to Criteria	1 Below Expectations	0 No Expectation
Research and Area of Slices	Records the diameters and crust widths of two pizzas each from two different pizza places and finds the area of each slice.	Records the diameters and crust widths of two pizzas each from two different pizza places and finds the area of each slice with minor errors.	Records the diameters and crust widths of two pizzas each from two different pizza places and finds the area of each slice with significant errors.	Records only part of the necessary data or does not find areas.	Does not attempt task.
Topping Areas	Correctly calculates each of the four areas of pizza slices covered by toppings; shows work.	Calculates each of the four areas of pizza slices covered by toppings with minor errors; shows work.	Calculates each of the four areas of pizza slices covered by toppings with significant errors or does not show sufficient work.	Attempts calculations, but they are incomplete or no work is shown.	Does not attempt task.
Crust Areas	Correctly calculates each of the four areas of crust on the pizza slices; shows work.	Calculates each of the four areas of crust on the pizza slices with minor errors; shows work.	Calculates each of the four areas of crust on the pizza slices with significant errors or does not show sufficient work.	Attempts calculations, but they are incomplete or no work is shown.	Does not attempt task.
Ratios and Conclusions	Determines the ratios of crust area to topping area for each of the four pizzas and makes a reasonable conclusion.	Determines the ratios of crust area to topping area for each of the four pizzas with minor errors and makes a reasonable conclusion.	Determines the ratios of crust area to topping area for each of the four pizzas with significant errors or does not make a reasonable conclusion.	Attempts to determine ratios, but they are incorrect or incomplete.	Does not attempt task.

UNIT 7: The Geometry of Circles

7.3 Investigate and Interpret Circle Equations

photo: Getty Images

Lesson Objectives

- Derive the equation of a circle in a coordinate plane.
- Write the equation of a circle that passes through particular coordinates.
- Identify the center, the radius, and points on the circle using an equation.

Essential Question

- How can you represent a circle in the coordinate plane algebraically?

photo: Getty Images

Investigations

Circles in the Coordinate Plane

Plan concentric circles where 500 trailers can park for a reunion.

More Formations

Can an equation define a circle on the coordinate plane, no matter where its center is?

Equivalent Circle Equations

Graphing form, expanded form—when does each have its purpose?

Key Vocabulary

completing the square, expanded form, radian

Discover

As you complete Engage and the investigations, record the most important ideas you've learned.

Engage	
Investigation 1	**Investigation 2**

Name ______________________ Date ____________

Check for Understanding

Investigate and Interpret Circle Equations: Investigation 1

1. The endpoints of a radius of a circle are (4, 6), which is the circle's center, and (7, 10), which is a point on the circle. Suppose a right triangle is drawn with this radius as the hypotenuse in order to find the radius' length. *Fill in the blanks in the sentence below.*

   The length of the shorter leg of the right triangle is ______ units, and the length of the longer leg is ______ units, so the length of the circle's radius is ______ units.

2. Which is an equation that represents all the points on the circle in the following figure?

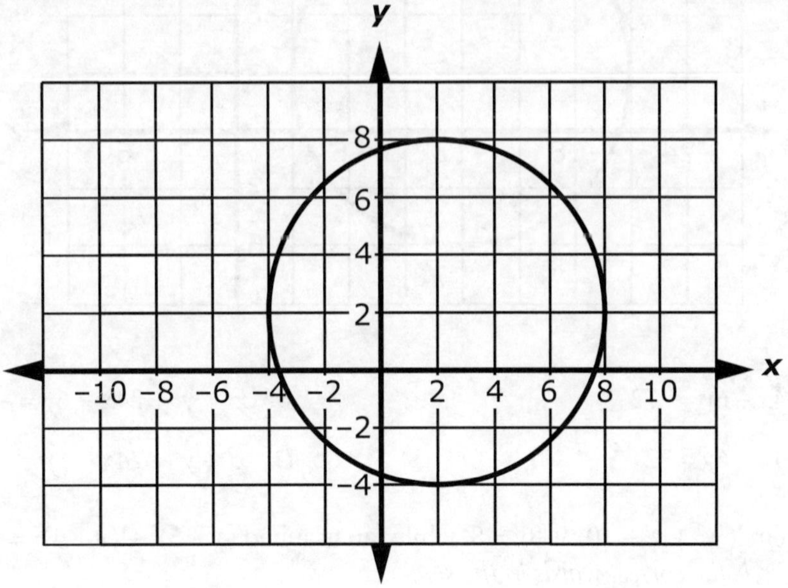

   A. $(x-2)^2+(y-2)^2=6$
   B. $(x+2)^2+(y+2)^2=6$
   C. $(x-2)^2+(y-2)^2=36$
   D. $(x+2)^2+(y+2)^2=36$

3. Consider the circle with the equation $(x+5)^2+(y-1)^2=100$. *Circle the correct words.*

   The points on the circle [do / do not] include (3, 7) and [do / do not] include (4, 5).

4. A circle has a diameter with the endpoints (−8, 9) and (22, −3). What is an equation that represents all the points on the circle? Explain how you got your answer.

Name ______________________________ Date ____________

Check for Understanding

Investigate and Interpret Circle Equations: Investigation 2

1. Which is the expanded form of the equation of the circle graphed in the figure below?

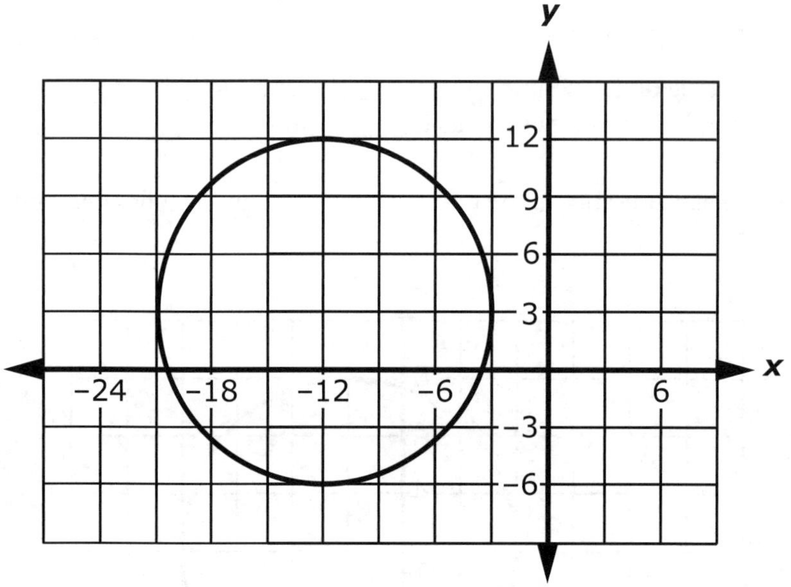

A. $x^2 + y^2 - 24x + 6y = -72$

B. $x^2 + y^2 + 24x - 6y = -72$

C. $x^2 + y^2 - 24x + 6y = 72$

D. $x^2 + y^2 + 24x - 6y = 72$

2. Amy graphed the circle $x^2 + y^2 - 10x + 8y = 8$, while Ian graphed $(x - 5)^2 + (y + 4)^2 = 49$. How do the circles compare? *Circle the correct words in the sentence.*

   Ian's circle has [the same / a different] center and [the same / a different] radius.

3. Seymour wants to find the center and radius of the circle represented by the equation $x^2 + y^2 + 6x - 12y = 16$, so he tried rewriting the equation in graphing form as shown.

$$x^2 + y^2 + 6x - 12y = 16$$
$$x^2 + 6x + y^2 - 12y = 16$$
$$x^2 + 6x + 36 + y^2 - 12y + 144 = 16 + 36 + 144$$
$$(x + 6)^2 + (y - 12)^2 = 196$$

   Seymour concluded that the center of the circle is at the point (−6, 12), while the radius is 14 units. However, Seymour's teacher told him that this is not correct. Identify Seymour's mistake(s), and give the circle's correct center and radius. Explain.

Summary

Before you attempt the Practice Exercises, review what you've learned.

Equation of a Circle

The equation of a circle in graphing form, with center (h, k) and radius r, is $(x - h)^2 + (y - k)^2 = r^2$.

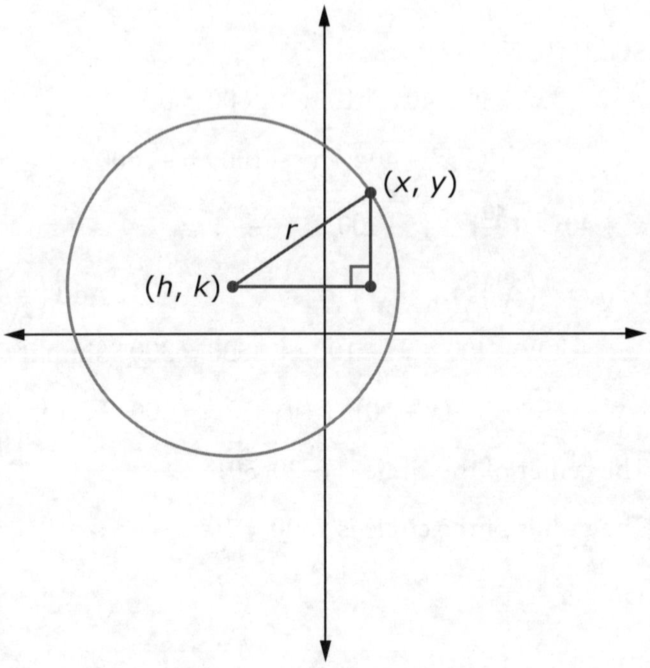

If (h, k) is the origin, then the equation is $x^2 + y^2 = r^2$.

The equation of a circle in graphing form can be used to identify its center, radius, and points on the circle.

EXAMPLE

Use the equation to identify the center and radius of each circle.

a.) $(x - 3)^2 + (y - 5)^2 = 81$

b.) $(x + 2)^2 + (y - 6)^2 = 3$

SOLUTION:

a.) The center is (3, 5), and $r^2 = 81$, so the radius is $\sqrt{81} = 9$.

b.) The equation can be written as:

$$(x - (-2))^2 + (y - 6)^2 = 3$$

The center is (−2, 6) and $r^2 = 3$, so the radius is $\sqrt{3}$.

EXAMPLE

a.) Write the equation of a circle with radius of 5 and center at (2, 3).

b.) Determine if (6, 9) and (5, 7) are points that lie on this circle.

SOLUTION:

a.) Substitute 2 for h, 3 for k, and 5 for r. The equation of the circle is $(x - 2)^2 + (y - 3)^2 = 25$.

b.) To determine if a point lies on the circle, substitute the point into the equation.

Substitute the point $(x, y) = (6, 9)$ into the equation.

$$(x - 2)^2 + (y - 3)^2 = 25$$
$$(6 - 2)^2 + (9 - 3)^2 \stackrel{?}{=} 25$$
$$4^2 + 6^2 \stackrel{?}{=} 25$$
$$52 \neq 25$$

The statement is false, so (6, 9) does not lie on the circle.

My Notes

Summary *(continued)*

Substitute the point $(x, y) = (5, 7)$ into the equation.

$$(x - 2)^2 + (y - 3)^2 = 25$$

$$(5 - 2)^2 + (7 - 3)^2 \stackrel{?}{=} 25$$

$$3^2 + 4^2 \stackrel{?}{=} 25$$

$$25 = 25$$

The statement is true, so (5, 7) lies on the circle.

Transferring between Representations

The equation of a circle in expanded form can be written in graphing form by completing the square, or by graphing the equation to find the center and radius.

Expanded form: $ax^2 + by^2 + cx + dy = e$

Graphing form: $(x - h)^2 + (y - k)^2 = r^2$

EXAMPLE

Write the equation of the circle in graphing form. Then, identify the center and radius of the circle.

$$x^2 + y^2 + 40x - 100y + 2600 = 0$$

SOLUTION:

$$x^2 + y^2 + 40x - 100y + 2600 = 0$$

$$x^2 + 40x + y^2 - 100y = -2600$$

$$x^2 + 40x + \left(\frac{40}{2}\right)^2 + y^2 - 100y + \left(-\frac{100}{2}\right)^2 =$$

$$-2600 + \left(\frac{40}{2}\right)^2 + \left(-\frac{100}{2}\right)^2 \quad \text{Add } \left(\frac{b}{2}\right)^2.$$

$$x^2 + 40x + 400 + y^2 - 100y + 2500 = 300 \quad \text{Simplify.}$$

$$(x + 20)^2 + (y - 50)^2 = 300 \quad \text{Factor.}$$

The center of the circle is (−20, 50).

The radius of the circle is $\sqrt{300} = 10\sqrt{3}$.

My Notes

Practice Exercises

Review what you've learned using these practice problems. For practice problems with feedback, try the Coach and Play items in the Practice section online.

1. Complete the statements regarding the equation of a circle.

A circle with equation $(x-4)^2 + (y-6)^2 = 25$ has a radius of ______ units.

The center point is ______.

One point on the circle is (______, 3).

2. Gertie the goat is tethered with an 8-foot rope to a stake at (2, −3) in the coordinate system on the ground of a large grassy enclosure.

What is the equation of the circle around the boundary of the area where Gertie can graze? *Write the entire equation.*

$(x -$ ______$)^2 + (y -$ ______$)^2 = ($______$)^2$

3. A circle centered at the origin passes through (−6, 4).

What is the equation of the circle?

The equation of the circle is $x^2 + y^2 =$ ______.

4. Point $P(-4, 4)$ is on the circle given by the equation $(x-2)^2 + (y-k)^2 = 100$.

Which coordinates could be the center of the circle? *Select all that apply.*

A. (−2, 2)

B. (−2, 6)

C. (2, −4)

D. (2, 4)

E. (2, 12)

5. A planetary model shows the circular orbit of a planet around a star in a coordinate plane. The plane is observed passing through the coordinates (2, 7), (12, −3), and (2, −13), which form the vertices of a right triangle.

What is the equation of the circle that describes the orbital path of the planet?

A. $(x+2)^2 + (y-3)^2 = 8^2$

B. $(x+2)^2 + (y-3)^2 = 10^2$

C. $(x-2)^2 + (y+3)^2 = 10$

D. $(x-2)^2 + (y+3)^2 = 10^2$

E. $(x+2)^2 + (y-3)^2 = 8$

6. The equation for a circle in the coordinate plane is $x^2 - 4x + y^2 + 6y - 3 = 0$.

The center of the circle is (2, −3). What is the equation for this circle in graphing form?

A. $(x+2)^2 + (y-3)^2 = 4^2$

B. $(x+2)^2 + (y-3)^2 = 4$

C. $(x-2)^2 = (y+3)^2 + 4^2$

D. $(x-2)^2 + (y+3)^2 = 4^2$

E. $(x-h)^2 + (y-k)^2 = r^2$

Practice Exercises *(continued)*

7. Steven and Mariko are playing a game. One rolls number cubes to get the parameters of a circle, and the other gives the equation of the circle.

Mariko has just rolled three number cubes: the one naming the x-coordinate of the center reads 3, the one naming the y-coordinate is 4, and the one for the radius is 2.

What does Steven give as the equation for the circle?

A. $(x - 3)^2 + (y - 4)^2 = 2^2$

B. $(x - 3)^2 + (y - 4)^2 = 2$

C. $(x + 3)^2 + (y + 4)^2 = 2^2$

D. $(x + 3)^2 + (y + 4)^2 = 2$

E. $(x - 4)^2 + (y - 3)^2 = 5^2$

8. Julia and Ian are practicing for their geometry test by drawing circles from equations. Julia gives Ian the following equation: $x^2 + 6x + y^2 - 10y - 15 = 0$.

Before Ian can draw this circle, he must find the coordinates of the center point and the length of the radius.

The circle is centered at ______ with a radius of ______ units.

9. Which of the following points lie on a circle with equation $(x - 1)^2 + (y - 3)^2 = 34$?

Select all that apply.

A. (1, 3)

B. (−4, 0)

C. (6, 0)

D. (4, 8)

E. (−3, 0)

10. Draw the circle described by the equation $x^2 + y^2 - 4x + 10y = -13$.

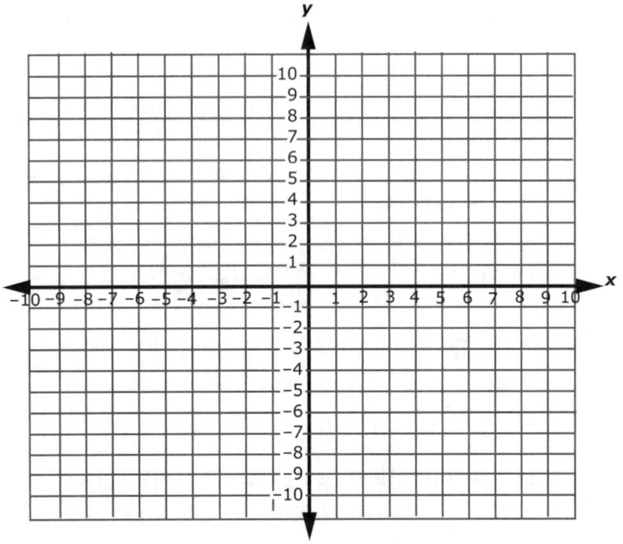

11. A square of side 8 has a vertex at the origin.

What is the equation of he circle centered at the origin that goes through the vertex of the square farthest from the origin?

The equation of the circle is $x^2 + y^2 =$ ______.

Apply

Geometry Detective: What Is a Reuleaux Triangle?

The Reuleaux triangle, shown here on the U.S. Bike Route sign, is named for a 19th-century German engineer, Franz Reuleaux. It is not a typical polygon; in fact, it may share more properties with circles than with triangles. How do you think Reuleaux triangles are constructed? What properties do they have that make them different from ordinary triangles?

Can a bicycle roll with triangular wheels?

The special properties of Reuleaux triangles can be extended to any regular polygon with an odd number of sides. To help you understand the construction of the triangle, study these images of a Reuleaux pentagon.

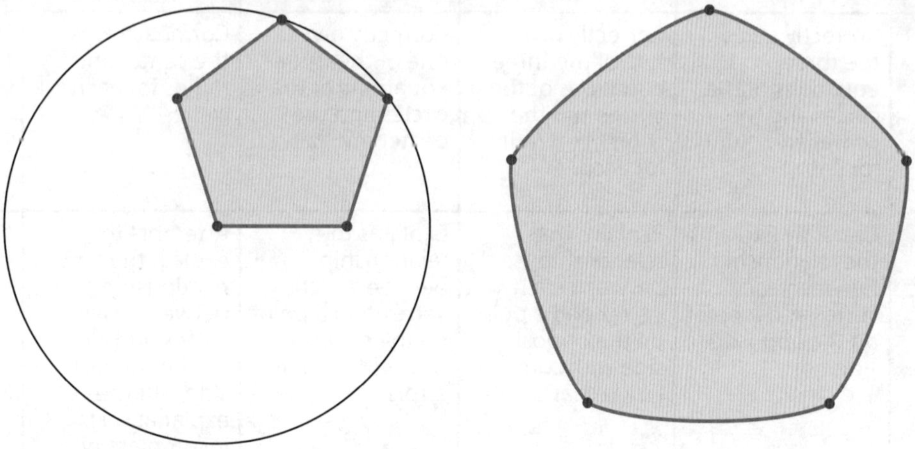

Using a compass and graph paper or an online geometry program:

- Construct a Reuleaux triangle.
- Identify the equation, center, and radius of each circle used.
- Determine the distance from each vertex to any point on the opposite side.
- Explain the results of the previous step using your knowledge of geometry.
- Research applications of the Reuleaux triangle, and explain the attribute that makes the application possible.

Show what you've learned by completing the other performance tasks in the online Apply section.

Apply *(continued)*

Your answer to Apply will be assessed on the following criteria:

1. Constructing a Reuleaux triangle
2. Identifying the circles by center, radius, and equation
3. Determining and explaining the distances from vertex to opposite side
4. Describing an application of a Reuleaux triangle

Criteria \ Scale	4	3	2	1	0
	Exceeds Criteria	**Meets Criteria**	**Progressing to Criteria**	**Below Expectations**	**No Expectation**
Construction	Correctly constructs a Reuleaux triangle.	Constructs a Reuleaux triangle with minor errors.	Constructs a Reuleaux triangle with significant errors.	Construction is not a Reuleaux triangle.	Does not attempt task.
Equations	Correctly gives the three equations of the circles and the center and radius for each.	Correctly gives two of the three equations of the circles and the center and radius for each.	Correctly gives one of the three equations of the circles and the center and radius.	Correctly gives the center and radius for each circle.	Does not attempt task.
Distance	Correctly explains the relationship between each vertex and a point on the opposite side.	Explains the relationship between each vertex and a point on the opposite side with some lack of clarity.	Explains the relationship between each vertex and a point on the opposite side with minor errors.	Attempts to explain the relationship between each vertex and a point on the opposite side, but the explanation is incomplete or incorrect.	Does not attempt task.
Application	Describes in detail an application of the Reuleaux triangle.	Describes an application of the Reuleaux triangle with some lack of clarity.	Gives an application of the Reuleaux triangle without any detail.	Gives an incorrect application of Reuleaux triangles.	Does not attempt task.

UNIT 8: Triangles and Circles

8.1 Investigate Concurrency in Triangles

photo: Getty Images

Lesson Objectives

- Identify, describe, and apply the relationship between the centroid and the medians of a triangle, the orthocenter and the altitudes of a triangle, the vertices of a triangle and the circumcenter, and the sides of a triangle and the incenter.
- Construct inscribed and circumscribed circles of triangles, and understand their relationship to the points of concurrency.

Essential Question

- What are the points of concurrency of a triangle, and how are they used to solve problems involving triangles?

photo: Getty Images

Investigations

City Centers

Where is the center of a triangle?

Location, Location, Location

A triangle's center can be based on the sides or the vertices. Use constructions to find these centers.

How Did They Do That?

Balance requires finding the center of mass. And where is that, exactly?

A Closer Look at Medians

What's the median's relationship to the centroid?

Concurrency in Different Types of Triangles

What do the points of concurrency tell you about a triangle?

Key Vocabulary

altitude, centroid, circumcenter of a triangle, concurrent, equidistant, Euler line, median (triangle), orthocenter of a triangle

Discover

As you complete Engage and the investigations, record the most important ideas you've learned.

Engage

Investigation 1	**Investigation 2**
Investigation 3	**Investigation 4**

Name ______________________________ Date ____________

Check for Understanding

Investigate Concurrency in Triangles: Investigation 1

1. Which of the following statements is correct about the circumcenter of a triangle?

A. It is the intersection of the bisectors of the triangle's three angles and is equidistant from the triangle's three sides.

B. It is the intersection of the bisectors of the triangle's three angles and is equidistant from the triangle's three vertices.

C. It is the intersection of the perpendicular bisectors of the triangle's three sides and is equidistant from the triangle's three sides.

D. It is the intersection of the perpendicular bisectors of the triangle's three sides and is equidistant from the triangle's three vertices.

2. Suppose point M is the circumcenter of $\triangle ABC$, while point N is the incenter of $\triangle ABC$. Which of the following statements could be correct? *Select all that apply.*

A. M and N are both inside $\triangle ABC$.

B. M is on $\triangle ABC$, and N is not.

C. N is outside $\triangle ABC$, and M is not.

D. M and N are both outside $\triangle ABC$.

E. M and N are both on $\triangle ABC$.

F. N is inside $\triangle ABC$, and M is not.

3. In the following figure, $\odot O$ is inscribed in $\triangle XYZ$. Use the fact that the sum of the areas of $\triangle XOY$, $\triangle XOZ$, and $\triangle YOZ$ is equal to the area of $\triangle XYZ$ to help determine the radius of $\odot O$. Based on what you found, what is the area of $\odot O$? Explain your answer.

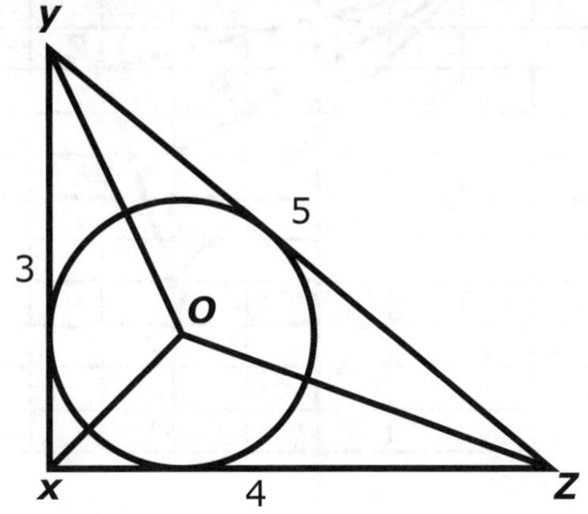

Name ______________________________ Date ____________

Check for Understanding

Investigate Concurrency in Triangles: Investigation 2

1. A crane operator is trying to lift a large, flat triangular piece of steel. How many of the medians of the triangular piece of steel must pass through its center of mass?

A. 0 **B.** 1 **C.** 2 **D.** 3

2. Megan constructed the three medians of $\triangle ABC$, which divided it into six smaller triangles. Which of these statements must be correct? *Select all that apply.*

A. All six of the smaller triangles are congruent.

B. All six of the smaller triangles have the same area.

C. All six of the smaller triangles are similar to $\triangle ABC$.

D. At least two of the smaller triangles have the same height.

E. At least two of the smaller triangles have congruent bases.

3. Pedro drew $\triangle XYZ$ as shown in the figure below. He also drew one of the medians of the triangle from vertex X to point W. On $\overline{XW}$ he plotted point P, which he believes is the centroid of the triangle. How can he use slopes to verify whether or not he plotted the correct centroid? Is Pedro's centroid correct? Explain how you got your answer.

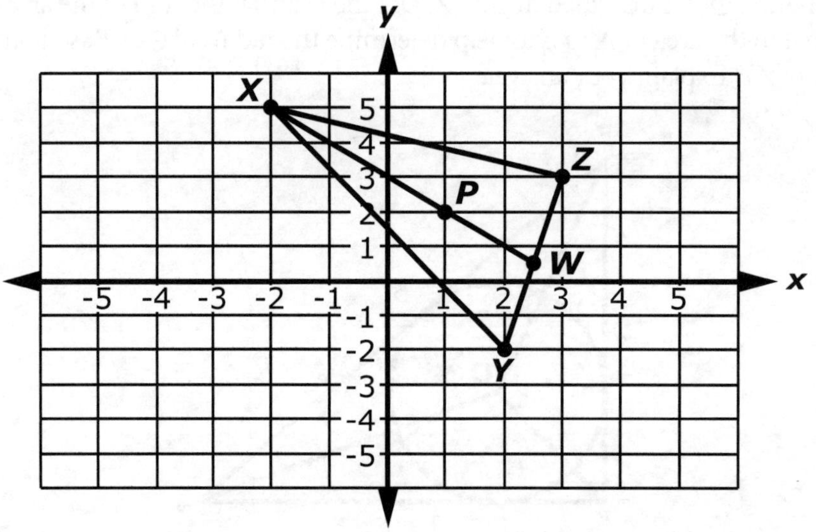

Name ______________________________ Date __________

Check for Understanding

Investigate Concurrency in Triangles: Investigation 3

1. Monique constructed $\overline{AD}$, which is a median of $\triangle ABC$. The centroid of $\triangle ABC$ is point X. Which of the statements correctly compares the lengths of two line segments?

A. The length of $\overline{AD}$ is two times the length of $\overline{AX}$.

B. The length of $\overline{AX}$ is two times the length of $\overline{XD}$.

C. The length of $\overline{AD}$ is three times the length of $\overline{AX}$.

D. The length of $\overline{AX}$ is three times the length of $\overline{XD}$.

2. If you draw $\triangle FGH$, with midsegment $\overline{JK}$, and medians $\overline{FJ}$ and $\overline{GK}$ that intersect at point X, which of the following must be true statements? *Select all that apply.*

A. $FX = GX$ **B.** $2KX = FX$ **C.** $\triangle FXG \sim \triangle JXK$ **D.** $FJ = 3JX$

E. $\triangle FXG \cong \triangle JXK$ **F.** $JX - KX$ **G.** $0.5GX = KX$ **H.** $FK = GJ$

3. In the figure shown, the three medians of $\triangle PQR$, which are $\overline{PM}$, $\overline{QN}$, and $\overline{RL}$, intersect at point X. The lengths of the sides of $\triangle PQR$ are $PR = 2$, $QR = 2$, and $PQ = 2\sqrt{2}$, while the lengths of its medians are $PM = \sqrt{5}$, $QN = \sqrt{5}$, and $RL = \sqrt{2}$. What are the perimeters of each of the six smaller triangles into which $\triangle PQR$ is divided? Explain.

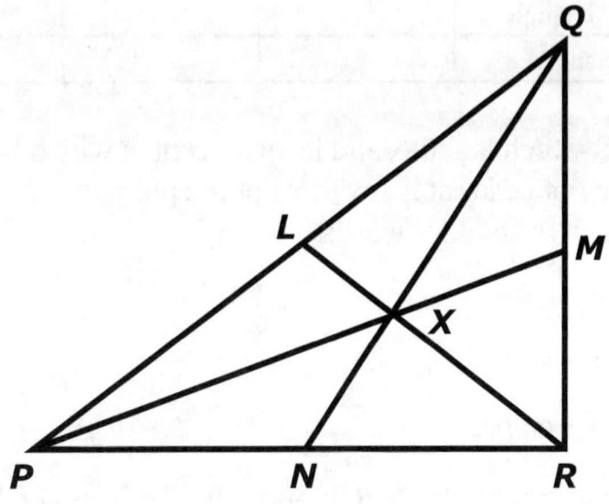

Name ______________________________ Date ____________

Check for Understanding

Investigate Concurrency in Triangles: Investigation 4

1. Which points of concurrency coincide with the orthocenter in an equilateral triangle?

A. only the centroid and incenter

B. only the centroid and circumcenter

C. only the circumcenter and incenter

D. the centroid, circumcenter, and incenter

2. For each of the points of concurrency given in the table, decide if it is inside, on, or outside the triangle. Then, put a checkmark in the appropriate column.

Point of Concurrency	Inside	On	Outside
Orthocenter of Acute Triangle			
Centroid of Acute Triangle			
Circumcenter of Acute Triangle			
Incenter of Acute Triangle			
Orthocenter of Right Triangle			
Centroid of Right Triangle			
Circumcenter of Right Triangle			
Incenter of Right Triangle			
Orthocenter of Obtuse Triangle			
Centroid of Obtuse Triangle			
Circumcenter of Obtuse Triangle			
Incenter of Obtuse Triangle			

3. Raymond constructed $\triangle ABC$, which is acute, and its orthocenter, which he labeled point P. Then he drew $\triangle ABP$, $\triangle ACP$, and $\triangle BCP$. He noticed that if any three of the points A, B, C, and P form a triangle, then the remaining point is the orthocenter. In other words:

- point C is the orthocenter of $\triangle ABP$
- point B is the orthocenter of $\triangle ACP$
- point A is the orthocenter of $\triangle BCP$

Based on this information, how can $\triangle ABP$, $\triangle ACP$, and $\triangle BCP$ be classified? Explain.

Summary

Before you attempt the Practice Exercises, review what you've learned.

> Three or more lines or curves that intersect at a single point are called **concurrent**, and the point of intersection is called a **point of concurrency**.

Orthocenter

> The **orthocenter** is the intersection of the altitudes of a triangle.

In an acute triangle, the orthocenter is inside the triangle.	In a right triangle, the orthocenter is the vertex of the right angle.	In an obtuse triangle, the orthocenter is outside the triangle.
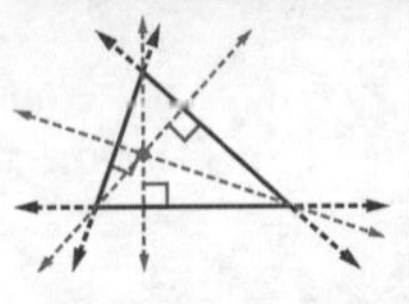	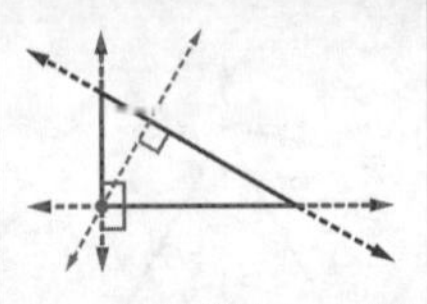	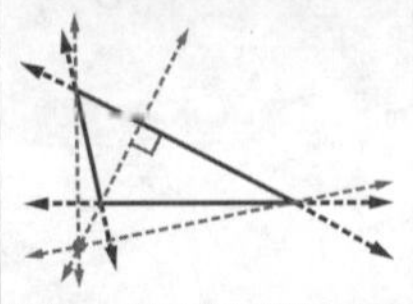

Circumcenter

> The **circumcenter** is the intersection of the perpendicular bisectors of the sides of a triangle.

The circumcenter is equidistant from the vertices of the triangle, so it is also the center of the circumscribed circle.

In an acute triangle, the circumcenter is inside the triangle.	In a right triangle, the circumcenter is the mid-point of the hypotenuse.	In an obtuse triangle, the circumcenter is outside the triangle.
	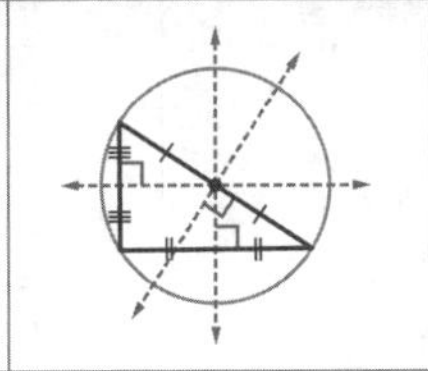	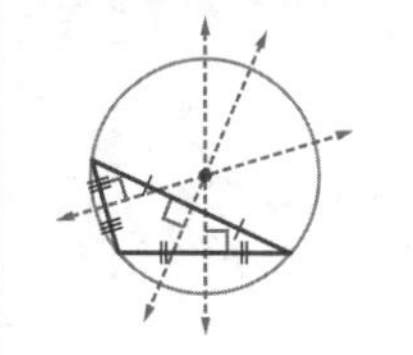

Incenter

> The **incenter** is the intersection of the angle bisectors of the angles of a triangle.

The incenter is equidistant from the sides of the triangle, so it is also the center of the inscribed circle. The incenter is always inside a triangle.

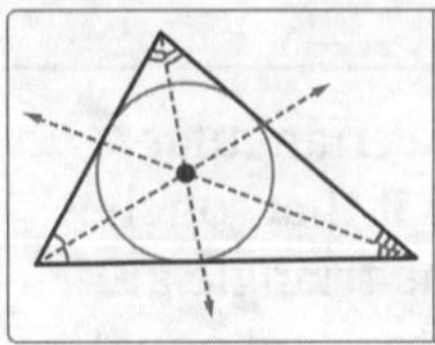
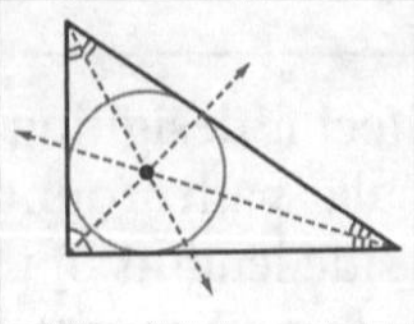
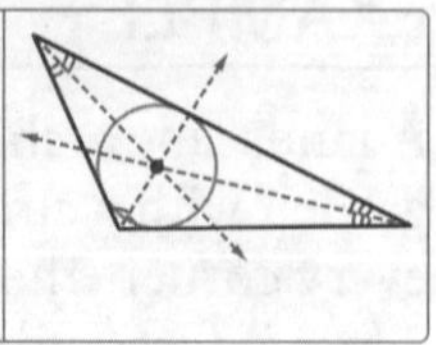

Centroid

> The **centroid** is the intersection of the medians of a triangle. The distance from a triangle's vertex to the centroid is $\frac{2}{3}$ the length of the median.

The centroid is always inside the triangle.

My Notes

Summary *(continued)*

Each median divides the original triangle into two triangles with equal areas. The three medians divide the triangle into six triangles with equal areas, with the centroid at the common vertex. The centroid is the center of mass, or balance point of a triangle.

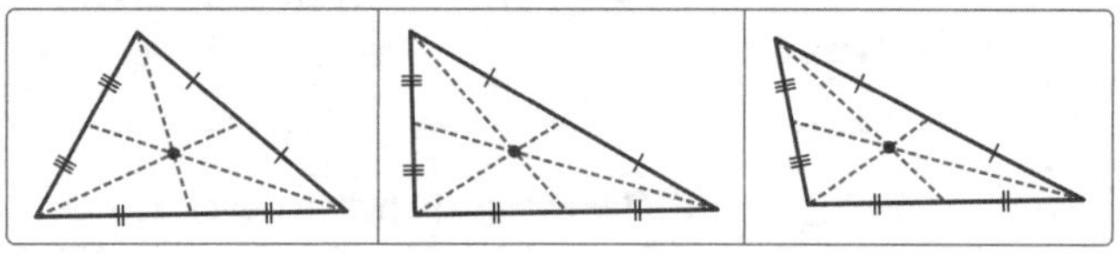

EXAMPLE

A landscape architect is designing a triangular garden with a circular path around it that touches every corner. The side lengths of the triangle are 3 feet, 4 feet, and 5 feet. She wants to plant equal areas of six different flowers and place a birdbath where the flower beds intersect. Draw a possible sketch of the garden. Describe the sketch, including how each point was located.

SOLUTION:

The sides of the triangle are a Pythagorean triple, so it is a right triangle. The path around it will be a circumscribed circle, so the center will be the midpoint of the hypotenuse, *C*.

The three medians will divide the triangle into six equal areas with the birdbath, *B*, at the centroid.

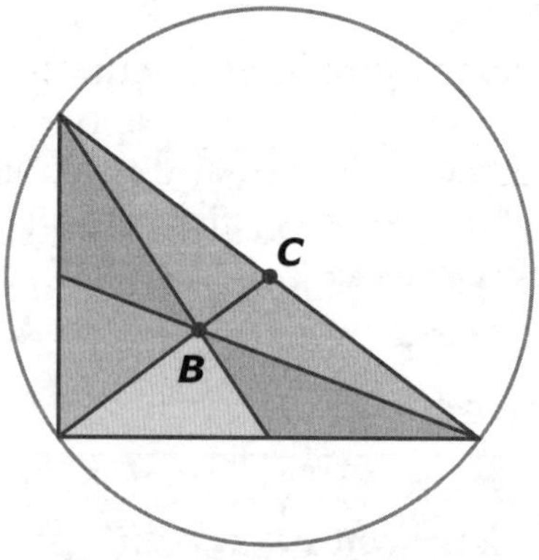

Equilateral and Isosceles Triangles

In an equilateral triangle, the orthocenter, circumcenter, incenter, and centroid all coincide. This point is also the center of rotational symmetry for the triangle.

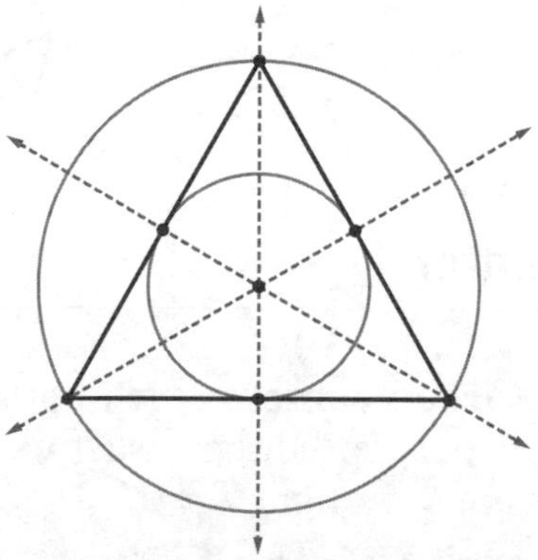

In an isosceles triangle, the orthocenter, circumcenter, incenter, and centroid are all on the line of symmetry of the triangle.

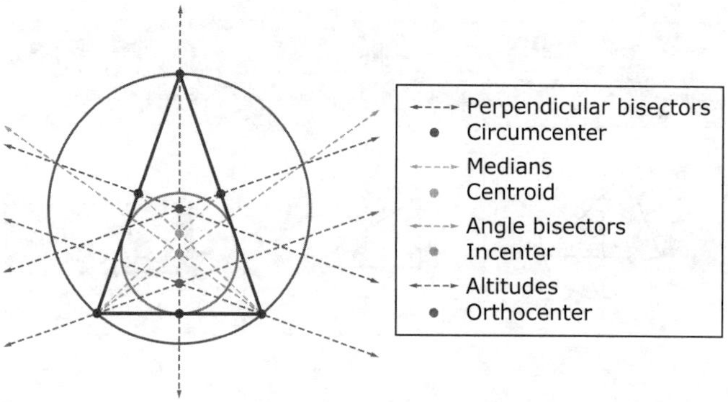

My Notes

Practice Exercises

Review what you've learned using these practice problems. For practice problems with feedback, try the Coach and Play items in the Practice section online.

1. The centroid of $\triangle ABC$ occurs at point M. Point Q is the midpoint of $\overline{BC}$, $\overline{AQ}$ is a median of the triangle, and $AM = 8$. What is AQ?

   A. 4

   B. 8

   C. 12

   D. 16

2. Michael constructs the incenter of a triangle.

   What does he find is true of the incenter?

   A. The incenter is inside the triangle only when it is acute.

   B. The incenter is outside the triangle when it is obtuse.

   C. The incenter is at the vertex of the right angle of a right triangle.

   D. The incenter is always inside the triangle.

3. Sanjay has just constructed a triangle, and finds that the orthocenter (the intersection of the three altitudes of the triangle) will lie outside of the triangle.

   What is the best explanation for Sanjay's observation?

   A. He has made a mistake; the orthocenter should be inside the triangle.

   B. He is correct; he has drawn an acute triangle.

   C. He is correct; he has drawn a right triangle.

   D. He is correct; he has drawn an obtuse triangle.

   E. He is correct; the orthocenter is always outside the triangle.

4. Cheyenne creates a triangle with a graphing program and then draws the median from each vertex. She calculates that the area of $\triangle ADG$ is 10 square units.

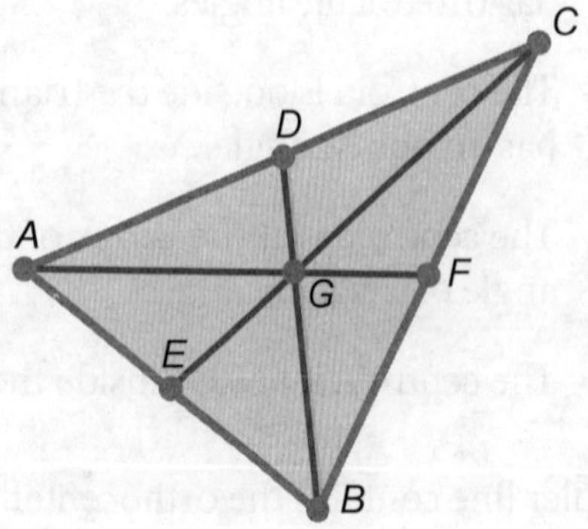

   What is the area of $\triangle ABC$?

   The area of $\triangle ABC$ is ________ square units.

5. A telecommunications company wants to put up a new cell phone tower to provide service to the three towns shown in the following diagram. The cell tower will be placed so that it is equidistant from all three towns, so that they all receive the same quality of service.

   The company starts by making a triangle with the three towns as the vertices. The cell tower will be placed at a certain point relative to the triangle.

   What is this point called?

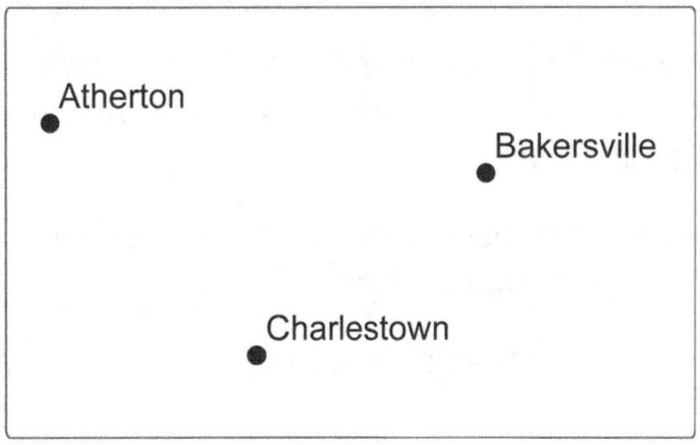

   A. incenter

   B. circumcenter

   C. orthocenter

   D. centroid

Practice Exercises *(continued)*

6. Some points of concurrency always lie inside the triangle, and others can sometimes lie outside.

Which is true of the centroid?

A. The centroid is never inside the triangle.

B. The centroid is inside the triangle only if it has three acute angles.

C. The centroid is outside the triangle if it has an obtuse angle.

D. The centroid is at the vertex of the right angle of a right triangle.

E. The centroid is always inside the triangle.

7. The Euler line contains the orthocenter, circumcenter, and centroid, among other significant points within a triangle.

What is true when the triangle is a right triangle?

A. The Euler line is collinear with the hypotenuse of the triangle.

B. The Euler line is a perpendicular bisector of the hypotenuse.

C. The Euler line is an altitude of the triangle.

D. The Euler line is an angle bisector of the triangle.

E. The Euler line is a median of the triangle.

8. A given triangle has its centroid in the same location as its orthocenter; that is, its medians are also altitudes.

Which statement is the best (most precise) description of the triangle?

A. The triangle is obtuse.

B. The triangle is scalene.

C. The triangle is acute.

D. The triangle is isosceles.

E. The triangle is equilateral.

9. Laura draws a triangle. Then, she finds the point that is equidistant from all three vertices.

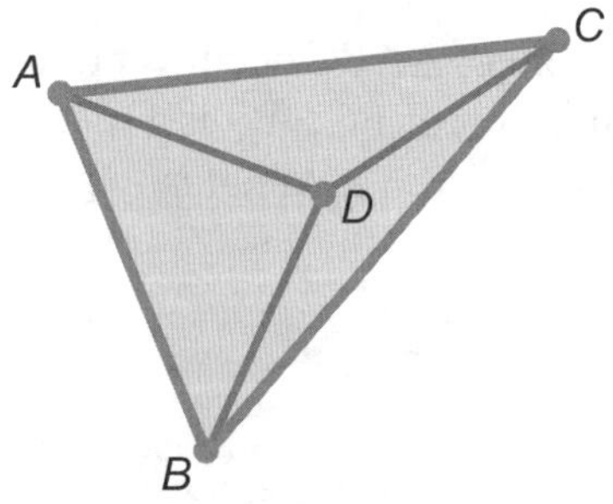

What is this point called?

A. centroid

B. circumcenter

C. incenter

D. orthocenter

E. Schiffler point

10. Right $\triangle ACE$ is subdivided by medians $\overline{AD}$, $\overline{BE}$, and $\overline{CF}$ as shown in the diagram.

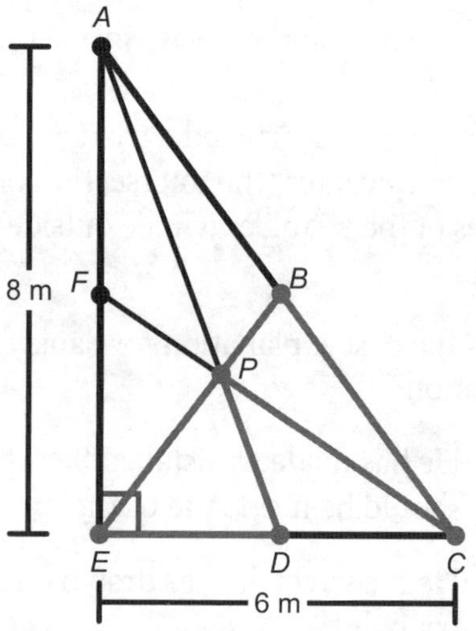

The lengths of the medians are $AD \approx 8.5$ meters, $BE = 5$ meters, and $CF \approx 7.2$ meters.

What is the combined perimeter of triangles $\triangle BCP$ and $\triangle DEP$?

Round your answer to the nearest tenth.

The combined perimeter of triangles $\triangle BCP$ and $\triangle DEP$ is ____________ meters.

Apply

Where Should the New Hospital Be Located?

Three cities are working together to select a central location for a regional hospital. If possible, they would like a site that is equally convenient for each city's residents.

A Complex Choice: View this video segment to learn about some of the factors that go into choosing the location of a hospital.

Besides being centrally located to Connorburgh, Browntown, and Scottsville, the hospital must be built on an existing road.

- Where should the new hospital be built?
- Why is this location the best place to put a hospital?
- Describe which triangle center works well in this scenario and how the existing roads impact the location.
- Include a snapshot of the map showing the suggested site and triangle center you constructed.
- Comment on the reasoning you used (including mathematics) to justify your recommendation.

Show what you've learned by completing the other performance tasks in the online Apply section.

Apply *(continued)*

Your answer to Apply will be assessed on the following criteria:

1. Identifying an appropriate location for the hospital and constructing the chosen triangle center on the map
2. Justifying which triangle center was used and why
3. Describing how existing roads impacted the decision on where to locate the hospital
4. Explaining why this is the best location for the hospital, using mathematics to justify the recommendation

Scale / Criteria	4	3	2	1	0
	Exceeds Criteria	**Meets Criteria**	**Progressing to Criteria**	**Below Expectations**	**No Expectation**
Location and Center	Correctly identifies an appropriate location based on a center of the triangle and constructs the center on the map.	Correctly identifies an appropriate location based on a center of the triangle and constructs the center with minor errors on the map.	Correctly identifies an appropriate location based on a center of the triangle, but does not construct the center on the map.	Does not identify an appropriate location based on the center of the triangle.	Does not identify a location.
Justification	Correctly justifies the center chosen and appropriately explains that choice of center.	Correctly justifies the center chosen and appropriately explains that choice of center with minor errors.	Justifies the center chosen and explains that choice of center, but one is incorrect.	Incorrectly justifies the center chosen and incorrectly explains that choice of center.	Does not justify center or explain choice.
Road Decision	Correctly discusses how the roads helped determine the best location for the hospital.	Discusses how the roads helped determine the best location for the hospital with minor errors in reasoning.	Discusses how the roads on the map helped determine the best location for the hospital with major errors in reasoning.	Only mentions briefly that roads were a concern in the location of the hospital.	Does not describe how the roads impact the decision.
Explanation	Correctly explains why location is best, using evidence from work done.	Correctly explains why location is best, using evidence from work done with minor errors.	Correctly explains why location is best, with little to no evidence used from work done.	Incorrectly explains why location is best.	Does not include an explanation of location.

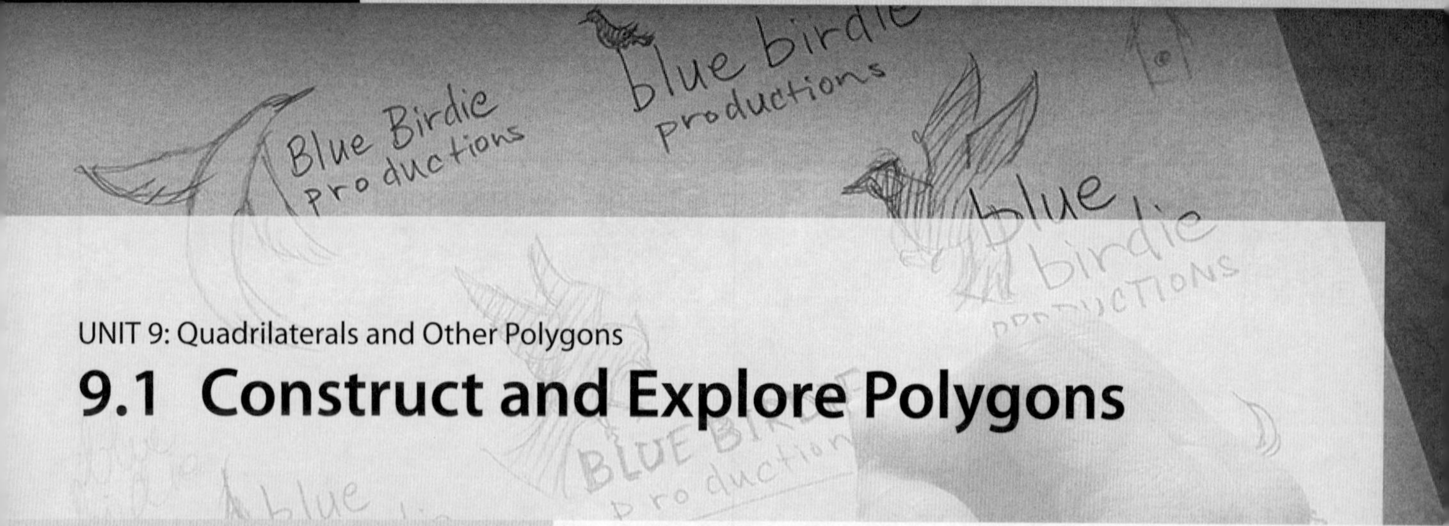

UNIT 9: Quadrilaterals and Other Polygons

9.1 Construct and Explore Polygons

Lesson Objectives

- Construct equilateral triangles, squares, and regular hexagons inscribed in a circle.
- Discover relationships between angles of a quadrilateral inscribed in a circle.
- Verify constructions both manually and with technology tools.
- Derive and apply the formula for the area of a regular polygon.

Essential Question

- How are geometric constructions used to inscribe a figure in a circle, and what properties of geometric figures are evidenced in the construction?

photo: Getty Images

Investigations

Designs and Prototypes

Design a logo to meet a client's expectations. Keep the polygon in the circle.

Inscribed Constructions

What are your choices for the logo? Construct inscribed regular polygons to see a few.

Common Ground

What do you know about an inscribed quadrilateral? Make and verify conjectures with constructions.

GOOOAAAALLL!

Find the surface area of a soccer ball. But first, find the area of regular pentagons and hexagons.

Key Vocabulary

diagonal, polygon, quadrilateral, square

Discover

As you complete Engage and the investigations, record the most important ideas you've learned.

Engage	Investigation 1
Investigation 2	Investigation 3

Name ____________________ Date __________

Check for Understanding

Construct and Explore Polygons: Investigation 1

1. After inscribing regular hexagon *ABCDEF* in a circle, Ethan used the hexagon to construct an equilateral triangle inscribed in the same circle. Which of these could be the equilateral triangle that Ethan constructed? *Select all that apply.*

   A. $\triangle BCD$ **B.** $\triangle ADF$ **C.** $\triangle DEF$ **D.** $\triangle ACE$

   E. $\triangle CDE$ **F.** $\triangle BEF$ **G.** $\triangle ABC$ **H.** $\triangle BDF$

2. A regular octagon and a square are inscribed in the same circle. Suppose a radius of the circle is drawn that intersects the circle at a point that is a vertex of the octagon, but not a vertex of the square. Which of the following statements about the radius must always be false?

   A. It is perpendicular to a side of the square.

   B. It passes through the midpoint of a side of the square.

   C. It is longer than half the length of a diagonal of the square.

   D. It intersects the circle at a point equidistant from the two adjacent vertices of the square.

3. Fran constructed regular pentagon *PQRST* inscribed in a circle, and then she drew a five-pointed star inside the pentagon as shown. The angles forming the points of the star are $\angle PRT$, $\angle QSP$, $\angle RTQ$, $\angle SPR$, and $\angle TQS$. What are the angles' measures? Explain.

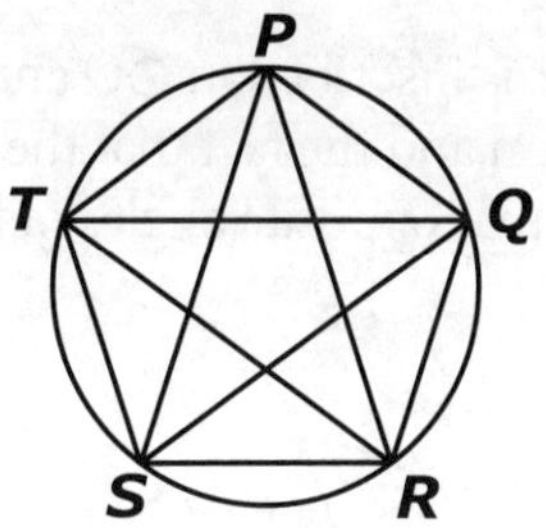

Name ______________________ Date ____________

Check for Understanding

Construct and Explore Polygons: Investigation 2

1. Suppose that for each side of a quadrilateral, you constructed a line segment that is a perpendicular bisector of the side and that passes through the quadrilateral. Is it possible for the quadrilateral to be inscribed in a circle?

   A. only if the perpendicular bisectors are all congruent

   B. only if the perpendicular bisectors are not all congruent

   C. only if the perpendicular bisectors all intersect at a single point

   D. only if the perpendicular bisectors do not all intersect at a single point

2. Russell inscribed quadrilateral *FGHK* in a circle. Which of these are possible measures of each of the four angles of the quadrilateral? *Select all that apply.*

   A. $m\angle F = 44°$; $m\angle G = 136°$; $m\angle H = 91°$; $m\angle K = 89°$

   B. $m\angle F = 53°$; $m\angle G = 85°$; $m\angle H = 127°$; $m\angle K = 95°$

   C. $m\angle F = 67°$; $m\angle G = 113°$; $m\angle H = 39°$; $m\angle K = 141°$

   D. $m\angle F = 76°$; $m\angle G = 18°$; $m\angle H = 104°$; $m\angle K = 162°$

   E. $m\angle F = 102°$; $m\angle G = 27°$; $m\angle H = 153°$; $m\angle K = 78°$

3. In the figure below, quadrilateral *WXYZ* is inscribed in $\odot O$ and divided into four isosceles triangles. How can you use the isosceles triangle theorem and the fact that the sum of the measures of the four central angles in the circle is 360° to help show that opposite angles of the quadrilateral are supplementary? Explain your answer.

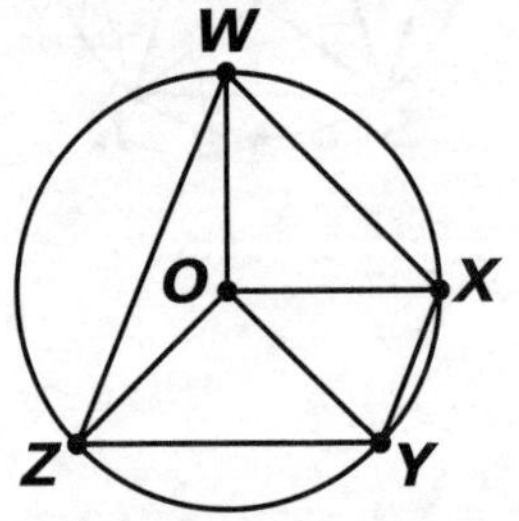

Name ______________________________ Date __________

Check for Understanding

Construct and Explore Polygons: Investigation 3

1. Each of the rows in the following table gives some information about a regular quadrilateral, or square. *Fill in the missing values in the table for each of the squares.*

Side Length (cm)	Apothem Length (cm)	Area (cm^2)
14		
		256
	12	
		324
20		
	6	

2. In the figure shown below, a regular hexagon has been inscribed in a circle with a radius of 10 units. Approximately how many square units is the area of the shaded region?

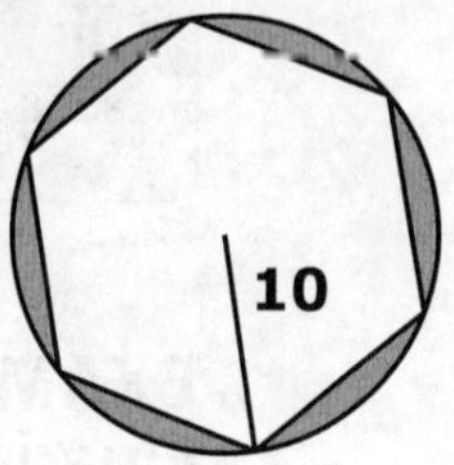

A. 54 **B.** 150 **C.** 164 **D.** 260

3. Dana used the formula for the area of a regular polygon to correctly derive a formula for the area of a regular octagon in terms of only *a*, the length of the octagon's apothem. How could she have done so, and what formula did she get? Explain your answer.

Summary

Before you attempt the Practice Exercises, review what you've learned.

If all the vertices of a polygon lie on a circle, then the figure is inscribed in the circle. Any regular polygon can be **inscribed** in a circle. A compass and straightedge or the Construction Tool can be used to construct inscribed polygons.

EXAMPLE: Construct an Inscribed Regular Dodecagon

SOLUTION:

First, construct an inscribed regular hexagon by using the radius of the circle to mark off six equal lengths around the circle. Connect the vertices to form the hexagon.

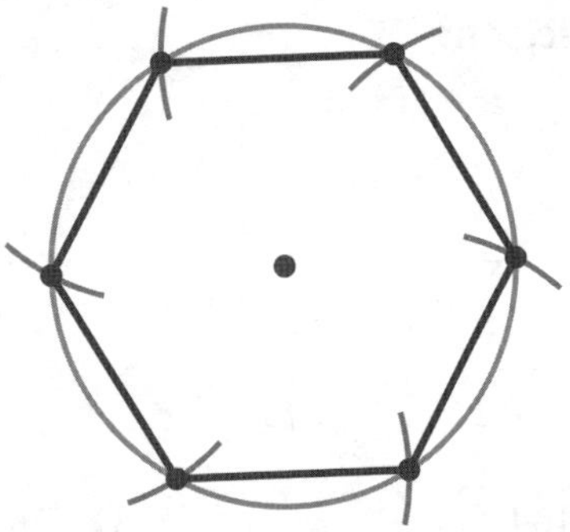

Construct diameters through the midpoints of opposite sides to locate the remaining vertices. Connect the vertices in order around the circle to form the dodecagon.

Inscribed Quadrilaterals

If a quadrilateral is inscribed in a circle, then both pairs of opposite angles are supplementary.

EXAMPLE: Use Properties of Inscribed Quadrilaterals

Prove that *KLMN* is a kite. (Note: A kite has at least two pairs of adjacent congruent sides.)

As shown, *KLMN* is inscribed in the given circle.

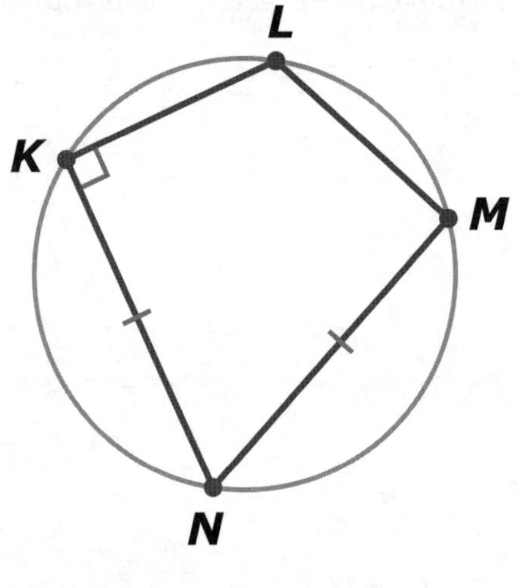

Therefore, $m\angle K + m\angle M = 180°$ and $m\angle L + m\angle N = 180°$.

My Notes

Summary *(continued)*

SOLUTION:

Given: Quadrilateral $KLMN$ inscribed in a circle with $\angle K$ a right angle and $\overline{KN} \cong \overline{MN}$.

Prove: $KLMN$ is a kite.

Proof: Because $KLMN$ is inscribed in a circle, $\angle K$ and $\angle M$ are supplementary. So, $m\angle K + m\angle M = 180°$. Angle K is a right angle, and $90° + m\angle M = 180°$. Therefore, $m\angle M = 90°$. Also, because $\angle K$ is a right angle, it is inscribed in a semicircle. This means $\overline{LN}$ is a diameter. $\triangle KLN$ and $\triangle MLN$ are right triangles with hypotenuse $\overline{LN}$ and $\overline{KN} \cong \overline{MN}$. Therefore, by the HL congruence theorem, $\triangle KLN \cong \triangle MLN$. Since the corresponding parts are congruent, we can conclude $\overline{KL} \cong \overline{ML}$. Quadrilateral $KLMN$ has two pairs of adjacent congruent sides and by definition, it is a kite.

Area of a Regular Polygon

If a regular polygon has a perimeter of P units and an apothem of a units, then the area A can be calculated using the formula:

$$\text{Area} = \frac{1}{2}aP$$

EXAMPLE: Calculate the Area of a Regular Octagon with a Perimeter of 56 cm

SOLUTION:

The perimeter of the regular octagon is 56 centimeters, so the length of one side is 7 centimeters. The apothem bisects the side of the octagon creating a right triangle. The hypotenuse of the triangle is the radius of the octagon. Because the apothem bisects the side of the octagon, the measure of one leg of the right triangle is 3.5 centimeters.

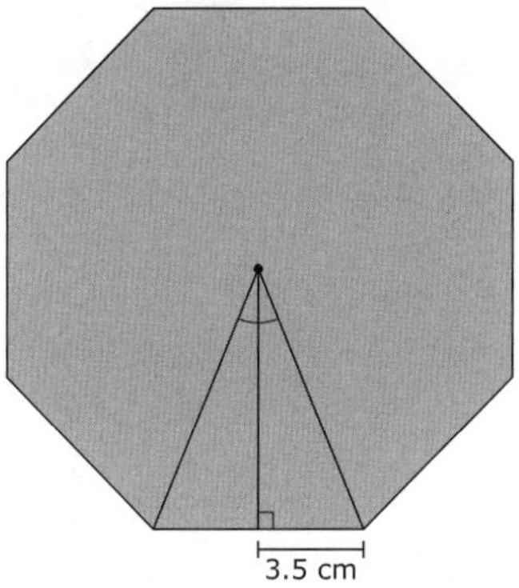

The angle formed by the apothem and the radius is $\left(\frac{1}{2}\right)\left(\frac{360°}{8}\right) = 22.5°$. We can use trigonometric ratios to calculate the measure of the apothem:

$$\tan(22.5°) = \frac{3.5}{a}$$

$$a\tan(22.5°) = 3.5$$

$$a = \frac{3.5}{\tan(22.5°)} \approx 8.45\ cm$$

Then, the area of the octagon is Area $\approx \frac{1}{2}(8.45)(56) \approx 236.6\text{ cm}^2$.

My Notes

Summary *(continued)*

Interior and Exterior Angles of Polygons

The sum of the interior angles of a convex, n-gon is Sum = $180°(n - 2)$.

The sum of the exterior angles of a convex, n-gon is 360°.

EXAMPLE

The measures of the interior angles of a pentagon are $(2x + 15)°$, $(x - 5)°$, $(1.5x + 7)°$, $(0.5x + 10)°$, and $(1.5x + 6)°$. Calculate the value of x.

SOLUTION:

$$(2x + 15)° + (x - 5)° + (1.5x + 7)° + (0.5x + 10)° + (1.5x + 6)° = 180°(5 - 2)$$
$$(6.5x + 33)° = 540°$$
$$(6.5x)° = 507°$$
$$x = 78°$$

EXAMPLE

For the pentagon in the previous example, what is the measure of the exterior angle with a corresponding interior angle that is $(2x + 15)°$?

SOLUTION:

$$(2x + 15)° + m\angle A = 180°$$
$$(2(78) + 15)° + m\angle A = 180°$$
$$171° + m\angle A = 180°$$
$$m\angle A = 9°$$

My Notes

Practice Exercises

Review what you've learned using these practice problems. For practice problems with feedback, try the Coach and Play items in the Practice section online.

1. A production crew is marking up a circular stage to identify where the actors will be standing as shown in the diagram.

The tape markings are joined by line segments that form specific angles, the measures of two of which are given in the diagram.

What are the measures of angles *A* and *B* where the tape lines meet?

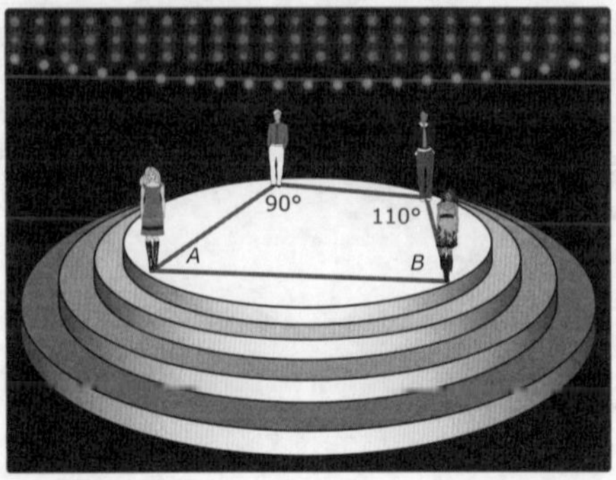

$m\angle A =$ __________ °

$m\angle B =$ __________ °

2. The measure of the exterior angles of a pentagon are $(3x + 17)°$, $(5x - 2)°$, $(7x + 4)°$, $(x + 21)°$, and $(2x + 32)°$.

Complete the statement. The value of $x =$ ______.

3. A circle with a diameter of 45 units contains an inscribed hexagon. You are to establish that the hexagon is regular by measuring the length of each side and comparing the results to the expected side length.

What is the value of the expected side length? *Express your answer as a decimal.*

Side length = ____________ units

4. Roger knows that adding the interior angle measures of a triangle yields a sum of 180°. But what sum does he get when he adds the measures of the three exterior angles *D*, *E*, and *F* shown on the triangle below?

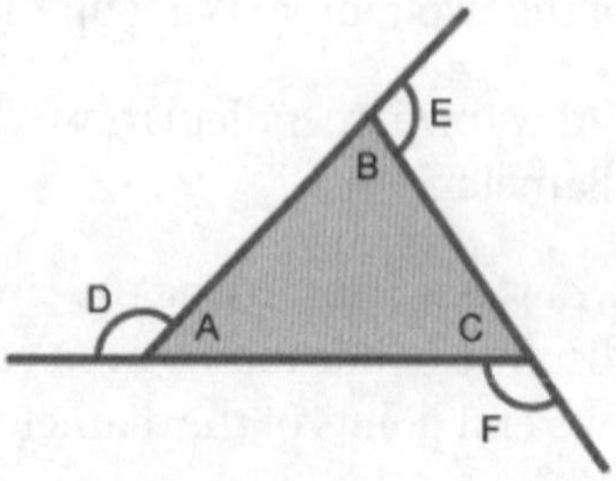

A. 180°

B. 270°

C. 360°

D. 540°

5. Talk about a cool apartment! An old lighthouse has been renovated, and the former caretaker's quarters are for rent. The ground floor living space is in the shape of a regular heptagon. The diagram below, taken from the lighthouse blueprints, shows the distance from *A* to *C* is 7.25 meters.

If *P* is the center of the heptagon and the circumscribed circle, apply appropriate strategies to calculate each measurement.

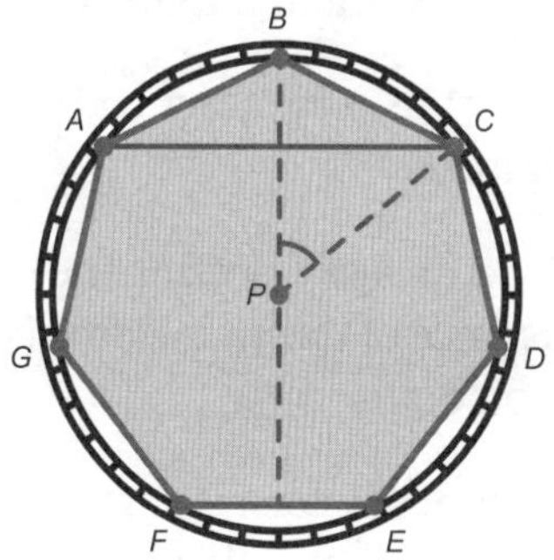

$m\angle BPC \approx$ __________ °

$PC \approx$ __________ m

The area of the inscribed regular heptagon is approximately __________ m^2.

Practice Exercises *(continued)*

6. You are given a circle with a diameter drawn between two points on the circle.

Which of the following procedures could be used to identify additional points on the circle needed to define an inscribed regular hexagon? *Select all that apply.*

A. Draw the perpendicular bisector of each of the two radii making up the diameter.

B. Draw the perpendicular bisector of the diameter.

C. Draw arcs intersecting the circle a distance of one radius from each of the two end points of the diameter.

D. Draw an arc intersecting the circle a distance of one diameter from one of the two end points of the diameter.

E. Draw arcs intersecting the circle a distance of one diameter from each of the two end points of the diameter.

7. You are given a circle with a diameter drawn between two points on the circle.

Which of the following procedures could be used to identify additional points on the circle needed to define an inscribed square?

A. Draw the perpendicular bisector of one of the two radii making up the diameter.

B. Draw the perpendicular bisector of each of the two radii making up the diameter.

C. Draw the perpendicular bisector of the diameter.

D. Draw arcs intersecting the circle a distance of one radius from each of the two points.

E. Draw arcs intersecting the circle a distance of one diameter from each of the two points.

8. The figure shows a circle along with an inscribed quadrilateral.

Two angles shown in the diagram have measures of 30° and 70°.

Write the values x and y for the measures of the other two angles indicated in the figure.

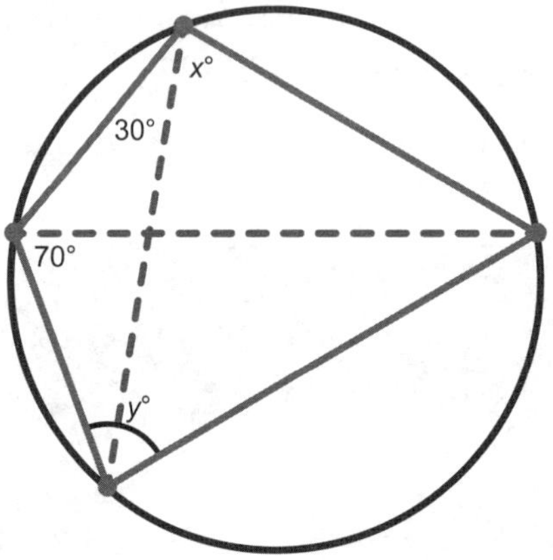

$x =$ __________

$y =$ __________

9. The measure of one interior angle of a regular polygon is 156°. How many sides does the polygon have?

A. 13

B. 15

C. 17

D. 24

10. Armand is playing with a math tiles set that includes various colored wooden shapes. He collects all the green equilateral triangles of the same size and starts arranging them into regular polygons.

The regular polygon that Armand can make is a ____________.

Apply

What Will You Make Your Pasta Look Like?

Have you ever seen how pasta is made? Types, such as spaghetti or penne with a fixed cross-section, are made by pushing the dough through a die. Watch the video to find out more about dies and extrusions.

How can you ensure precision when making dies?

Dies and Extrusions: View this video segment about molding equipment.

Imagine you are part of a team opening a new restaurant. You want to offer fresh pasta cut into unique shapes to stand out from the competition. Design two new dies for the pasta extruder. Both should be polygons—one convex and one concave. Use geometric constructions to draw the dies. Depending on the polygon, place either some or all of the vertices on the circle.

In order to have the die made to your standards, you will need to record the steps on how to create each die. How would you use measurements to verify the precision of each resulting die?

Show what you've learned by completing the other performance tasks in the online Apply section.

Apply *(continued)*

Your answer to Apply will be assessed on the following criteria:

1. Designing a convex die (a polygon inscribed in a circle)
2. Designing a concave die (a polygon inscribed in a circle)
3. Recording the steps necessary to create each die
4. Recording how to use measurements to verify the precision of each replicate

Scale / Criteria	4 Exceeds Criteria	3 Meets Criteria	2 Progressing to Criteria	1 Below Expectations	0 No Expectation
Convex Die	Correctly designs a die that is a convex polygon inscribed in a circle.	Designs a die that is a convex polygon inscribed in a circle with minor errors.	Designs a die that is a convex polygon inscribed in a circle with significant errors.	Designs a simple die that is not a convex polygon inscribed in a circle.	Does not design a convex die.
Concave Die	Correctly designs a die that is a concave polygon inscribed in a circle.	Designs a die that is a concave polygon inscribed in a circle with minor errors.	Designs a die that is a concave polygon inscribed in a circle with significant errors.	Designs a simple die that is not a concave polygon inscribed in a circle.	Does not design a concave die.
Steps	Records clear steps to replicate the die.	Records clear steps to replicate the die with minor errors.	Records steps to replicate the die with significant errors.	Records steps that do not replicate the die.	Does not record the steps.
Precision	Records how to use measurements to verify precision.	Records how to use measurements to verify precision with minor errors.	Records how to use measurements to verify precision with significant errors.	Records steps that will not verify precision.	Does not record how to verify precision.

photo: Getty Images

UNIT 9: Quadrilaterals and Other Polygons

9.2 Prove and Apply Theorems about Quadrilaterals

Lesson Objectives

- Prove theorems about parallelograms, rectangles, kites, and trapezoids.
- Apply theorems about quadrilaterals to solve real-world and mathematical problems.

Essential Question

- How can relationships among the sides, angles, and diagonals of a quadrilateral be used to analyze, describe, and classify quadrilaterals?

photo: Getty Images

Investigations

How Do They Work?

Analyze design. How do geometric features make things work?

Curious Quadrilaterals

Investigate the differences in quadrilaterals. What makes a trapezoid special?

Puzzling Parallelograms

Use what you know about parallel lines to identify properties of parallelograms.

Popcorn Anyone?

How are isosceles trapezoids and kites related? Are they so different from other quadrilaterals?

Classification of Quadrilaterals

Sort quadrilateral jewels and classify them by their properties.

Quadrilaterals in the Coordinate Plane

Add another level of precision to quadrilateral classification on the coordinate plane.

Key Vocabulary

concave polygon, converse, convex polygon, diagonal, isosceles trapezoid, kite, parallelogram, polygon, quadrilateral, rectangle, rhombus, square, trapezoid

Discover

As you complete Engage and the investigations, record the most important ideas you've learned.

Engage	**Investigation 1**
Investigation 2	**Investigation 3**
Investigation 4	**Investigation 5**

Name ______________________________ Date ____________

Check for Understanding

Prove and Apply Theorems about Quadrilaterals: Investigation 1

1. Three of the interior angles of a convex quadrilateral measure 98°, 64°, and 116°, respectively. *Fill in the blanks in the sentence with the correct numbers of degrees.*

The sum of the measures of all the interior angles of the quadrilateral is ____________, so the measure of the quadrilateral's fourth interior angle must be ____________.

2. Which of the following statements is correct about rhombuses and squares?

A. All rhombuses are squares, and all squares are rhombuses.

B. All rhombuses are squares, but not all squares are rhombuses.

C. Not all rhombuses are squares, but all squares are rhombuses.

D. Not all rhombuses are squares, and not all squares are rhombuses.

3. Given that an isosceles trapezoid is defined as a quadrilateral that has at least one pair of parallel sides and base angles that are congruent, circle the correct words below.

A parallelogram is [always / sometimes / never] an isosceles trapezoid, and a rectangle is [always / sometimes / never] an isosceles trapezoid.

4. Angela drew trapezoid $WXYZ$, where $\overline{WX} \parallel \overline{ZY}$. Then she constructed $\overline{MN}$, where M is the midpoint of $\overline{WZ}$, and N is the midpoint of $\overline{XY}$, as shown in the figure. If the height of trapezoid $WXYZ$ is h units, and the lengths of its bases are b_1 and b_2 units, what is the area of trapezoid $MNYZ$? Explain how you got your answer.

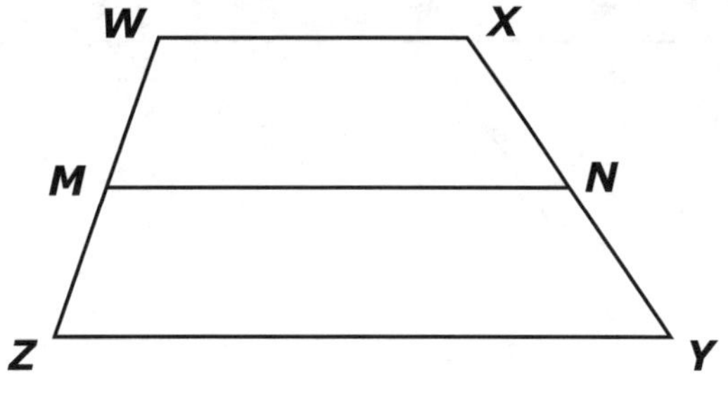

Name ______________________________ Date ____________

Check for Understanding

Prove and Apply Theorems about Quadrilaterals: Investigation 2

1. Jacqueline drew parallelogram $ABCD$. Which of these could be its side lengths?

   A. $AB = 18$ cm; $BC = 18$ cm; $CD = 9$ cm; $AD = 9$ cm

   B. $AB = 21$ cm; $BC = 7$ cm; $CD = 7$ cm; $AD = 21$ cm

   C. $AB = 15$ cm; $BC = 3$ cm; $CD = 15$ cm; $AD = 3$ cm

   D. $AB = 5$ cm; $BC = 5$ cm; $CD = 25$ cm; $AD = 25$ cm

2. The diagonals of parallelogram $FGHK$ intersect at point P, with $FP = 8$ and $GK = 6$. *Fill in the blanks with the correct lengths.*

   $HP =$ ______ $FH =$ ______ $GP =$ ______ $KP =$ ______

3. Wynn and Melanie were each given information about a different parallelogram. Wynn was given the measure of one of his parallelogram's interior angles, while Melanie was given the length of one of her parallelogram's sides. *Circle the correct words below.*

   With only the information given, Wynn [can / cannot] determine the measures of his parallelogram's other three interior angles, and Melanie [can / cannot] determine the lengths of her parallelogram's other three sides.

4. In the figure shown, quadrilateral $WXYZ$ is a parallelogram, with $\overline{ZM}$ bisecting $\angle WZY$ and $\overline{YM}$ bisecting $\angle XYZ$. Is it possible to know the exact measure of $\angle YMZ$? If so, what is the measure of the angle? Explain how you got your answer.

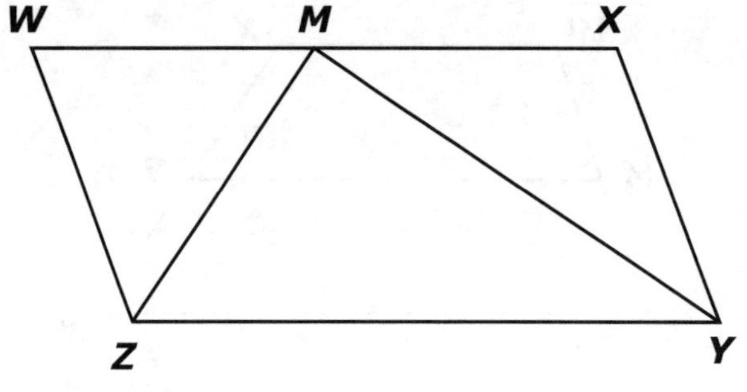

Name ______________________________ Date __________

Check for Understanding

Prove and Apply Theorems about Quadrilaterals: Investigation 3

1. Vera was given isosceles trapezoid *FGHK*, with $\overline{FK} \cong \overline{GH}$. She drew its diagonals, $\overline{FH}$ and $\overline{GK}$, and proved they are congruent by showing that $\triangle FHK \cong \triangle GKH$. If she finished her proof in as few steps as possible, what was her reason when stating that $\triangle FHK \cong \triangle GKH$?

   A. AAS **B.** ASA **C.** SAS **D.** SSS

2. Bill found the sum of the lengths of the two diagonals of a square to be 26 cm, April found the sum of the lengths of the two diagonals of a rhombus to be 22 cm, and Jay found the sum of the lengths of the two diagonals of a rectangle to be 28 cm. Which of the following statements could be true? *Select all that apply.*

   A. One diagonal of Bill's square is longer than each diagonal of Jay's rectangle.

   B. One diagonal of April's rhombus is longer than each diagonal of Bill's square.

   C. One diagonal of Bill's square is shorter than each diagonal of April's rhombus.

   D. One diagonal of April's rhombus is shorter than each diagonal of Jay's rectangle.

   E. One diagonal of Jay's rectangle is shorter than each diagonal of April's rhombus.

3. In the figure shown below, quadrilateral *QRST* is a kite, with diagonals $\overline{QS}$ and $\overline{RT}$ intersecting at point *P*. If the length of $\overline{QS}$ is *a* units, and the length of $\overline{RT}$ is *b* units, what is the area of the kite? Explain how you got your answer.

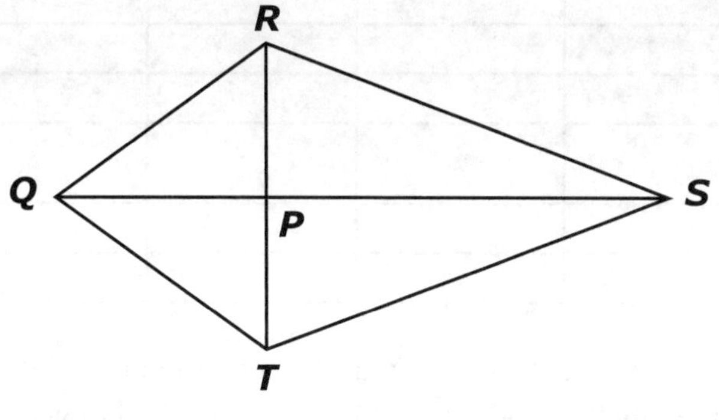

Name ______________________________ Date ______________

Check for Understanding

Prove and Apply Theorems about Quadrilaterals: Investigation 4

1. Consider the meanings of *equilateral* and *equiangular*, and then circle the correct words.

A quadrilateral that must be equilateral, but that does not have to be equiangular, is a [rectangle / rhombus / square], while a quadrilateral that must be equiangular, but that does not have to be equilateral, is a [rectangle / rhombus / square].

2. Which of the following polygons do not have to have at least two pairs of adjacent supplementary angles? *Select all that apply.*

A. quadrilateral
B. parallelogram
C. square
D. rhombus
E. isosceles trapezoid
F. trapezoid
G. kite
H. rectangle

3. For each of the polygons listed in the table, decide if it must have at least one pair of each of the attributes shown and put a checkmark in the appropriate column(s).

Polygon	Parallel Opposite Sides	Congruent Opposite Sides	Congruent Adjacent Sides	Congruent Opposite Angles	Congruent Adjacent Angles
Quadrilateral					
Parallelogram					
Square					
Rhombus					
Kite					
Isosceles Trapezoid					
Trapezoid					
Rectangle					

4. Kurt drew a quadrilateral whose diagonals are congruent but not perpendicular, while Athena drew a quadrilateral whose diagonals are perpendicular but not congruent. For each person's quadrilateral, what is the minimum number of diagonals that must be bisected by the other? Explain how you arrived at your answer.

Name ______________________________ Date __________

Check for Understanding

Prove and Apply Theorems about Quadrilaterals: Investigation 5

1. Which of the following groups of coordinates could represent the vertices of a quadrilateral that can be classified as a rectangle? *Select all that apply.*

   A. (–7, 8), (3, 8), (8, –7), (8, 3)
   B. (–9, 6), (1, 6), (1, –5), (–9, –5)
   C. (–5, 4), (5, 4), (0, 10), (0, –1)
   D. (–2, 11), (8, 9), (6, –1), (–4, 1)
   E. (–3, 9), (12, 9), (–5, –2), (10, –2)
   F. (–7, 6), (8, 4), (6, –3), (–9, –1)

2. Parallelogram *FGHK* is in Quadrant I of a coordinate plane. *Fill in the coordinates for* G.

   F(3, 11) *G*(____, ____) *H*(12, 1) *K*(1, 4)

3. Isosceles trapezoid *QRST* is shown in the figure below. Suppose the midpoint of $\overline{QR}$ is *W*, the midpoint of $\overline{RS}$ is *X*, the midpoint of $\overline{ST}$ is *Y*, and the midpoint of $\overline{QT}$ is *Z*. After line segments joining the midpoints of adjacent sides of the trapezoid are drawn, what would be the most precise term for quadrilateral *WXYZ*? Explain your answer.

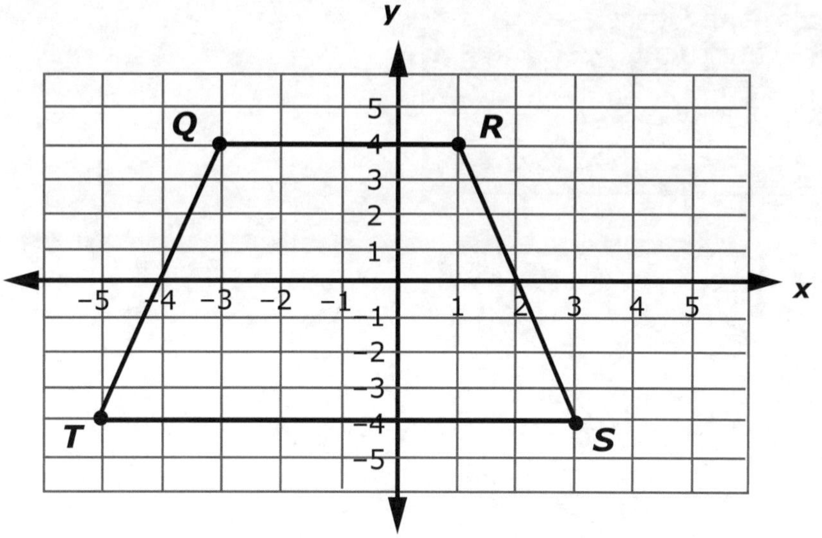

Summary

Before you attempt the Practice Exercises, review what you've learned.

Properties of Quadrilaterals

In the investigations, you explored the following properties of quadrilaterals.

Properties of Parallelograms

- The opposite angles of a parallelogram are congruent.
- The consecutive angles of a parallelogram are supplementary.
- The opposite sides of a parallelogram are congruent.
- The diagonals of a parallelogram bisect each other.

Properties of Other Quadrilaterals

- The diagonals of an isosceles trapezoid are congruent.
- The diagonals of a kite are perpendicular.

EXAMPLE

Prove that the opposite angles of a parallelogram are congruent.

SOLUTION:

Given: Parallelogram $ABCD$

Prove: $\angle A \cong \angle C$, $\angle B \cong \angle D$.

Proof:

By the definition of a parallelogram, $\overline{AD} \parallel \overline{CB}$ and $\overline{AB} \parallel \overline{CD}$. $\angle A$ and $\angle B$ are consecutive angles of parallel lines, so they are supplementary. Also, $\angle B$ and $\angle C$ are consecutive angles of parallel lines, so they are supplementary. By the congruent supplements theorem, $\angle A \cong \angle C$. $\angle C$ and $\angle D$ are also consecutive angles of parallel lines, so they are supplementary. So by the congruent supplements theorem, $\angle B \cong \angle D$.

My Notes

Summary *(continued)*

EXAMPLE

Show that the diagonals of figure *PQRS* are perpendicular.

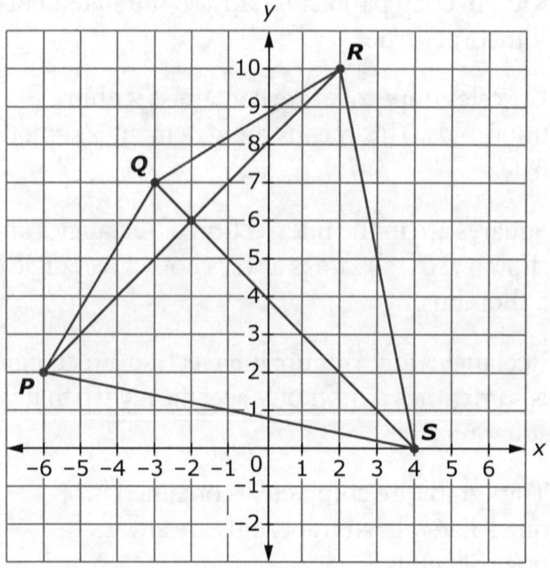

SOLUTION:

The slope of diagonal $\overline{RP}$ is: $\frac{10-2}{2-(-6)} = \frac{8}{8} = 1$.

The slope of diagonal $\overline{QS}$ is: $\frac{7-0}{-3-4} = \frac{7}{-7} = -1$.

The product of the slopes is −1, so $\overline{RP} \perp \overline{QS}$.

Hierarchy of Quadrilaterals

The diagram shows the relationship among quadrilaterals.

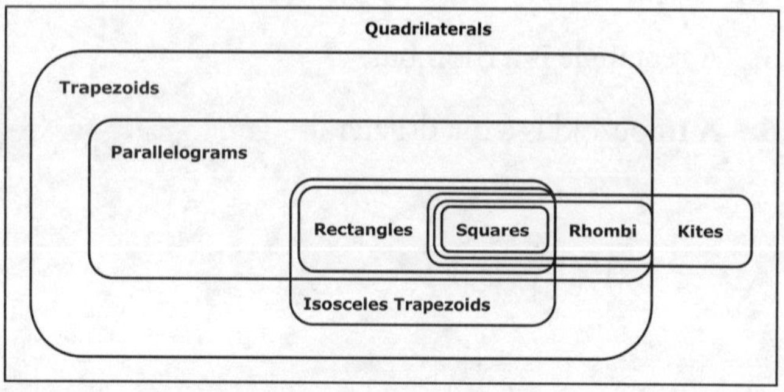

You can use the diagram to help you understand the properties of quadrilaterals.

My Notes

Summary *(continued)*

EXAMPLE

Identify each statement as always, sometimes, or never true.

1. A square is a parallelogram.
2. A kite is a parallelogram.
3. A trapezoid has congruent legs.
4. A square is both a rectangle and a rhombus.
5. A rectangle is a rhombus.
6. A trapezoid is a quadrilateral.

SOLUTION:

1. Squares are completely contained in parallelograms, so the statement is always true.
2. Kites overlap parallelograms, so the statement is sometimes true.
3. Isosceles trapezoids are contained within trapezoids. This means the statement is sometimes true.
4. Squares are in the intersection of rectangles and rhombi, so a square is always both a rectangle and a rhombus.
5. Rectangles and rhombi intersect, so a rectangle is sometimes a rhombus, specifically when it is a square.
6. Trapezoids are completely contained in quadrilaterals, so trapezoids are always quadrilaterals.

My Notes

Practice Exercises

Review what you've learned using these practice problems. For practice problems with feedback, try the Coach and Play items in the Practice section online.

1. You are given a circle with an inscribed quadrilateral.

Which of the following conditions will guarantee the quadrilateral is a square?

A. All four interior angles have the same measure.

B. All four sides have the same length.

C. Both pairs of opposite sides are parallel.

D. The diagonals intersect at the center of the circle.

E. The perpendicular bisectors of the sides intersect at the center of the circle.

F. The diagonals are perpendicular to each other.

G. The perpendicular bisectors of the sides are perpendicular to each other.

2. Quadrilateral *PQRS* has the following properties:

a) Opposite sides are parallel and equal.

b) Diagonals are equal in length.

c) Diagonals do not intersect at right angles.

Which of the following best describes quadrilateral *PQRS*?

A. *PQRS* is a square.

B. *PQRS* is a rectangle.

C. *PQRS* is a rhombus.

D. *PQRS* is an isosceles trapezoid.

3. The figure shows two rhombuses that are similar.

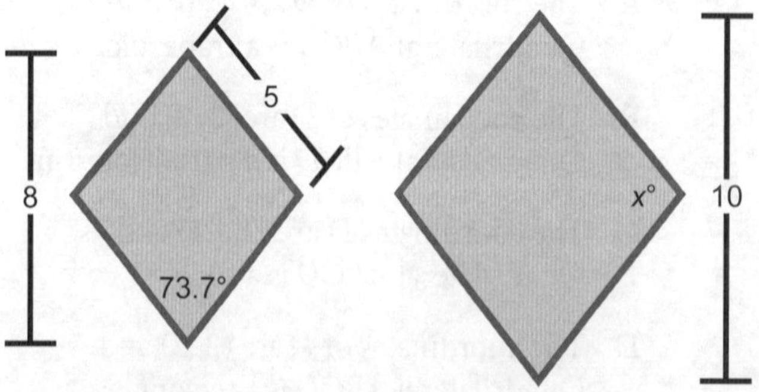

In the spaces provided below, write the values of the perimeter and x for the larger rhombus on the right.

Round values to the nearest tenth.

Perimeter = ____________ units

x = ____________ degrees

4. Two opposite vertices of a rhombus, $A(-5, 7)$ and $C(3, -5)$, such that $\overline{AC}$ forms a diagonal of the rhombus are given in the coordinate plane. The rhombus has two sides that are parallel to the x-axis.

Use this information to construct the rhombus by drawing the following five lines: the line containing the other diagonal of the rhombus and the four lines containing the four sides of the rhombus.

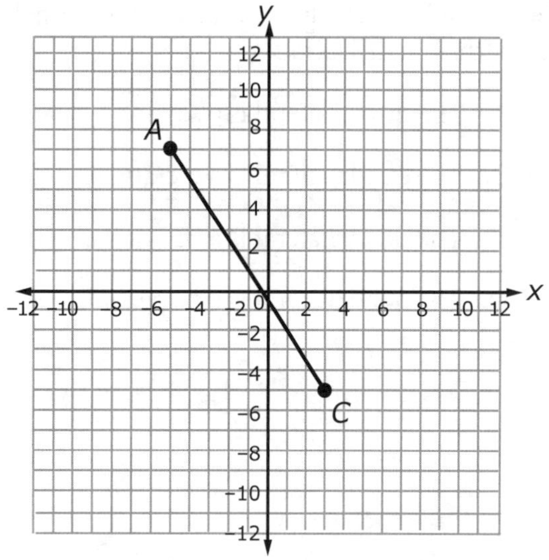

Practice Exercises *(continued)*

5. Quadrilateral $ABCD$ has three disclosed vertices $A(-2, 3)$; $B(2, 6)$; $C(6, 3)$.

Which of the following could be true?

A. The coordinates of D are (1, 0) and Quadrilateral $ABCD$ is a trapezoid.

B. The coordinates of D are (2, 0) and Quadrilateral $ABCD$ is a parallelogram.

C. The coordinates D are (2, −1) and Quadrilateral $ABCD$ is a square.

D. The coordinates of D are (2, 2) and Quadrilateral $ABCD$ is a rectangle.

6. One diagonal of a rhombus measures 15 centimeters. The other diagonal measures 24 centimeters. Apply the properties of a rhombus to calculate the perimeter and area of this rhombus.

The perimeter of the rhombus is ______ centimeters.

The area of the rhombus is ______ square centimeters.

7. Quadrilateral $ABCD$ has the following properties:

a) Opposite sides are parallel

b) Opposite sides are equal

c) Diagonals are equal and bisect each other at right angles

Based on these conditions, what can be concluded about the quadrilateral?

A. The quadrilateral is a square.

B. The quadrilateral is a rectangle.

C. The quadrilateral is a rhombus.

D. The quadrilateral is a trapezoid.

8. When plotted in a standard coordinate plane, the coordinates (1, −1), (−3, −3), (5, −3), and (1, −5) form a geometric shape. Which of the following types of quadrilaterals are formed when they are connected with lines?

I. parallelogram

II. rectangle

III. rhombus

IV. square

A. I and II only

B. I and III only

C. II only

D. II and IV only

9. Parallelogram $JKLM$ has the following vertices: $J(-8, 1)$, $K(2, 5)$, $L(4, 0)$, $M(-6, -4)$. Determine if this parallelogram is a rectangle.

The first two answers are numerical.

JL = ______

KM = ______

Does JL equal KM? Write *yes* or *no*. ______

Is parallelogram $JKLM$ a rectangle? Write *yes* or *no*. ______

10. Quadrilateral $ABCD$ is a rhombus. Therefore, it can be assumed that ______.

A. $\triangle ABC \cong \triangle ABD$ because of the Side-Side-Side congruent triangle rule.

B. $\triangle ABC \cong \triangle DBC$ because of the Side-Angle-Side congruent triangle rule.

C. $\triangle ABC \cong \triangle CDA$ because of the Side-Side-Side congruent triangle rule.

D. A relationship between the triangles within Quadrilateral $ABCD$ cannot be determined.

Apply

How Would You Organize a County Fair?

Imagine you are the organizer of a local county fair. You want to create a setup that is fun and easy for visitors to find their way around. To make an accurate map of the fairgrounds, you might survey the land and determine the shape and size of any structures involved. Watch this video to get started!

photo: Pixabay

Surveying a Fairground: View this video segment about organizing a county fair.

The midway of a fair is a long, straight stretch of land with activities on each side. Suppose you are to design the midway. On one side, the booths must be congruent isosceles trapezoids; on the other side, they must be congruent parallelograms. You can only use isosceles trapezoids and parallelograms that are not rectangles. Every booth needs to be covered by a tarp supported by a vertical post at each corner and by struts (narrow beams) along the edges and the diagonals.

The longest side of each booth must be 20 feet. Determine reasonable sizes for the other dimensions. Decide how many booths you want on each side of your midway.

- Make a plan that minimizes the number of different-sized struts so assembly is easier.
- Choose a scale, calculate the dimensions, and make a drawing of your planned booths.
- Calculate the number and lengths of the posts and struts you would need to hang tarps over all of the booths.
- Generalize the number of different posts and struts you would need if you had n booths on each side of the midway. Write a rule for the number of each type and size, in terms of n.

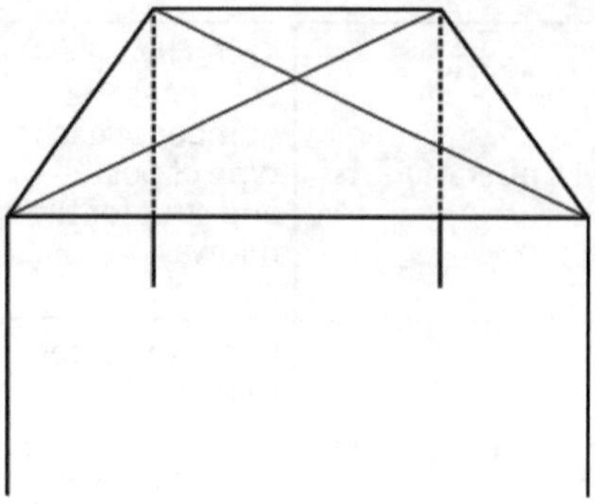

How many of each do you need for your midway?

Show what you've learned by completing the other performance tasks in the online Apply section.

Apply *(continued)*

Your answer to Apply will be assessed on the following criteria:

1. Creating a plan that minimizes different parts and making a scale drawing
2. Calculating the lengths of all of the struts, showing your reasoning
3. Determining the number of each type of post and strut for your midway
4. Writing rules for the number of each type of post and strut for *n* of your booths per side

Criteria \ Scale	4	3	2	1	0
	Exceeds Criteria	**Meets Criteria**	**Progressing to Criteria**	**Below Expectations**	**No Expectation**
Plan and Drawing	Creates a plan that meets criteria and uses up to five different lengths of materials.	Creates a plan that meets criteria and uses up to six different lengths of materials.	Creates a plan that uses up to seven different lengths of materials and does not meet one criterion.	Creates a plan that uses more than seven lengths of material and does not meet more than one criterion.	Does not attempt task.
Lengths of Struts	Calculates the lengths of all of the struts correctly.	Calculates the lengths of all of the struts with one error.	Calculates the lengths of all of the struts with two or three errors.	Calculates the lengths of only one or two of the struts or calculates more than three of the lengths incorrectly.	Does not attempt task.
Number of Parts	Correctly calculates the number of each type of post and strut for the midway.	Calculates the number of each type of post and strut for the midway with a minor error.	Calculates the number of each type of post and strut for the midway with several errors.	Calculates the number of each type of post and strut for the midway incorrectly or incompletely.	Does not attempt task.
Rules for General Case	Correctly writes rules for the number of each type of post and strut for *n* of the booths per side.	Writes rules for the number of each type of post and strut for *n* of the booths per side with lack of clarity.	Writes rules for the number of each type of post and strut for *n* of the booths per side with minor errors.	Writes rules that are incomplete or incorrect.	Does not attempt task.

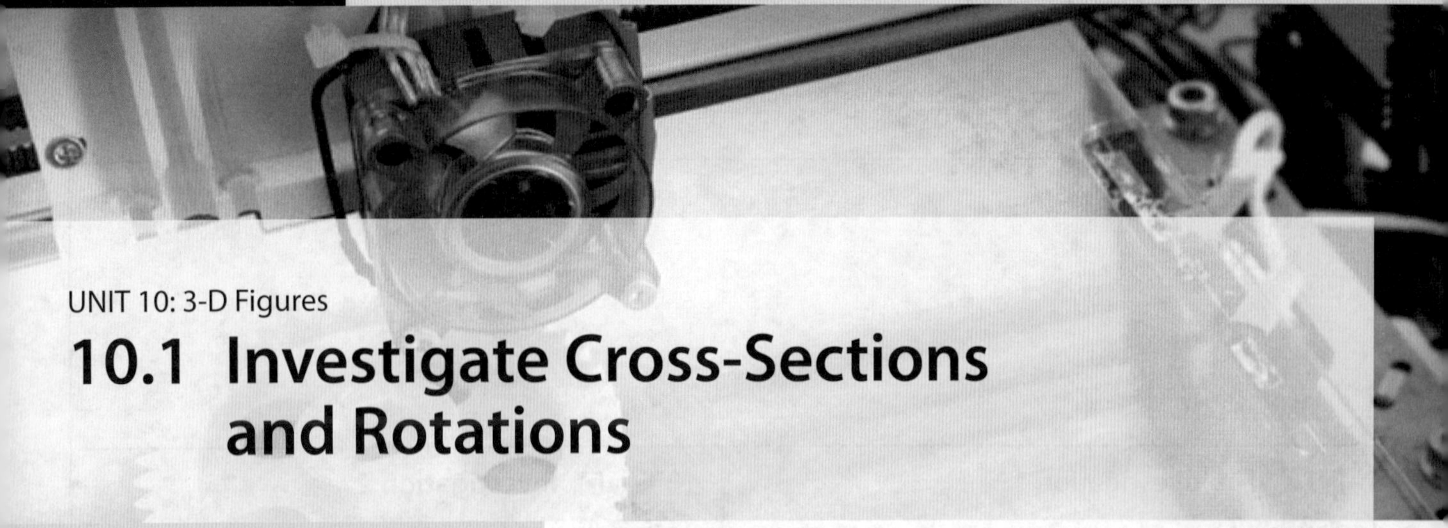

UNIT 10: 3-D Figures

10.1 Investigate Cross-Sections and Rotations

photo: Getty Images

Lesson Objectives

- Visualize and model cross-sections of three-dimensional figures.
- Visualize and describe three-dimensional objects created by rotations of two-dimensional figures.
- Visualize and represent spheres, cylinders, cones, pyramids, and prisms as an accumulation of two-dimensional cross-sections.
- Approximate real-world objects as combinations of geometric shapes to make scale models and explore properties.

Essential Question

- How can you visualize and represent three-dimensional objects?

photo: Getty Images

Investigations

3-D Printers

Simulate the process of a 3-D printer. Create a three-dimensional object in layers.

Cross-Sections

Dig into a three-dimensional shape. What other shapes are inside?

Making Faces

Turn a 2-D shape into 3-D. Rotate a cross-section. What does it form?

Key Vocabulary

cone, cross-section, cylinder, isometric drawing, orthographic drawing, perspective drawing, pyramid, sphere

Discover

As you complete Engage and the investigations, record the most important ideas you've learned.

Engage	
Investigation 1	Investigation 2

Name ______________________ Date __________

Check for Understanding

Investigate Cross-Sections and Rotations: Investigation 1

1. Which of these can be a cross-section of a cylinder? *Select all that apply.*

   A. scalene triangle **B.** pentagon **C.** isosceles triangle

   D. circle **E.** equilateral triangle **F.** rectangle

   G. irregular quadrilateral **H.** hexagon

2. Vladimir sliced a cube with a plane to produce the cross-section shown in the figure below. Crystal produced the exact same cross-section by slicing a square pyramid with a plane. Which of these statements is true about the plane Crystal used?

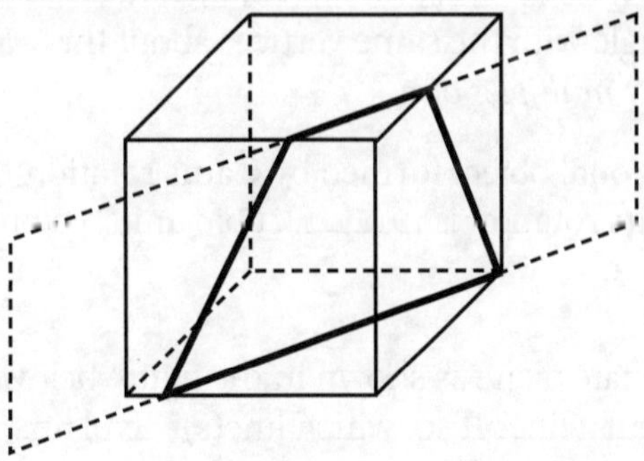

   A. It must have been parallel to the base of the pyramid.

   B. It must not have been parallel to the base of the pyramid.

   C. It must have been perpendicular to the base of the pyramid.

   D. It must not have been perpendicular to the base of the pyramid.

3. Lori found a cross-section of a 3-D shape to be a square. *Circle the correct words.*

   Lori's shape [could / could not] be a cone and [could / could not] be a tetrahedron.

4. Mr. Epstein taught his class that the cross-sections of a sphere are circles that have varying radii less than or equal to the radius of the sphere. To demonstrate, he created two cross-sections of the same sphere by slicing the sphere with two parallel planes. The first cross-section is a circle with a radius of 37 cm, which is also the radius of the sphere. The second cross-section is a circle with a radius of 35 cm. How far apart were the two planes that Mr. Epstein used to create the cross-sections? Explain your answer.

Name ____________________ Date ____________

Check for Understanding

Investigate Cross-Sections and Rotations: Investigation 2

1. Rotating which of these two-dimensional shapes about the x-axis in a coordinate plane will produce a cylinder with a height of 6 units and a radius of 8 units?

   A. a triangle with vertices at (–4, 0), (4, 0), and (4, 6)

   B. a triangle with vertices at (–3, 0), (3, 0), and (3, 8)

   C. a rectangle with vertices at (–4, 0), (–4, 6), (4, 0), and (4, 6)

   D. a rectangle with vertices at (–3, 0), (–3, 8), (3, 0), and (3, 8)

2. Cam rotated a rectangle with vertices at (0, 0), (0, 7), (10, 0), and (10, 7) about the x-axis in a coordinate plane, while Peg rotated a rectangle with the same vertices about the y-axis. *Fill in the blanks in the following sentence with the correct multiples of* π.

   The volume of the three-dimensional object formed by Cam's rotation is ______ cubic units, and the volume of the one formed by Peg's rotation is ______ cubic units, so one object's volume is ______ cubic units greater than that of the other.

3. Mayumi drew $\triangle RST$ in a coordinate plane as shown in the figure below. Is it possible for her to produce a cone by rotating the triangle about a line? If so, which line(s)? Explain.

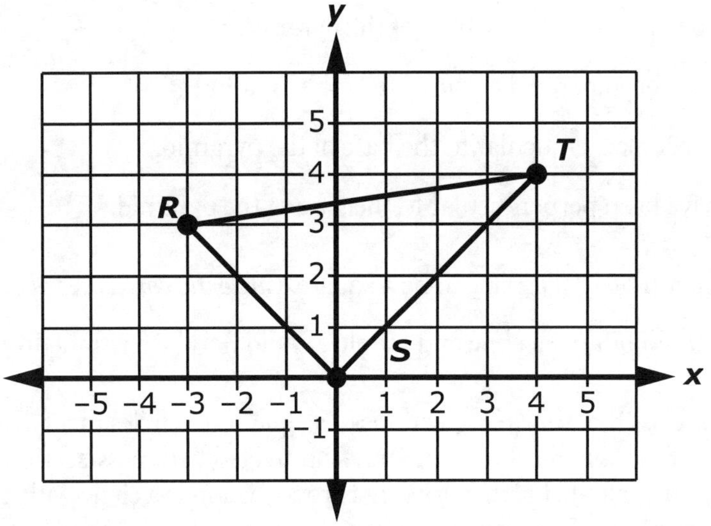

Summary

Before you attempt the Practice Exercises, review what you've learned.

Cross-Sections

A **cross-section** is the intersection of a plane and a three-dimensional object. The shape of the cross-section will vary depending on the location and angle.

The table shows some examples of cross-sections of common three-dimensional objects.

Object	Location of Plane	Cross-Section
Sphere	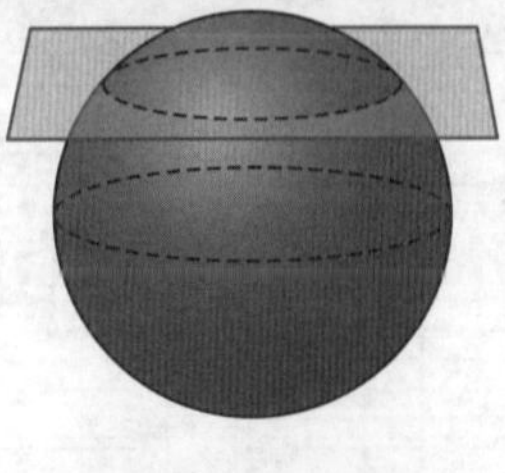	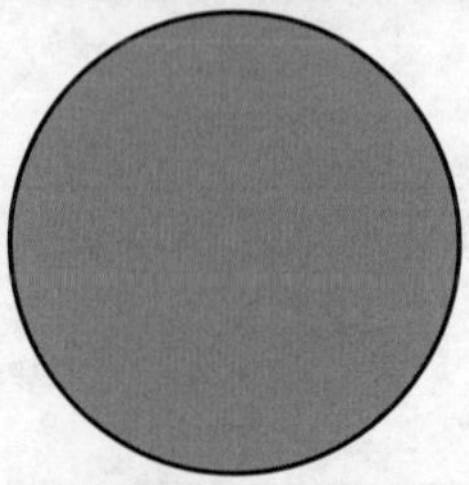
Cylinder	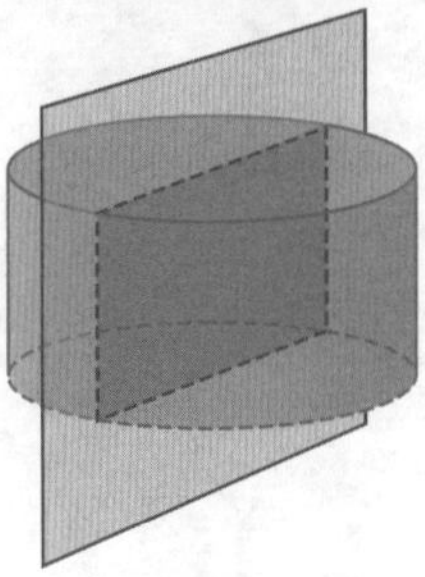	
Cone	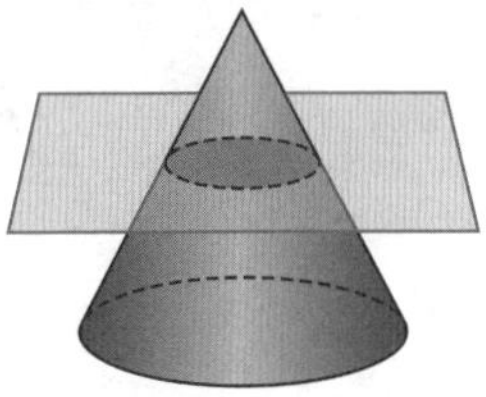	
Pyramid	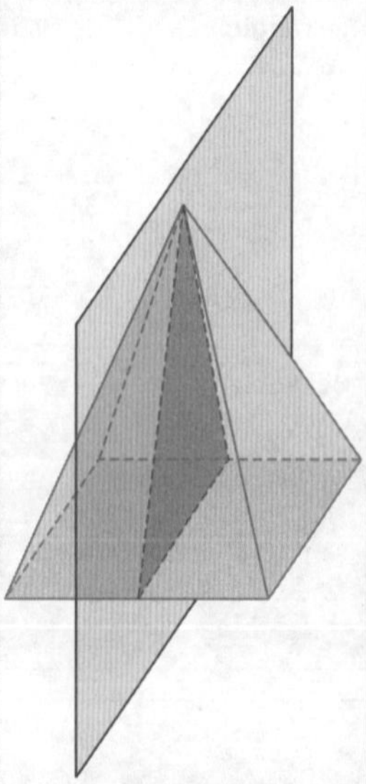	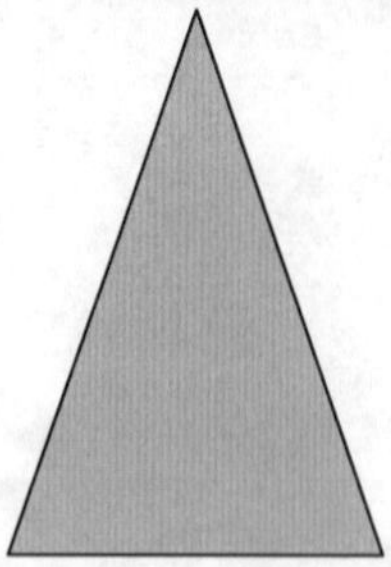
Prism	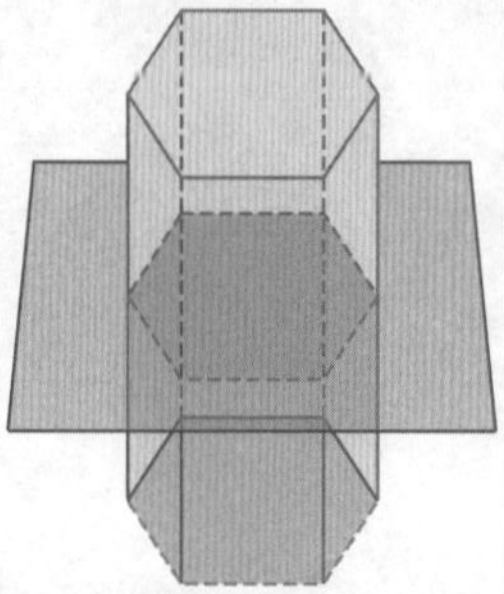	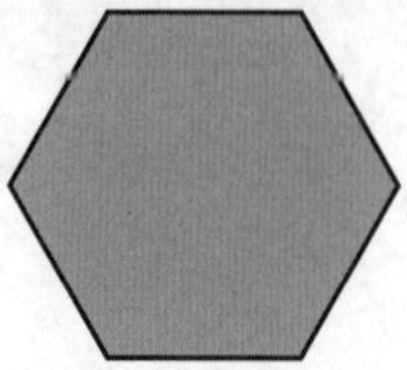

My Notes

Summary *(continued)*

The location of the plane determines the shape of each cross-section. For example, a cube can produce each of the following cross-sections:

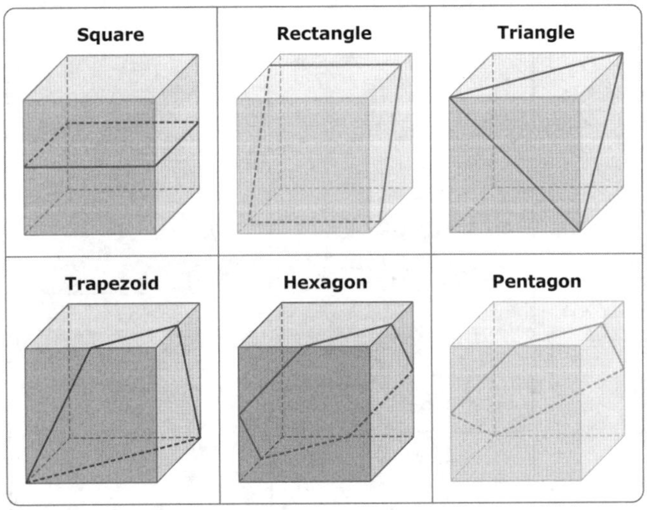

Rotations of 2-D Figures

The process for drawing a three-dimensional object resulting from the rotation of a two-dimensional region in the coordinate plane about an axis is as follows:

1. Sketch the two-dimensional object.
2. Draw its reflection across the axis of rotation.
3. Draw ellipses to indicate three dimensions.

The following steps show how to draw the three-dimensional object that results from rotating a triangle with vertices at (2, 0), (5, 0), and (2, 3) about the y-axis.

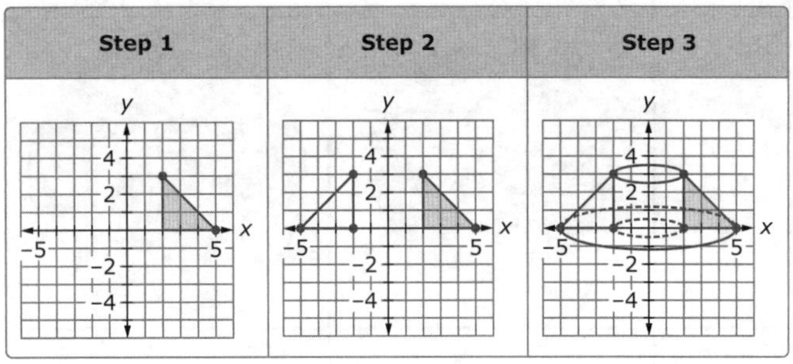

My Notes

Practice Exercises

Review what you've learned using these practice problems. For practice problems with feedback, try the Coach and Play items in the Practice section online.

1. Mariam is mailing a framed photo of herself and her sister to her grandfather for his birthday. The picture, including the frame, is a rectangle that measures 10 inches by 12.5 inches, and is very flat—less than 0.25 inch.

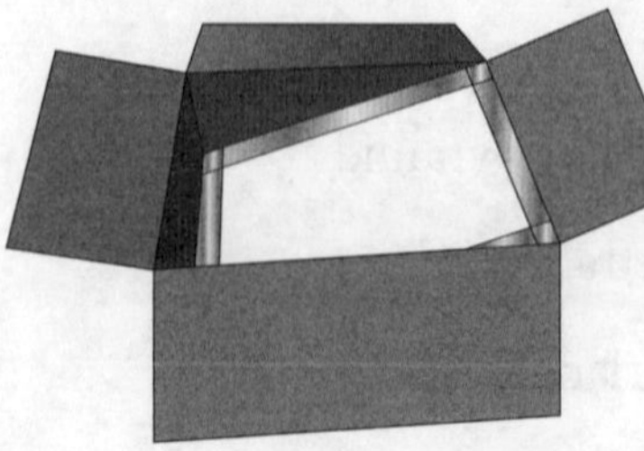

The dimensions of available packing boxes, all of which are rectangular prisms, are listed below.

Which boxes could she use to pack the gift, making sure she has a little extra room in the box for packing material? *Select all that apply.*

   A. 12 in. × 10.5 in. × 5 in.

   B. 12 in. × 11 in. × 2 in.

   C. 13 in. × 8 in. × 6 in.

   D. 13 in. × 10 in. × 4 in.

   E. 15 in. × 9 in. × 3 in.

2. Which of the following could be cross-sections of a right cone having base radius 5 feet and height 11 feet? *Select all that apply.*

   A. a circle of radius 1 foot

   B. a circle of radius 5.5 feet

   C. an equilateral triangle

   D. an isosceles triangle

   E. an ellipse

3. Marcus creates a two-dimensional cross-section by slicing a cube parallel to its base. If the volume of the cube is 4,096 cubic inches, what is the area of the cross-section?

The area of the cross-section is ____________ square inches.

4. What is the resulting cross-sectional figure created by the intersection of a horizontal plane with a vertical cylinder?

   A. circle

   B. ellipse

   C. square

   D. rectangle

5. Kym shined a flashlight on a particular solid. When she shined the flashlight from the front, the solid cast a triangular shadow on the wall. When Kym shined the flashlight from the side, the solid cast a rectangular shadow on the wall. When she shined the flashlight from the top, the solid cast a rectangular shadow on the floor.

What is the name of this mystery solid?

   A. a triangular pyramid

   B. a rectangular-based pyramid

   C. a triangular prism

   D. a square-based pyramid

Practice Exercises *(continued)*

6. What shape is the cross-section of a cone cut by a plane that is parallel to its base?

A. a circle

B. a triangle

C. a parabola

D. a hyperbola

7. A right square pyramid has a height of 8 inches and a square base measuring 12 inches on each side. A vertical plane cuts the pyramid so that the line of intersection with the square base is parallel to one of the sides of the base.

Which of the following are possible cross-sections that result from the cut of the pyramid?

A. a square with each side measuring 4 inches

B. a trapezoid

C. an isosceles triangle with height 5 inches

D. a scalene triangle

E. a rectangle that is not a square

F. an isosceles triangle with height 8 inches

G. a pentagon

8. A cone has a cross-section through the vertex and a diameter of the base. What figure will be the result?

A. circle

B. ellipse

C. triangle

D. trapezoid

9. Which of the following 3-D figures can have a 2-D cross-section that is a trapezoid? *Select all that apply.*

A. cube

B. cylinder

C. rectangular prism

D. sphere

E. square pyramid

F. tetrahedron

G. triangular prism

10. Each of the two-dimensional objects can be modeled by stacking a series of geometrically similar shapes.

Match each object to the possible cross-sectional shape or shapes that could be used to model it.

For each object, select all the cross-sectional shapes that apply.

____

____

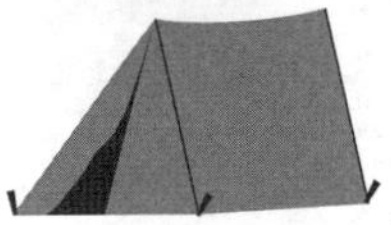
____

A. circle

B. triangle

C. rectangle or square

Apply

What Would a Vase Based on Your Profile Look Like?

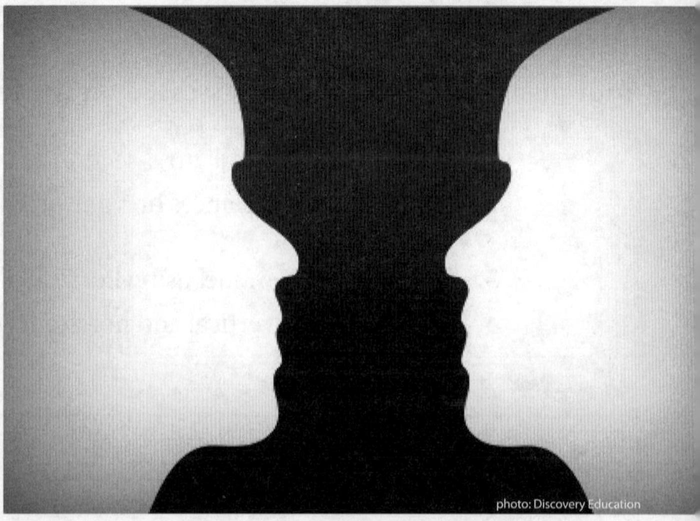

This common illusion, known as Rubin's vase, is named after its inventor, Edgar Rubin.

How can you use your own profile to make a personal version of Rubin's vase?

- Start by printing a photo of your profile.
- Draw a vertical line about an inch away from the tip of your nose to act as the axis of rotation.
- Draw the vertical cross-section of the solid formed by rotating your profile around this line.
- Measure the distance from your profile to the vertical line at different points.
- Using the measurements and the thickness of the materials as a guide, cut circles to represent cross-sections of your vase.
- Thread the circles onto a piece of string or glue them together to complete your model.
- Describe the vertical and horizontal cross-sections of the model.

Show what you've learned by completing the other performance tasks in the online Apply section.

Apply *(continued)*

Your answer to Apply will be assessed on the following criteria:

1. Drawing the vertical cross-section of your Rubin's vase and describing your process
2. Measuring the distances on your profile to create an appropriate number of horizontal cross-sections
3. Building a 3-D model using horizontal cross-sections
4. Describing the vertical and horizontal cross-sections of your model

Scale / Criteria	4	3	2	1	0
	Exceeds Criteria	**Meets Criteria**	**Progressing to Criteria**	**Below Expectations**	**No Expectation**
Cross-Section Drawing	Submits an accurate drawing of a vertical cross-section and describes the process used.	Submits a drawing of a vertical cross-section, with minor errors, and describes the process used.	Submits a drawing of a vertical cross-section and describes the process used, with major errors.	Submits a drawing of a vertical cross-section in which the personal profile image is not correctly reflected and does not describe the process used.	Does not submit a cross-section drawing.
Measured Distances	Correctly reports multiple distances from the edge of the vertical cross-section to the line of symmetry.	Reports multiple distances from the edge of the vertical cross-section to the line of symmetry, with minor errors.	Reports multiple distances from one edge of the vertical cross-section to the other edge.	Reports an inadequate number of distances from the edge of vertical cross-section to the line of symmetry.	Does not report measured distances.
3-D Model	Correctly cuts out disks based on the measurements, and builds an approximate model of the vase.	Cuts out disks based on the measurements and builds an approximate model of the vase found, but there are minor errors.	Cuts out disks based on the measurements, and builds an approximate model of the vase, but there are major errors.	Cuts out shapes other than circles based on the measurements, and builds an approximate model of the vase.	Does not build a model of the vase.
Cross-Section Descriptions	Correctly describes the vertical and horizontal cross-sections of the model.	Describes the vertical and horizontal cross-sections of the model, with minor errors.	Describes either the vertical or horizontal cross-sections of the model, but not both.	Incorrectly describes both the vertical and horizontal cross-sections of the model.	Does not describe the vertical or horizontal cross-sections.

UNIT 10: 3-D Figures

10.2 Develop and Apply Volume Formulas

photo: Getty Images

Lesson Objectives

- Develop volume formulas using informal arguments about sums of areas.
- Solve problems involving the area and volume of spheres, cones, cylinders, and pyramids.
- Use 3-D figures to model and solve real-world problems involving area and volume.

Essential Question

- How can decomposition and recomposition of three-dimensional objects be used to develop and explain volume formulas?

photo: Getty Images

Investigations

Sculptures

Estimate the volume of the 3-D object you created in layers.

Cavalieri's Principle

How are area and volume related? Develop a volume formula for all prisms and cylinders.

Pyramids and Cones

Estimate the volume of a pyramid. What's the formula? Does it work for a cone too?

Spherical Sculptures

How are cylinders, cones, and spheres related? How can you find the volume of a sphere?

Tree Sculptures

Put the volume formulas to work solving real-world design problems.

Geometric Art

Create nets of 3-D figures to find surface area. Is there a formula?

Key Vocabulary

cone, cross-section, cylinder, pyramid, prism, sphere

Discover

As you complete Engage and the investigations, record the most important ideas you've learned.

Engage	**Investigation 1**
Investigation 2	**Investigation 3**
Investigation 4	**Investigation 5**

Name ______________________________ Date __________

Check for Understanding

Develop and Apply Volume Formulas: Investigation 1

1. Two prisms are shown in the figure below. Do the prisms have the same volume?

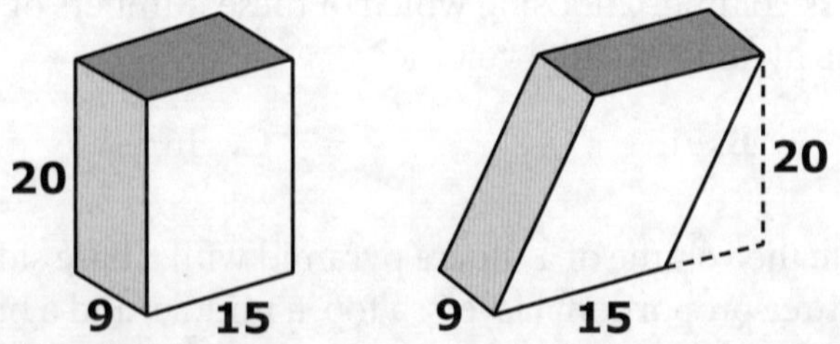

   A. Yes, because they are similar.

   B. No, because they are not similar.

   C. Yes, because they have the same height and the same cross-sectional area at every level.

   D. No, because they do not have the same height and do not have the same cross-sectional area at every level.

2. Kathy is passing out fliers that are rectangular and measure 5 inches by 7 inches. She began with a stack of 4,000 that formed a prism with a volume of 280 in.^{3} So far, she has passed out 800, and her remaining fliers are no longer neatly stacked. *Fill in the blanks in the following sentences.*

   Kathy's original stack of fliers was ______ inches high, with each flier having a thickness of ______ inches. Each of Kathy's remaining fliers has an area of ______ in.^{2}, so even though they are no longer neatly stacked, their total volume is ______ in.^{3}

3. Mike and Lydia each have a cylinder. Mike's cylinder has a radius of 6 cm and a volume of 324π cm^3, while Lydia's has a radius of 6 cm and a height of 9 cm. In Mike's cylinder the bases are aligned, with one directly above the other. *Circle the correct words in the sentences below.*

   If Lydia's cylinder has bases that are aligned, with one directly above the other, then its volume [is / is not] equal to that of Mike's. If Lydia's cylinder does not have bases that are aligned, then its volume [is / is not] equal to that of Mike's.

4. What is the converse of Cavalieri's principle? Is the converse true? Explain your answer.

Name ____ Date ____________

Check for Understanding

Develop and Apply Volume Formulas: Investigation 2

1. Tia will model a 48-cm-high square pyramid with proportional layers and find the sum of their volumes to approximate the pyramid's volume. Choosing which of these numbers of layers will give her an approximation closest to the pyramid's actual volume?

A. 6 **B.** 12 **C.** 16 **D.** 24

2. Suppose that to approximate the volume of a square pyramid with a base side length of 6 in. and a height of 3 in., you model it with three proportional layers: a top, a middle, and a bottom layer. *Fill in the blanks in the table and in the sentence that follows.*

Layer	Base Side Length (in.)	Height (in.)	Volume (in.^{3})
Top			
Middle			
Bottom			

The pyramid's approximate volume is ______ in.^{3}, and its actual volume is ______ in.^{3}

3. Willard approximated the volume, V, of a square pyramid with a base side length of 5 cm and a height of 4 cm by using the formula $V \approx b^2h\left(\frac{1}{3} + \frac{1}{2n} + \frac{1}{6n^2}\right)$, where b is the base side length, h is the height, and n is the number of proportional layers that model the pyramid. First, he found an approximate volume of 62.5 cm^3 when the pyramid is modeled with 2 layers, and then he found an approximate volume of 44 cm^3 when it is modeled with 5 layers. According to his calculations, the pyramid's approximate volume decreased by $\frac{62.5 - 44}{62.5} = 0.296 = 29.6\%$ when he added 3 layers. He believes that the approximate volume will continue to decrease by 29.6% for every 3 layers added. Is Willard correct in his reasoning? Explain your answer without using the formula to calculate the pyramid's approximate volume when it is modeled with 8 layers.

Name ______________________ **Date** ____________

Check for Understanding

Develop and Apply Volume Formulas: Investigation 3

1. Which of these is equal to the volume of a sphere with a radius of r units?

   A. the volume of a cylinder with a radius of r units and a height of r units minus the volumes of two cones, each with a radius of r units and a height of r units

   B. the volume of a cylinder with a radius of r units and a height of $2r$ units minus the volumes of two cones, each with a radius of r units and a height of r units

   C. the volume of a cylinder with a radius of $2r$ units and a height of r units minus the volumes of two cones, each with a radius of r units and a height of r units

   D. the volume of a cylinder with a radius of $2r$ units and a height of $2r$ units minus the volumes of two cones, each with a radius of r units and a height of r units

2. A cone, a cylinder, and a sphere each have a radius of 9 cm, and the height of the cone is 18 cm. The volume of the sphere is the average of the volume of the cone and the volume of the cylinder. *Fill in the blank in the following sentence.*

   According to the information given, the height of the cylinder must be _______ cm.

3. On the left in the figure below are a cylinder and a cone, both of radius r units, with the cone inside the cylinder. On the right is a hemisphere of radius r units. A horizontal cross-section of each is being taken a units above the bottoms of the cylinder and hemisphere. The radius of the cone's cross-section is a units, and the radius of the hemisphere's cross-section is b units. How can you use this information to determine the relationship between the volumes of the cylinder, cone, and hemisphere? Explain.

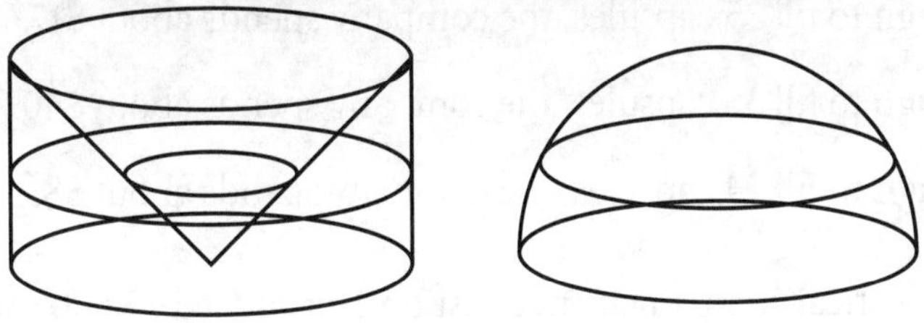

Name ______________________________ Date ______________

Check for Understanding

Develop and Apply Volume Formulas: Investigation 4

1. A pharmaceutical company is selling a new medicine in capsule form. Each capsule can be modeled by a cylinder with a hemisphere at each end as shown in the figure. The length of the cylinder used to model a capsule is 15 mm, and its diameter is 8 mm.

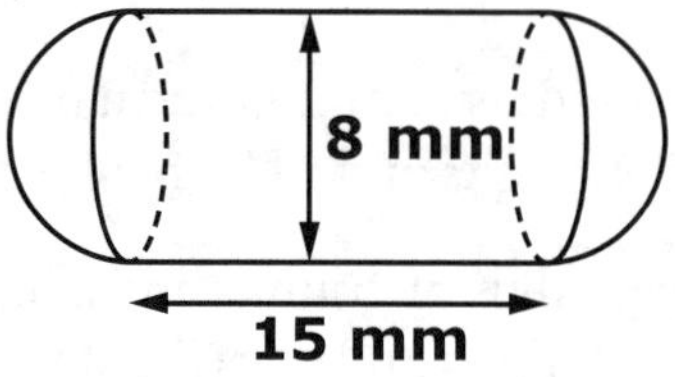

What is the approximate volume of each capsule?

A. 268 mm^3 **B.** 754 mm^3 **C.** 888 mm^3 **D.** 1,022 mm^3

2. The density of the new medicine that the pharmaceutical company from question 1 is selling is 1.468 mg per cubic millimeter. *Fill in the blank in the sentence below.*

To the nearest milligram, the amount of medicine in each capsule is ______ mg.

3. The pharmaceutical company from question 1 spends \$0.01 to produce 10 mg of the new medicine. Which of the following statements are correct? *Select all that apply.*

A. To produce enough to fill 46 capsules, the company spends about \$69.

B. To produce enough to fill 22 capsules, the company spends about \$24.

C. To produce enough to fill 15 capsules, the company spends about \$17.

D. To produce enough to fill 36 capsules, the company spends about \$40.

E. To produce enough to fill 54 capsules, the company spends about \$81.

4. Last month, the pharmaceutical company from question 1 produced 75,000 capsules of the new medicine. If the diameter of the cylinder used in the model were 6 mm instead of 8 mm, about how many more capsules could the company have produced? Explain.

Name ______________________________ Date ____________

Check for Understanding

Develop and Apply Volume Formulas: Investigation 5

1. Carmelo is a draftsman using computer-aided design software. He must meet specifications calling for an 18-foot-high cylinder with two circular bases and a surface area of between 400 ft^2 and 500 ft^2. *Circle the correct words in the following sentence.*

A radius of 3 ft for the cylinder [will / will not] allow Carmelo to meet the specifications, and one of 4 ft [will / will not] allow him to meet the specifications.

2. The net of a cone consists of a smaller circle with a radius of 7 cm and a sector of a larger circle with a radius of 25 cm. *Fill in the blanks in the sentence with multiples of π.*

The area of the smaller circle is ______ cm^2, and the area of the sector of the larger circle is ______ cm^2, so the surface area of the cone is ______ cm^2.

3. Rhonda sliced a sphere with a plane to create a circular cross-section. She claims that the area of the cross-section is $\frac{1}{4}$ the surface area of the sphere. Is her claim correct?

A. no

B. only if the plane passed through the sphere's center

C. yes

D. only if the plane did not pass through the sphere's center

4. The figure below shows the net of an aluminum cylinder of radius r units. The aluminum in only the rectangular portion of the net was melted and reshaped into a sphere of radius r units without having any aluminum left over. What is the ratio of the length of the base of the net's rectangular portion to its height? Explain how you got your answer.

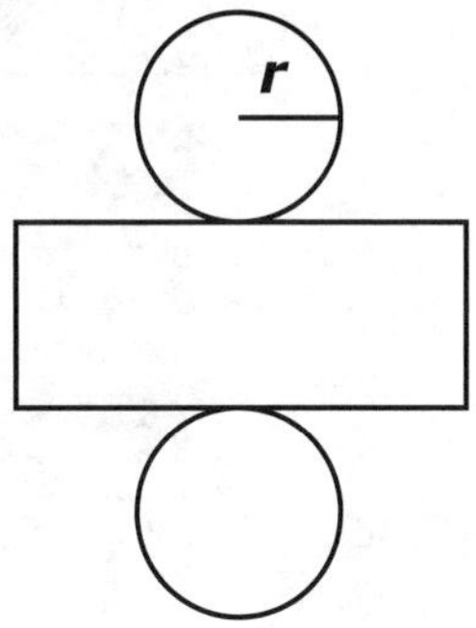

Summary

Before you attempt the Practice Exercises, review what you've learned.

Volume

Cavalieri's principle states that if two 3-D figures have the same height and the same cross-sectional area at every level, then their volumes are equal. This principle can be used to generalize the formula for rectangular prisms to irregular and oblique prisms and cylinders.

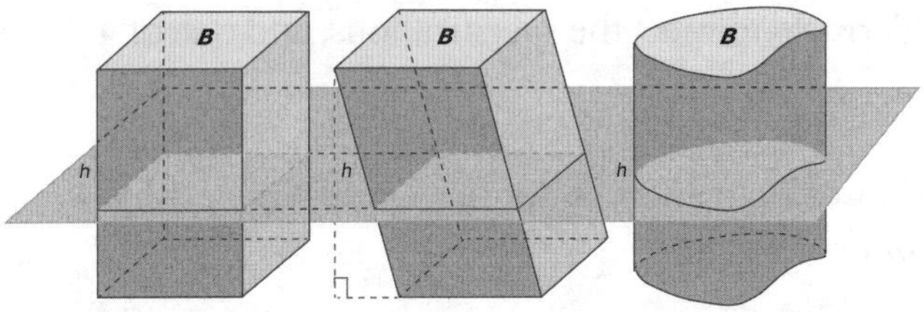

Volume of a Prism or Cylinder

For any prism or cylinder, $V = Bh$, where B is the area of the base and h is the height.

For a cylinder, the volume can be written as $V = \pi r^2 h$.

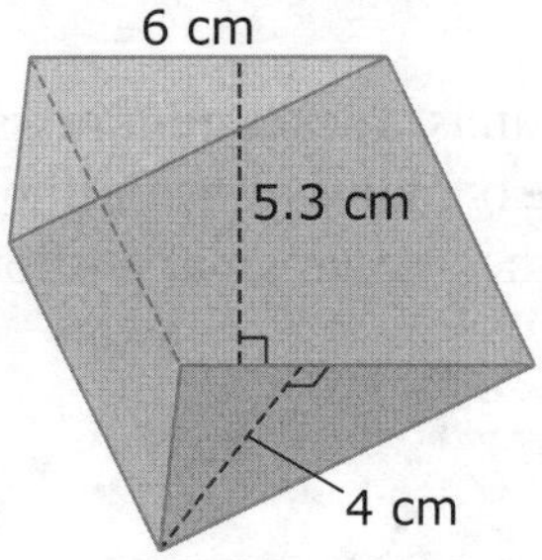

$V = Bh$

$V = \left(\frac{1}{2}(4)(6)\right)5.3$

$V = 63.6 \text{ cm.}^3$

Cavalieri's principle can also be used to extend the formula for a square pyramid to any pyramid or cone.

Volume of a Pyramid or Cone

For any pyramid or cone, $V = \frac{1}{3}Bh$, where B is the area of the base and h is the height.

For a cone, the volume can be written as $V = \frac{1}{3}\pi r^2 h$.

Volume of a Sphere

For any sphere, $V = \frac{4}{3}\pi r^3$, where r is the radius.

Surface Area

Surface Area of a Right Cylinder

For a right cylinder, $S = 2\pi r^2 + 2\pi rh$, where r is the radius and h is the height.

Surface Area of a Right Cone

For a right cone, $S = \pi r^2 + \pi rs$, where r is the radius and s is the slant height.

Surface Area of a Sphere

For a sphere, $S = 4\pi r^2$, where r is the radius.

Practice Exercises

Review what you've learned using these practice problems. For practice problems with feedback, try the Coach and Play items in the Practice section online.

1. Which of the following statements best explains the relationship between the volume of a cone and a volume of a cylinder with identical radii and heights?

A. The volume of a cone is half of the volume of a cylinder because the volume of a triangular cross-section of a cone is half of the volume of a rectangular cross-section of a cylinder.

B. The volume of a cone is one-third of the volume of a cylinder because the volume of a pyramid is one-third of the volume of a prism with identical base and height.

C. The volume of a cylinder is twice the volume of a cone because two cones can be placed vertex to base to create a cylinder.

D. The volume of a cylinder is triple the volume of a cone because a rectangular cross-section can be formed using three congruent triangular cross-sections from the vertex of the cone.

2. Tyrion creates a set of nesting shapes as a toy set for his baby nephew. The outer figure is a cube with an edge length of 20 centimeters. The middle figure is a square pyramid with an identical square base and the same height as the cube. The innermost figure is a cone with a diameter and height equal to the edge of the cube.

Calculate each difference in the volumes.

Round your answers to the nearest whole cubic centimeter.

Difference between cube and pyramid:

___________ cubic inches

Difference between pyramid and cone: ___________ cubic inches

Difference between cube and cone: ___________ cubic inches

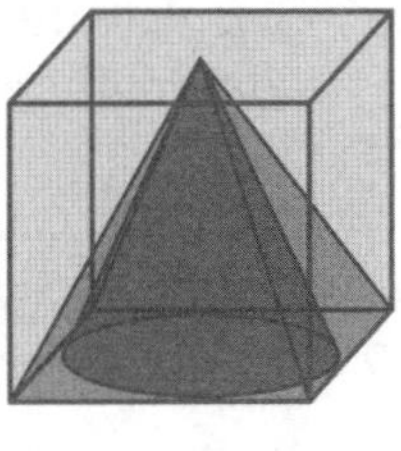

3. A stack of identical 8-centimeter-by-10-centimeter rectangular playing cards is sitting on a table. The stack is 1 centimeter high and has a volume of 80 cubic centimeters. The table is bumped and the stack of cards is shifted so that the stack is no longer a right rectangular prism, but all of the cards are still part of the stack.

Which of the following statements are true of the shifted stack of cards? *Select all that apply.*

A. The shifted stack of cards has a volume of 80 cubic centimeters because it has the same base area and height.

B. The area of a card at any location in the stack remains the same.

C. The width of the stack from the leftmost edge to the rightmost edge will remain 10 centimeters because the volume is the same.

D. Cavalieri's principle cannot be applied to the volume of the stack because the stack is a rectangular prism instead of a cylinder.

E. The area of the middle card in the stack is 40 square centimeters because it is half of 80.

4. A cylinder has a height equal to its radius. A cone has the same height and radius as the cylinder.

Which of the following statements is true?

A. The cone has half the volume of the cylinder.

B. The cylinder is 4 times the volume of the cone.

C. A sphere with an equal radius has a volume equal to the combined volumes of the cylinder and cone.

D. A sphere with an equal radius has a volume equal to half the combined volumes of the cylinder and cone.

Practice Exercises *(continued)*

5. A cone has a diameter of 10 centimeters and a height of 12 centimeters. A square pyramid has an edge length of 10 centimeters and a height of 12 centimeters.

Which of the following statements are true about the cone and square pyramid? *Select all that apply.*

A. The cone has less volume than the square pyramid.

B. The square pyramid has about 228 cubic centimeters less volume than the cone.

C. The cone has a greater surface area than the square pyramid.

D. The square pyramid has a little more than 77 square inches greater surface area than the cone.

E. The cone and square pyramid have equal surface area and equal volume because they have the same base length and height.

6. Nina needs to create a pond space that has a volume of 134 cubic feet and a depth of 4 feet. She proposes three possible pool designs:

- a prism with a square top
- an inverted cone with a circular opening
- a hemisphere with a circular opening

Calculate the area of exposed water for each of Nina's proposed pool designs.

The area of exposed water for the prism pool is

____________ square feet.

The area of exposed water for the inverted cone pool is ____________ square feet.

The area of exposed water for the hemisphere pool is approximately ____________ square feet.

7. A cone has a height of 12 centimeters and a volume of 314.16 cubic centimeters.

The formula for the volume of a cone is $V =$ ____________.

The radius of the cone is ____________ centimeters.

8. What is the radius of a sphere with a volume of 36π?

A. 2 **B.** 3

C. 4 **D.** 5

9. Mr. Litterski creates a ball from rubber bands. The circumference of the ball is 9.4 inches. What is the approximate surface area and volume of the ball?

Round your answers to the nearest unit.

The surface area of the ball is approximately

__________ square inches.

The volume of the ball is approximately

__________ cubic inches.

10. Corinne's Place specializes in hand-carved, wooden beads for bracelets and necklaces. One of Corinne's favorite designs is an octahedron, as shown.

An octahedron can be thought of as two square-based pyramids with the squares joined. All eight triangles of the octahedron are congruent equilateral triangles.

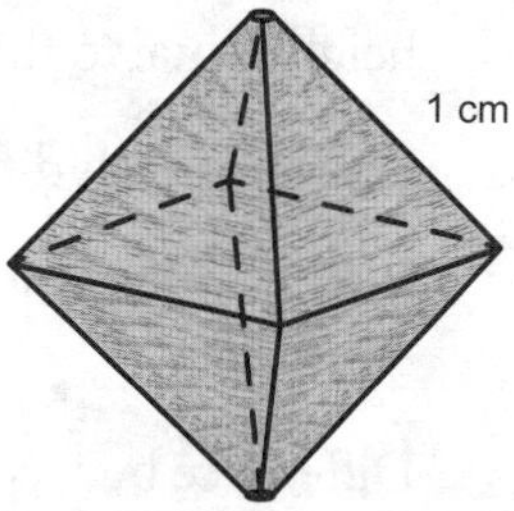

Calculate the volume of an octahedron bead if the side lengths are 1 centimeter.

The volume is approximately ____________ cubic centimeters.

Apply

How Much Would It Cost to "Print" a 3-D Game Piece?

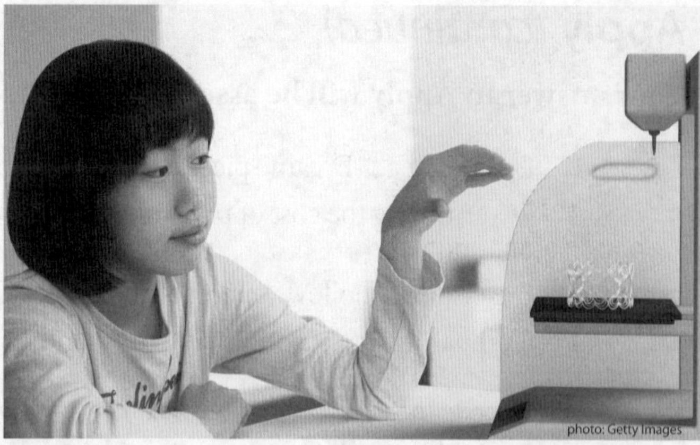
photo: Getty Images

Making ideas reality, one layer of filament at a time!

Imagine that you have invented a board game, and you want to use a 3-D printer to create customized game pieces. Design a game piece no larger than 4 cm tall that is a solid, composite, 3-D object, incorporating at least two different shapes. Draw and describe your game piece, and then calculate its volume, making reasonable approximations as needed.

Your 3-D printer uses 1.75 mm filament. The filament is a long cylinder of plastic with a diameter of 1.75 mm that is melted to form the layers of the printed object. A 1 kg spool has a length of about 330 m and costs $22.50.

- About how much would it cost for you to print one game piece?
- About how many pieces could you print using a 1 kg spool?

To save time and money, you consider making the game piece hollow. Describe the shape of the space inside the piece, and describe the shape of the piece's exterior. Compare the cost of printing the hollow version with the cost of the solid version and summarize your results.

Show what you've learned by completing the other performance tasks in the online Apply section.

Apply *(continued)*

Your answer to Apply will be assessed on the following criteria:

1. Designing a 3-D game piece to print and calculating the volume
2. Determining the cost of printing one solid piece and the number of pieces you could print using one spool
3. Describing a hollow version of the object and finding the cost of printing
4. Comparing and summarizing the printing process for the two versions of your game piece

Criteria \ Scale	4 Exceeds Criteria	3 Meets Criteria	2 Progressing to Criteria	1 Below Expectations	0 No Expectation
Object Design	Designs a game piece that meets the criteria and accurately calculates the volume.	Designs a game piece that meets the criteria and calculates the volume with minor errors.	Designs a game piece that meets the criteria, but does not accurately calculate the volume.	Designs a game piece that does not meet the criteria and does not accurately calculate the volume.	Does not submit.
Cost of Solid Object	Accurately calculates the cost of printing the solid game piece and the number of pieces that can be printed from one spool.	Calculates the cost of printing the solid game piece and the number of pieces that can be printed from one spool, but makes minor errors.	Accurately calculates either the cost of printing the solid game piece or the number of pieces that can be printed from one spool, but not both.	Does not accurately calculate the cost of printing the solid game piece or the number of pieces that can be printed from one spool.	Does not attempt any calculations.
Cost of Hollow Object	Describes a hollow version of the game piece and accurately calculates the cost of printing.	Describes a hollow version of the game piece and calculates the cost of printing, but makes minor errors.	Describes a hollow version of the game piece and calculates the cost of printing, but makes major errors.	Describes a hollow version of the game piece, but does not calculate the cost of printing.	Does not describe a hollow version of the game piece or the cost to print it.
Comparison	Provides a strong and accurate comparison of the costs of printing the solid and hollow pieces.	Provides a reasonable comparison of the costs of printing the solid and hollow pieces.	Provides a comparison of the costs of printing the solid and hollow pieces, but uses faulty logic in making conclusions.	Includes a weak comparison and summary of the costs of printing the solid and hollow pieces and uses faulty logic in making conclusions.	Does not compare the game pieces.

UNIT 11: Conic Sections

11.1 Analyze Graphs and Equations of Circles and Ellipses

photo: Getty Images

Lesson Objectives

- Extend understanding of the Pythagorean theorem to derive the equation of a circle.
- Translate between the equation of a circle and its graph.
- Explain the relationship between a circle and its center.
- Derive the equation of an ellipse.
- Explain the relationship between foci and an ellipse.
- Translate between the equation of an ellipse and its graph.

Essential Questions

- How are the center and radius of a circle or the foci of an ellipse related to their corresponding equations?
- How do the characteristics of the conic equations affect the key features of their graphs?

Investigations

Finding Conic Sections

Use the Conic Section Explorer to find the shapes that result from a plane intersecting a cone.

Circle Explorer

What does the Pythagorean theorem have to do with the equation of a circle?

Graphing the Form of a Circle

The equation of a circle can be written in different forms. One form makes key features apparent.

Ellipse Explorer

Extend a circle to an ellipse using focal points and Pythagoras.

Graphing Ellipses

Build the equation of an ellipse using key characteristics. But what, exactly, is key?

The Equation of an Ellipse

The equation of an ellipse can be written in different forms. Which form spotlights key characteristics?

Key Vocabulary

axes, center, circle, cone, conic section, ellipse, expanded form, focus (foci), major axis, minor axis, radius, vertex

Discover

As you complete Engage and the investigations, record the most important ideas you've learned.

Engage	**Investigation 1**
Investigation 2	**Investigation 3**
Investigation 4	**Investigation 5**

Name ____________________ Date __________

Check for Understanding

Analyze Graphs and Equations of Circles and Ellipses: Investigation 1

1. Dora drew a circle. Then, she drew a right triangle with a hypotenuse that has one endpoint at the center of the circle and one endpoint on the circle. Next, she used the Pythagorean theorem to derive the circle's equation. What does the hypotenuse of the right triangle represent?

   A. the radius of the circle
   B. the diameter of the circle
   C. the square of the circle's radius
   D. the square of the diameter of the circle

2. Examine the circle, and then fill in the blanks in the sentence that follows.

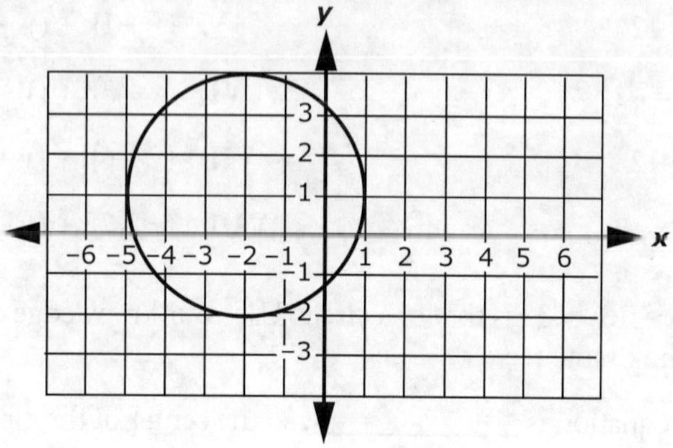

The center of the circle is at (______, ______), and the length of its radius is ______ units, so the equation of the circle is ____________.

3. Which of the circles have radii that are less than 5 units? *Select all that apply.*

   A. $(x + 3)^2 + (y - 4)^2 = 30$
   B. $(x - 6)^2 + (y - 11)^2 = 2$
   C. $(x + 12)^2 + (y + 5)^2 = 4$
   D. $(x - 1)^2 + (y + 8)^2 = 36$
   E. $(x - 9)^2 + (y - 10)^2 = 18$
   F. $(x + 7)^2 + (y + 2)^2 = 52$

4. Suppose you want to use a right triangle with a hypotenuse that has endpoints at the center of a circle and on the circle, along with the Pythagorean theorem, to derive the equation of the circle. How many possible right triangles could you draw? Explain.

Name ______________________________ Date ______________

Check for Understanding

Analyze Graphs and Equations of Circles and Ellipses: Investigation 2

1. Match each of the equations of a circle in the left column with the equivalent graphing form of the equation in the right column.

A. $x^2 + y^2 - 6x + 4y = 12$	**I.** $(x - 2)^2 + (y + 3)^2 = 25$
B. $x^2 + y^2 + 2x - 12y = -21$	**II.** $(x - 1)^2 + (y + 6)^2 = 16$
C. $x^2 + y^2 + 4x - 6y = 12$	**III.** $(x + 3)^2 + (y - 2)^2 = 25$
D. $x^2 + y^2 - 12x + 2y = -21$	**IV.** $(x + 6)^2 + (y - 1)^2 = 16$
E. $x^2 + y^2 - 4x + 6y = 12$	**V.** $(x + 1)^2 + (y - 6)^2 = 16$
F. $x^2 + y^2 - 2x + 12y = -21$	**VI.** $(x - 3)^2 + (y + 2)^2 = 25$
G. $x^2 + y^2 + 6x - 4y = 12$	**VII.** $(x - 6)^2 + (y + 1)^2 = 16$
H. $x^2 + y^2 + 12x - 2y = -21$	**VIII.** $(x + 2)^2 + (y - 3)^2 = 25$

2. The equation $x^2 + y^2 + 16x - 10y = 32$ is that of a circle. Use your knowledge of completing the square to fill in the blanks in the following sentence:

The graphing form of the equation is ____________, so the center of the circle that the equation represents is at the point (______, ______), and the length of the radius of the circle is ______ units.

3. Brad claims that converting the equations $x^2 + y^2 - 14x + 48y = 1$ and $x^2 + y^2 + 30x - 40y = 1$ to graphing form shows that the circles represented by the two equations have radii of the same length. Therefore, he has concluded that when two equations in the form $x^2 + y^2 + ax + by = c$ have the same value of c, the lengths of the radii of the circles represented by the equations must be the same, even if the values of a and b in the equations are different. Is Brad correct? Explain your reasoning.

Name ______________________________ Date ____________

Check for Understanding

Analyze Graphs and Equations of Circles and Ellipses: Investigation 3

1. A horizontal ellipse represented by the equation $\frac{x^2}{a^2} + \frac{y^2}{b^2} = 1$ has foci at the points $(-c, 0)$ and $(c, 0)$. The point (x, y) lies on the ellipse. If m is the distance from (x, y) to $(-c, 0)$, and n is the distance from (x, y) to $(c, 0)$, what is the value of $m + n$?

A. a **B.** b **C.** $2a$ **D.** $2b$

2. A vertical ellipse represented by the equation $\frac{x^2}{a^2} + \frac{y^2}{b^2} = 1$ has foci at the points $(0, c)$ and $(0, -c)$. In the blanks below, write a in terms of b and c, and b in terms of a and c.

$a =$ ____________ $b =$ ____________

3. Ross drew the graph of an ellipse represented by the equation $\frac{x^2}{a^2} + \frac{y^2}{b^2} = 1$. Which statements must be correct about the graph Ross drew? *Select all that apply.*

A. If $a > b$, the length of the major axis of the ellipse is $2b$.

B. If $b > a$, the foci of the ellipse lie on the ellipse's major axis.

C. If $a > b$, the foci of the ellipse lie on the ellipse's major axis.

D. If $b > a$, the major axis of the ellipse is a segment of the x-axis.

E. If $a > b$, the major axis of the ellipse is a segment of the y-axis.

F. If $b > a$, the length of the major axis of the ellipse is longer than $2a$.

G. If $a > b$, the length of the major axis of the ellipse is shorter than $2a$.

4. Consider the horizontal ellipse represented by the equation $\frac{x^2}{a^2} + \frac{y^2}{b^2} = 1$ that has foci at the points $(-c, 0)$ and $(c, 0)$. What can you say about the value of $\frac{c}{a}$? What happens to the ellipse as the value of $\frac{c}{a}$ gets smaller and smaller? Explain your reasoning.

Name ______________________________ Date ______________

Check for Understanding

Analyze Graphs and Equations of Circles and Ellipses: Investigation 4

1. The graph of an ellipse is shown. Fill in the blanks in the sentence that follows.

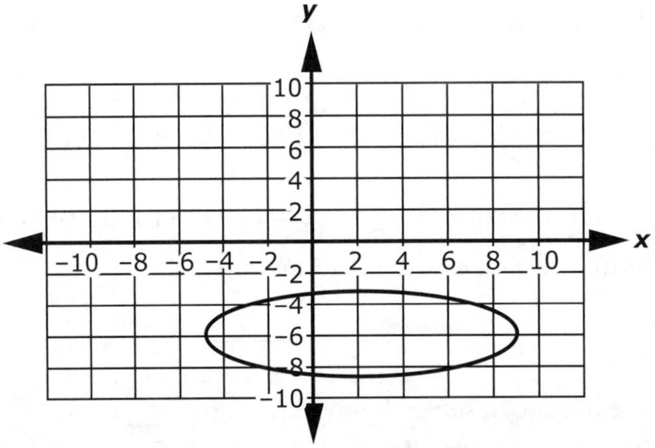

The center of the ellipse is at the point (_______, _______), half the length of the ellipse's horizontal axis is _______ units, and half the length of the ellipse's vertical axis is _______ units, so the equation of the ellipse must be _____________.

2. Which of the following equations represents an ellipse for which the sum of the lengths of the focal radii is 18 units?

A. $\frac{(x+8)^2}{4}+\frac{(y+10)^2}{9}=1$

B. $\frac{(x+20)^2}{4}+\frac{(y+2)^2}{36}=1$

C. $\frac{(x+3)^2}{4}+\frac{(y+6)^2}{81}=1$

D. $\frac{(x+18)^2}{4}+\frac{(y+1)^2}{324}=1$

3. An ellipse passing through the point (−5, 14) has a center at (−1, 8) and a vertex at (−9, 8). What is the equation of the ellipse? Explain how you got your answer.

Name ______________________________ Date __________

Check for Understanding

Analyze Graphs and Equations of Circles and Ellipses: Investigation 5

1. Match each of the equations of an ellipse in the left column with an equivalent equation in graphing form in the right column.

A. $4x^2 + 25y^2 - 16x + 150y = -141$ **I.** $\frac{(x+2)^2}{25} + \frac{(y-3)^2}{4} = 1$

B. $25x^2 + 4y^2 - 100x + 24y = -36$ **II.** $\frac{(x-3)^2}{4} + \frac{(y+2)^2}{25} = 1$

C. $4x^2 + 25y^2 - 24x + 100y = -36$ **III.** $\frac{(x+3)^2}{4} + \frac{(y-2)^2}{25} = 1$

D. $25x^2 + 4y^2 - 150x + 16y = -141$ **IV.** $\frac{(x-2)^2}{25} + \frac{(y+3)^2}{4} = 1$

E. $4x^2 + 25y^2 + 16x - 150y = -141$ **V.** $\frac{(x-3)^2}{25} + \frac{(y+2)^2}{4} = 1$

F. $25x^2 + 4y^2 + 150x - 16y = -141$ **VI.** $\frac{(x-2)^2}{4} + \frac{(y+3)^2}{25} = 1$

2. Consider the ellipse with the equation $64x^2 + 9y^2 - 128x + 90y = 287$. *Fill in the blanks.*

The coordinates of the foci are (_______, _______) and (_______, _______).

3. An elliptical pool table has an inner wall in the shape of an ellipse and a hole at one focus of the ellipse. If you place a ball at the other focus and bounce it anywhere off the inner wall, it will go into the hole. Suppose the inner wall can be modeled by the equation $3x^2 + 11y^2 - 24x = 18$, where the units are in feet and the hole is at the origin. How far away from the hole should a ball be placed so that it will go into the hole when it is bounced anywhere off the inner wall? Explain how you got your answer.

Summary

Before you attempt the Practice Exercises, review what you've learned.

A conic section is formed by the intersection of a double cone and a plane. Two possible types of conic sections are **circles** and **ellipses**.

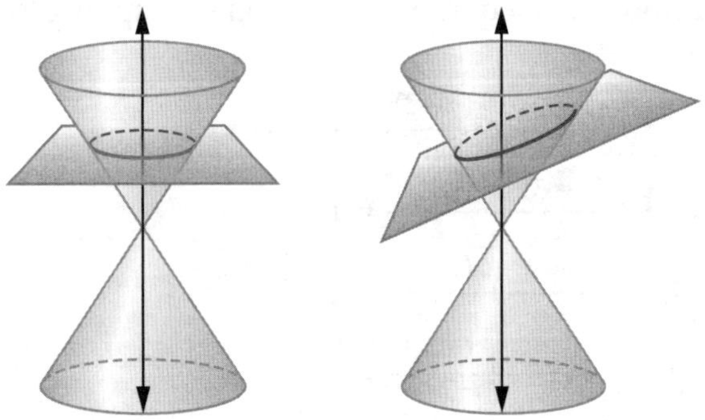

Circle
The plane is perpendicular to the axis of the cone.

Ellipse
The plane intersects one part of the cone but is not necessarily perpendicular to the axis.

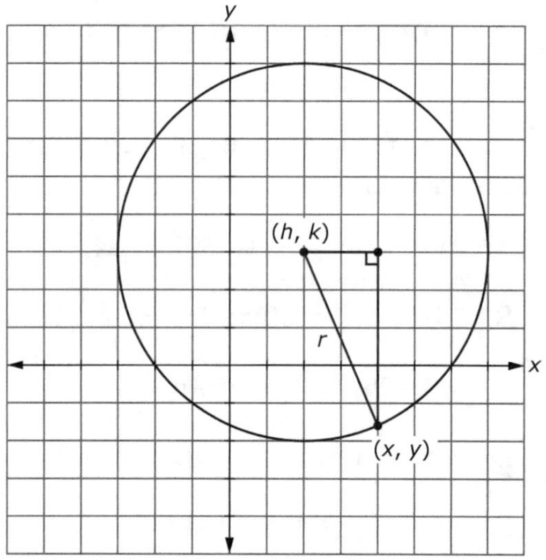

Circles and ellipsis can be represented in the coordinate plane using equations.

Circles

The equation of a **circle** with radius r centered at the point (h, k) is $(x - h)^2 + (y - k)^2 = r^2$.

My Notes

EXAMPLE

What is the equation of a circle with the endpoints of a diameter at (2, 7) and (−4, 3)?

The center of the circle is the midpoint of the diameter:

$\left(\frac{2 + (-4)}{2}, \frac{7 + 3}{2}\right) = (-1, 5)$, so $h = -1$ and $k = 5$.

Use the distance formula to find the radius of the circle, which is the distance from the center to either endpoint of the diameter:

$d = \sqrt{(-1 - 2)^2 + (5 - 7)^2} = \sqrt{13}$, so $r = \sqrt{13}$.

The equation of the circle is $(x + 1)^2 + (y - 5)^2 = 13$. The graph is shown.

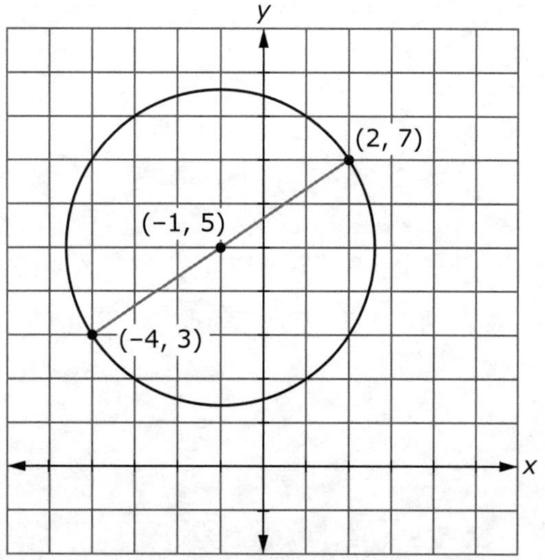

Summary *(continued)*

Ellipses

The equation of an ellipse centered at the point (0, 0) is $\frac{x^2}{a^2} + \frac{y^2}{b^2} = 1$, and the ellipse passes through the points $(-a, 0)$, $(a, 0)$, $(0, b)$, and $(0, -b)$.

If $a > b$, then the ellipse is oriented horizontally. The major axis has endpoints $(-a, 0)$ and $(a, 0)$, the foci are at $(-c, 0)$ and $a^2 = b^2 + c^2$.

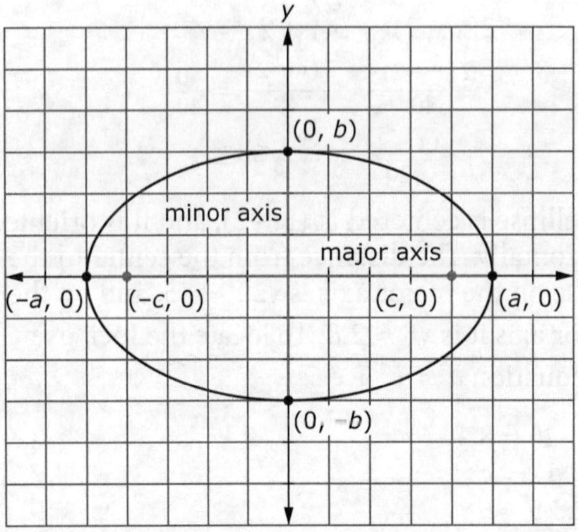

If $b > a$, then the ellipse is oriented vertically. The major axis has endpoints $(0, -b)$ and $(0, b)$, the foci are at $(0, -c)$ and $(0, c)$, and $b^2 = a^2 + c^2$.

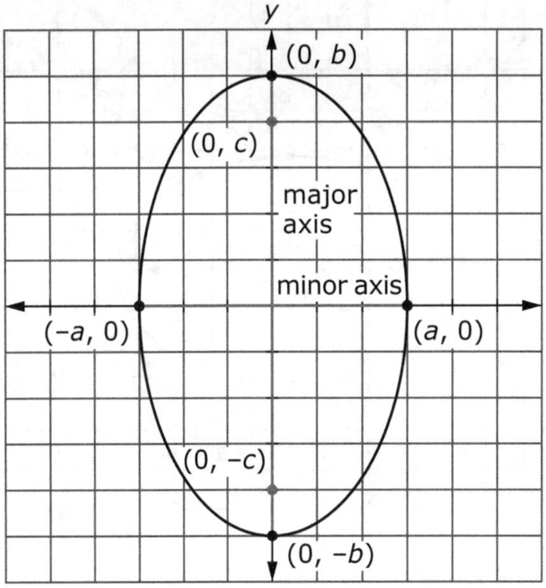

If the center is at (h, k), then the equation is $\frac{(x-h)^2}{a^2} + \frac{(y-k)^2}{b^2} = 1$.

EXAMPLE

What is the equation of an ellipse with foci at (1, 5) and (1, −3) that passes through the point (3, 1)?

First, plot the foci. The center is the midpoint of the segment connecting the foci:

$$\left(\frac{1+1}{2}, \frac{5+(-3)}{2}\right) = (1, 1), \text{ so } h = 1 \text{ and } k = 1.$$

If you plot the point (3, 1), you can see that it lies on the minor axis and that $b = 2$. The distance from the center to either of the foci is 4, so $c = 4$. The ellipse is oriented vertically, so $b^2 = a^2 + c^2$. Substitute the values of b and c and solve to find a^2.

$$2^2 = a^2 + 4^2$$
$$4 = a^2 + 16$$
$$20 = a^2$$

My Notes

Summary *(continued)*

The equation of the ellipse is $\frac{(x-1)^2}{4} + \frac{(y-1)^2}{20} = 1$. The graph is shown.

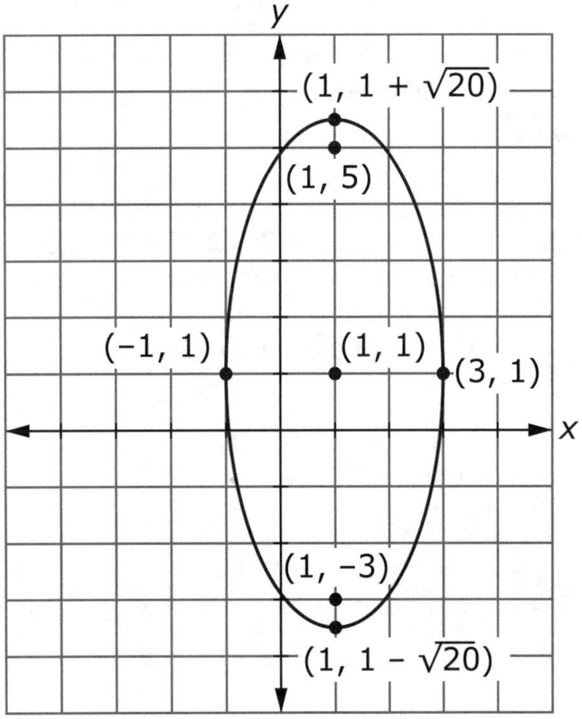

EXAMPLE

Graph $2x^2 + 5y^2 - 8x + 20y = 12$ and label the foci.

First, write the equation in graphing form.

$$2x^2 + 5y^2 - 8x + 20y = 12$$
$$(2x^2 - 8x) + (5y^2 + 20y) = 12$$
$$2(x^2 - 4x) + 5(y^2 + 4y) = 12$$
$$2(x^2 - 4x + 4) + 5(y^2 + 4y + 4) = 12 + 8 + 20$$
$$2(x-2)^2 + 5(y+2)^2 = 40$$
$$\frac{2(x-2)^2}{40} + \frac{5(y+2)^2}{40} = \frac{40}{40}$$
$$\frac{(x-2)^2}{20} + \frac{(y+2)^2}{8} = 1$$

The ellipse is centered at $(2, -2)$, and it is oriented horizontally. The distance from the center to the ellipse on the major axis is $\sqrt{20} \approx 4.5$, and on the minor axis it is $\sqrt{8} \approx 2.8$. To locate the foci, use the equation $a^2 = b^2 + c^2$.

$$20 = 8 + c^2$$
$$12 = c^2$$
$$\pm 3.5 \approx c$$

So, the foci are about 3.5 units to the left and right of the center. The graph is shown.

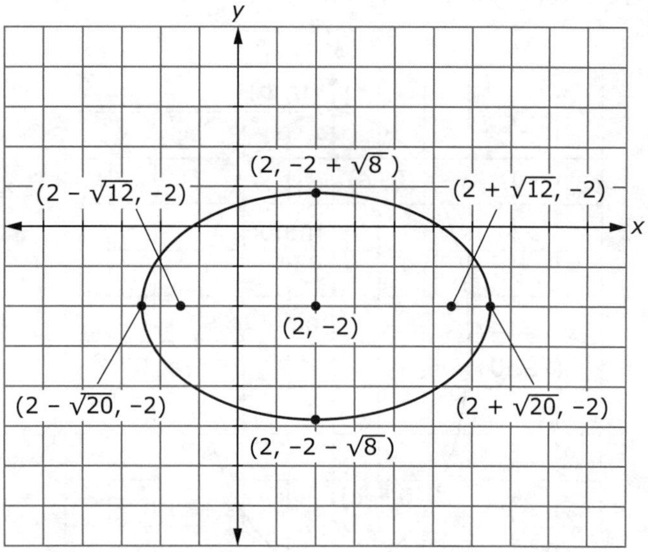

My Notes

Practice Exercises

Review what you've learned using these practice problems. For practice problems with feedback, try the Coach and Play items in the Practice section online.

1. An ellipse is defined by the equation $\frac{x^2}{25} + \frac{y^2}{9} = 1$.

Which of the following represents the equation of the same ellipse translated so that its foci are located at the points (–8, 10) and (0, 10)?

A. $\frac{(x-8)^2}{25} + \frac{(y+10)^2}{9} = 1$

B. $\frac{(x-4)^2}{25} + \frac{(y+10)^2}{9} = 1$

C. $\frac{(x+4)^2}{25} + \frac{(y-10)^2}{9} = 1$

D. $\frac{(x+8)^2}{25} + \frac{(y-10)^2}{9} = 1$

2. An ellipse has focal points (1, 8) and (1, 2). The sum of the distances from any point on the ellipse to the foci is 10 units.

Enter the coordinates of the center of the ellipse and the lengths of the longest horizontal line segment and longest vertical line segment in the ellipse.

Center point: (______, ______)

Longest horizontal line segment: ______

Longest vertical line segment: ______

3. An ellipse is constructed using the focal points (5, 8) and (11, 8), and the specification that the sum of the two focal radii is 12 units.

Write the coordinates of the center of the ellipse along with the lengths of the longest horizontal line segment and longest vertical line segment in the ellipse.

Center point: (______, ______)

Length of the longest horizontal line segment: ______

Length of the longest vertical line segment: ______

4. What is the equation of the circle centered at the point (–2, 3) and passing through the origin?

Write your answer in standard form, $Ax^2 + By^2 + Cx + Dy = E$.

$x^2 + y^2 +$ ______ $x +$ ______ $y =$ ______

5. Four points are plotted on a coordinate graph: $A(3, 0)$, $B(0, 3)$, $C(-3,0)$, and $D(0, -3)$. These points are to be used as the centers of circles.

Look at the four descriptions of circles. Then, match the equations to the corresponding descriptions.

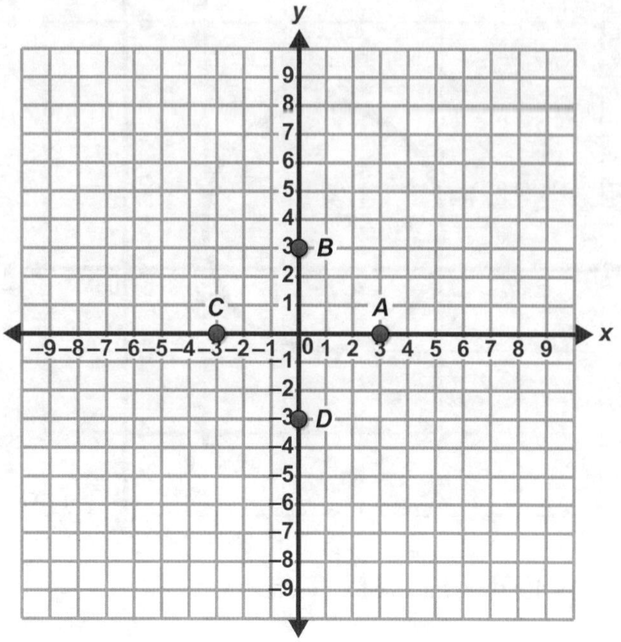

A. $x^2 + (y - 3)^2 = 36$

B. $(x - 3)^2 + y^2 = 36$

C. $x^2 + (y + 3)^2 = 36$

D. $(x + 3)^2 + y^2 = 36$

Centered at A, passing through C: ______

Centered at B, passing through D: ______

Centered at C, passing through A: ______

Centered at D, passing through B: ______

Practice Exercises *(continued)*

6. Suppose you are asked to graph the circle corresponding to the equation $x^2 - 10x + y^2 + 6y = -16$.

Write the coordinates of the center point (x_0, y_0) and the radius r that you use to graph the circle.

$x_0 =$ ______

$y_0 =$ ______

$r =$ ______

7. The circle in the figure corresponds to the equation $(x - x_0)^2 + (y - y_0)^2 = r^2$.

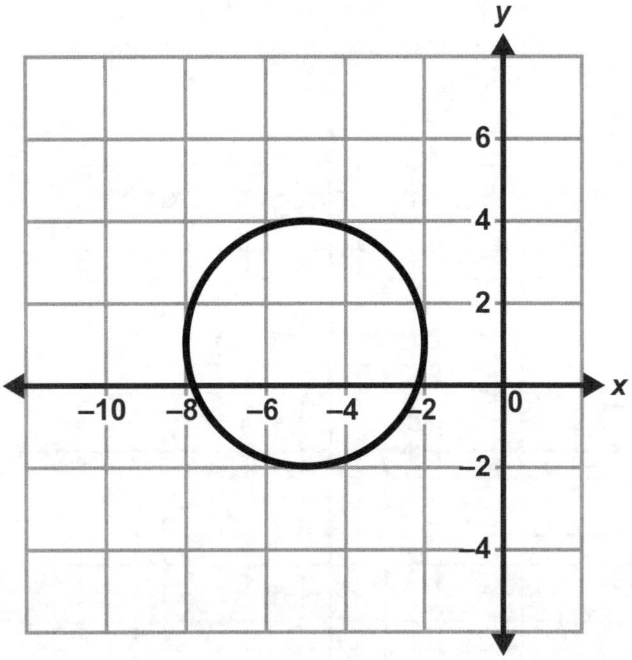

Write the values of x_0, y_0, and r.

$x_0 =$ ______

$y_0 =$ ______

$r =$ ______

8. An ellipse is defined by the equation $3x^2 - 6x + 4y^2 + 16y = -7$.

What are the center and radius of the largest circle that can be inscribed in that ellipse?

Center: (______, ______)

Radius: ______

9. Identify the conics which are circles. *Select all that apply.*

A. $4(x + 2)^2 + 2(y + 4)^2 = 20$

B. $\frac{(y + 1)^2}{10} - \frac{(x + 2)^2}{10} = 1$

C. $3x^2 + 7x - 5 = 3y^2 + 5y + 1$

D. $\frac{5(x - 1)^2}{40} + \frac{y^2}{8} = 1$

E. $x^2 + y^2 + 1 = 0$

F. $(2x - 3)^2 + 4y^2 = 20$

10. A circle is centered at the point (3, −2) and passes through the point (10, 8).

Fill in the missing values to write the equation of the circle.

$(x -$ ______$)^2 + (y -$ ______$)^2 = ($______$)^2$

11. An ellipse with one focus at the point (3, 0) and the other at the point $(3, y_0)$, where $0 < y_0 < 4$, is constructed by setting the sum of the focal radii equal to 8. Suppose y_0 is increased, while the sum of the focal radii remains equal to 8.

Which of the following statements accurately describes other changes that will occur as a result? *Select all that apply.*

A. The horizontal dimension of the ellipse will decrease.

B. The horizontal dimension of the ellipse will increase.

C. The vertical dimension of the ellipse will decrease.

D. The vertical dimension of the ellipse will increase.

E. The location of the center of the ellipse will move downward.

F. The location of the center of the ellipse will move upward.

G. The ellipse will become rounder.

H. The ellipse will become more elongated.

Apply

Where Am I?

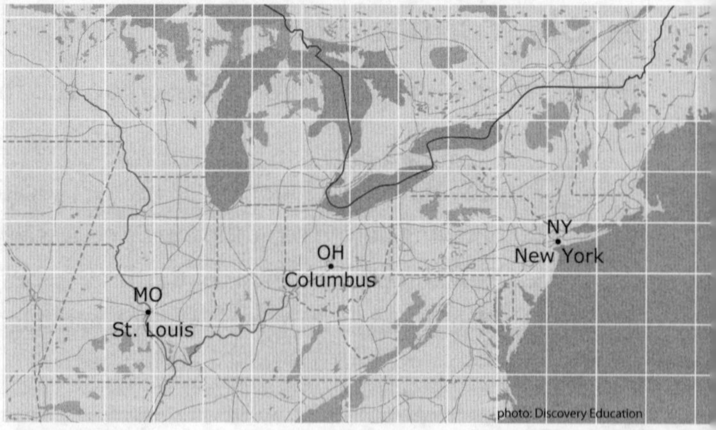

photo: Discovery Education

Triangulation is a way of determining something's location using the locations of other things.

Today you get to make a puzzle for your classmates. You will hide — and they will use triangulation to find you.

- On a map that includes a scale and an assigned origin, or on a full sheet of graph paper with the origin located in the center of the page, choose a secret location or a point; do not reveal it to anyone.
- Now choose three locations around it, plot the three locations, and find the three actual distances to your secret location.
- Write the puzzle clue. It only needs to include the three locations and their real distances from your mystery location.
- While paying attention to the scale, draw three circles with centers at your three locations and with radii the length of the three distances.
- Provide a solution that identifies your mystery location with algebraic and graphic evidence.

Show what you've learned by completing the other performance tasks in the online Apply section.

Apply *(continued)*

Your answer to Apply will be assessed on the following criteria:

1. Creating a puzzle that gives three points with their distances from the mystery location and providing the solution of the mystery location
2. Showing how to solve the puzzle on a scaled map or grid
3. Providing the equations of the three circles
4. Identifying your mystery location with algebraic and graphical evidence

Criteria \ Scale	4 Exceeds Criteria	3 Meets Criteria	2 Progressing to Criteria	1 Below Expectations	0 No Expectation
Original Puzzle	States three locations and their distances from the mystery location, and then states the mystery location.	States three locations from the mystery location, but one piece of information is missing, such as a distance or the mystery location.	States at least two locations, but two or three pieces of information are missing, such as distances or the mystery location.	States at least one location, but four or more pieces of information are missing, such as distances and the mystery location.	Does not state information to solve the puzzle or the solution.
Map of Solution	Provides a map and scale marked with the coordinates of the three locations, three circles drawn correctly, and the coordinates of the mystery location.	Provides a map, but one detail is missing or incorrect: scale for the map, coordinates of the three locations, three circles, or the coordinates of the mystery location.	Provides a map, but two or three details are missing or incorrect: scale for the map, coordinates of the three locations, three circles, or the coordinates of the mystery location.	Provides a map, but four or more details are missing or incorrect: scale for the map, coordinates of the three locations, three circles, or the coordinates of the mystery location.	Does not provide a map, or provides a map without any information on it.
Equations of Circles	Correctly states the equations of the three circles shown on the map, taking into account the scale factor from the map.	States the equations of the three circles shown on the map, but contains a minor error.	States the equations of the three circles shown on the map, but contains a major error, or only gives the correct equations for two circles.	States the equations of the three circles shown on the map, but contains a major error, or only gives the correct equations for two circles.	Does not provide equations for the circles.
Proof of Triangulation	Verifies the mystery location is located on all three circles through calculations.	Verifies the mystery location is located on all three circles through calculations, but with a minor error.	Verifies the mystery location is located on two of the circles through calculations, or one of the three calculations has a major error.	Verifies the mystery location is located on one of the circles through calculations, or two of the three calculations has a major error.	Does not attempt to prove the triangulation is correct.

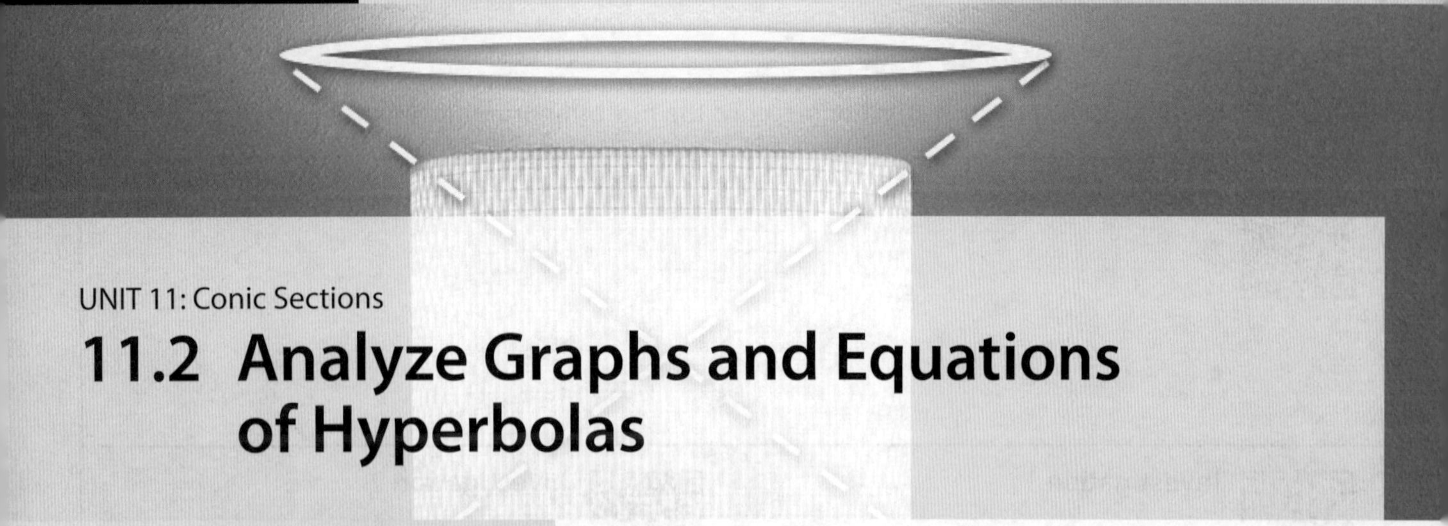

UNIT 11: Conic Sections

11.2 Analyze Graphs and Equations of Hyperbolas

photo: Getty Images

Lesson Objectives

- Derive the equation of a hyperbola.
- Translate between the equation of a hyperbola and its graph.
- Explain the relationship between foci and a hyperbola.

Essential Question

- How are the foci of a hyperbola related to its equation?

photo: Getty Images

Investigations

Hyperbolas

Take a closer look. What shapes form when a plane intersects a double cone?

Sketching Hyperbolas in the *x*-*y* Plane

What is a hyperbola? How do vertices and foci define one?

Graphing Hyperbolas

Explore graphs and equations of hyperbola. What key characteristics define the graph?

The Equation of a Hyperbola

Write equations for hyperbola. Use asymptotes as a guide.

Graphing Form of Ellipses and Hyperbolas

Analyze forms of hyperbola equations. Which form sheds more light on hyperbola properties?

Key Vocabulary

asymptote, center, conjugate axis, expanded form, focus (foci), standard form, transverse axis, vertex

Discover

As you complete Engage and the investigations, record the most important ideas you've learned.

Engage	
Investigation 1	**Investigation 2**
Investigation 3	**Investigation 4**

Name ____________________ Date __________

Check for Understanding

Analyze Graphs and Equations of Hyperbolas: Investigation 1

1. Which of these does the hyperbola shown below have? *Select all that apply.*

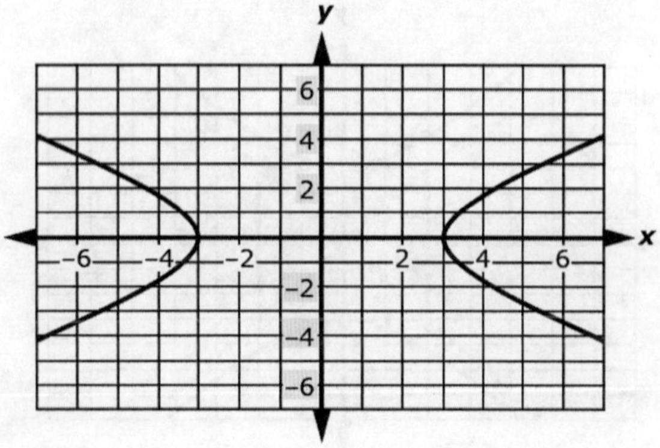

A. a center with an x-coordinate of –3

B. a center with a y-coordinate of 0

C. a focus with an x-coordinate of 0

D. a focus with a y-coordinate of –3

E. a vertex with an x-coordinate of 3

F. a vertex with a y-coordinate of 0

2. The points referenced in the following table each lie on a hyperbola and are all closer to focus 1 than they are to focus 2. Fill in the missing numbers of units in the table.

	Number of Units From Point to Focus 1	Number of Units From Point to Focus 2
Point 1	6	14
Point 2	13	
Point 3		18
Point 4	17	
Point 5		16
Point 6	3	

3. The distance between a vertex of a hyperbola and one of the hyperbola's foci is 2 units, while the distance between the same vertex and the hyperbola's other focus is 9 units. What can you say about the distance between the hyperbola's two vertices? Explain.

Name ______________________________ Date ______________

Check for Understanding

Analyze Graphs and Equations of Hyperbolas: Investigation 2

1. What is the equation of the hyperbola graphed?

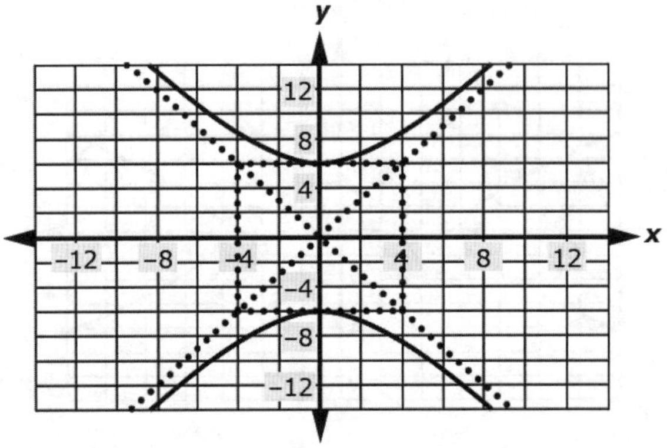

A. $\frac{x^2}{16} - \frac{y^2}{36} = 1$ **B.** $\frac{y^2}{16} - \frac{x^2}{36} = 1$ **C.** $\frac{x^2}{36} - \frac{y^2}{16} = 1$ **D.** $\frac{y^2}{36} - \frac{x^2}{16} = 1$

2. Which of these equations defining a hyperbola has a graph with foci at the points (−8, 0) and (8, 0)? *Select all that apply.*

A. $\frac{x^2}{43} - \frac{y^2}{21} = 1$ **B.** $\frac{y^2}{64} - \frac{x^2}{15} = 1$ **C.** $\frac{y^2}{13} - \frac{x^2}{51} = 1$

D. $\frac{x^2}{64} - \frac{y^2}{11} = 1$ **E.** $\frac{x^2}{29} - \frac{y^2}{35} = 1$ **F.** $\frac{y^2}{41} - \frac{x^2}{23} = 1$

3. Consider the hyperbola with the equation $\frac{x^2}{25} - \frac{y^2}{100} = 1$. *Fill in the blanks.*

The graph of the hyperbola has a center at the point (______, ______), vertices at the points (______, ______) and (______, ______), a transverse axis with a length of ______ units, a conjugate axis with a length of ______ units, and asymptotes with the equations $y =$ ______ and $y =$ ______.

4. Suppose you know the vertices of the auxiliary rectangle of a hyperbola. To graph the hyperbola, what additional information would you need to know at minimum? Explain.

Name ______________________________ Date ____________

Check for Understanding

Analyze Graphs and Equations of Hyperbolas: Investigation 3

1. The vertices of a hyperbola are at the points (–7, 3) and (3, 3), and one focus of the hyperbola is at the point (5, 3). What is the hyperbola's equation?

A. $\frac{(x-2)^2}{24} - \frac{(y+3)^2}{25} = 1$ **B.** $\frac{(x+2)^2}{24} - \frac{(y-3)^2}{25} = 1$

B. $\frac{(x-2)^2}{25} - \frac{(y+3)^2}{24} = 1$ **D.** $\frac{(x+2)^2}{25} - \frac{(y-3)^2}{24} = 1$

2. Below is the graph of a hyperbola with its vertices and a focus shown. *Fill in the blanks.*

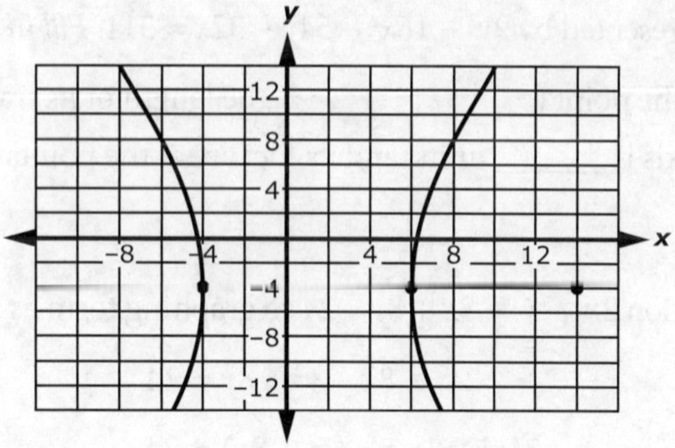

The equations of the asymptotes of the hyperbola are $y = \pm\frac{\square}{\square}(x - \square) - \square$.

3. The outline of a shadow on a wall is in the shape of one branch of a hyperbola. When a coordinate plane is transposed onto the wall, the vertex of this branch is at the point (–4, 3), and the asymptotes of the hyperbola are $y = \frac{3}{4}x + 3$ and $y = -\frac{3}{4}x - 3$. Explain how you can use a graph to help find the branch's equation solved for y in terms for x.

Name ______________________________ Date ______________

Check for Understanding

Analyze Graphs and Equations of Hyperbolas: Investigation 4

1. Which of these pairs of equations of a hyperbola are equivalent to each other?

   A. $9y^2 - x^2 - 18y - 8x = 88$ and $\frac{(y-1)^2}{9} - \frac{(x-4)^2}{81} = 1$

   B. $\frac{(y+7)^2}{49} - (x+3)^2 = 1$ and $y^2 - 49x^2 + 14y + 294x = 441$

   C. $4x^2 - 9y^2 - 64x + 36y = -76$ and $\frac{(x-8)^2}{36} - \frac{(y-2)^2}{16} = 1$

   D. $\frac{(x+5)^2}{4} - \frac{(y+6)^2}{64} = 1$ and $16x^2 - y^2 + 160x + 12y = -300$

2. Consider the hyperbola represented by $9y^2 - 16x^2 - 54y - 32x = 511$. *Fill in the blanks.*

   The hyperbola's center is at the point (______, ______), the length of its transverse axis is ______ units, the length of its conjugate axis is ______ units, and its foci are at the points (______, ______) and (______, ______).

3. Warren converted the equation $2x^2 - y^2 + 8x + 8y = 24$ to graphing form as follows.

$$2x^2 + 8x - y^2 + 8y = 24$$

$$2(x^2 + 4x) - (y^2 + 8y) = 24$$

$$2(x^2 + 4x + 4) - (y^2 + 8y + 16) = 24 + 4 + 16$$

$$2(x+2)^2 - (y+4)^2 = 44$$

$$\frac{(x+2)^2}{22} = \frac{(y+4)^2}{44} = 1$$

   What mistake(s) did Warren make when converting the equation? In your answer, explain how to fix his error(s), and give the correct graphing form of the equation.

Summary

Before you attempt the Practice Exercises, review what you've learned.

A **hyperbola** is a type of conic section.

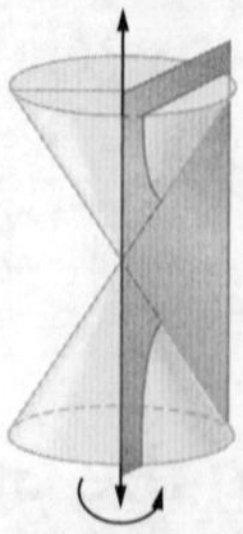

Hyperbola

Hyperbolas can be represented in the coordinate plane using equations.

Horizontal Hyperbolas

The equation of a hyperbola centered at the point (0, 0) that opens to the left and right is $\frac{x^2}{a^2} - \frac{y^2}{b^2} = 1$.

- The vertices are $(-a, 0)$ and $(a, 0)$, which are connected by a segment called the transverse axis.
- The points $(0, b)$ and $(0, -b)$ are connected by a segment called the conjugate axis.
- The rectangle with vertices at $(-a, -b)$, $(-a, b)$, (a, b), and $(a, -b)$ is called the auxiliary rectangle.
- The lines $y = -\frac{b}{a}x$ and $y = \frac{b}{a}x$ are the asymptotes of the hyperbola.
- The foci of the hyperbola are at $(-c, 0)$ and $(c, 0)$, and $a^2 + b^2 = c^2$.
- If the center is at (h, k), then the equation is $\frac{(x-h)^2}{a^2} - \frac{(y-k)^2}{b^2} = 1$.

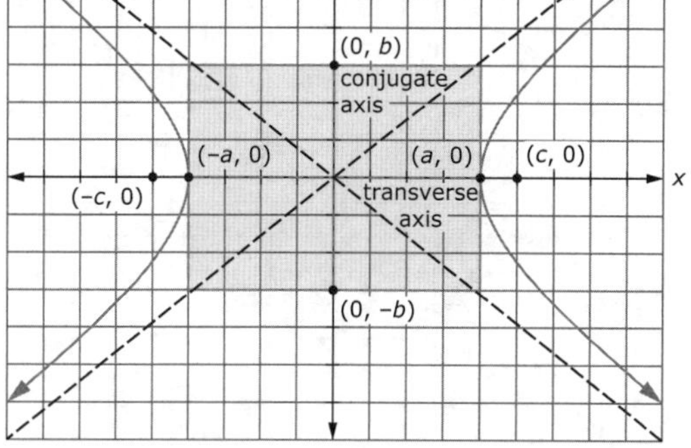

Vertical Hyperbolas

The equation of a hyperbola centered at the point (0, 0) that opens up and down is $\frac{y^2}{b^2} - \frac{x^2}{a^2} = 1$.

- The vertices are $(0, b)$ and $(0, -b)$, which are connected by the transverse axis.
- The points $(-a, 0)$ and $(a, 0)$ are connected by the conjugate axis.
- The auxiliary rectangle has vertices at $(-a, -b)$, $(-a, b)$, (a, b), and $(a, -b)$.
- The lines $y = -\frac{b}{a}x$ and $y = \frac{b}{a}x$ are the asymptotes of the hyperbola.
- The foci of the hyperbola are at $(0, -c)$ and $(0, c)$, and $a^2 + b^2 = c^2$.
- If the center is at (h, k), then the equation is $\frac{(y-k)^2}{b^2} - \frac{(x-h)^2}{a^2} = 1$.

My Notes

Summary *(continued)*

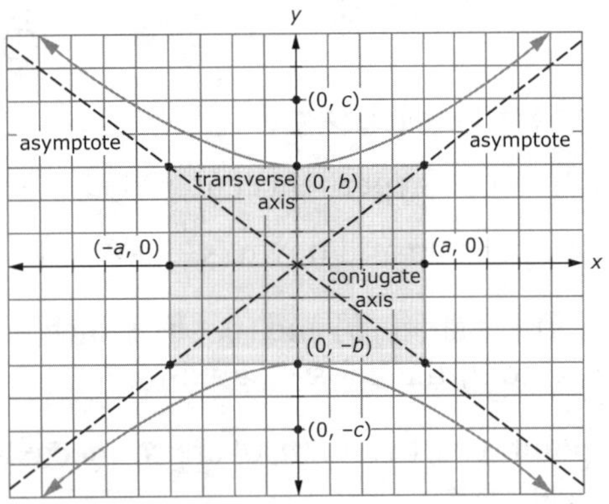

EXAMPLE: Using Key Information to Write an Equation

What is the equation of a hyperbola with vertices at (2, 3) and (6, 3) with an asymptote of $y = x - 1$?

SOLUTION:

First, plot the vertices. The center is the midpoint of the segment connecting the vertices:

$$\left(\frac{2+6}{2}, \frac{3+3}{2}\right) = (4, 3), \text{ so } h = 4 \text{ and } k = 3.$$

If you graph the asymptote, you can see that the auxiliary rectangle is a square with sides of length 4, so $a = 2$ and $b = 2$.

The equation of the hyperbola is $\frac{(x-4)^2}{4} - \frac{(y-3)^2}{4} = 1$.

The graph is shown.

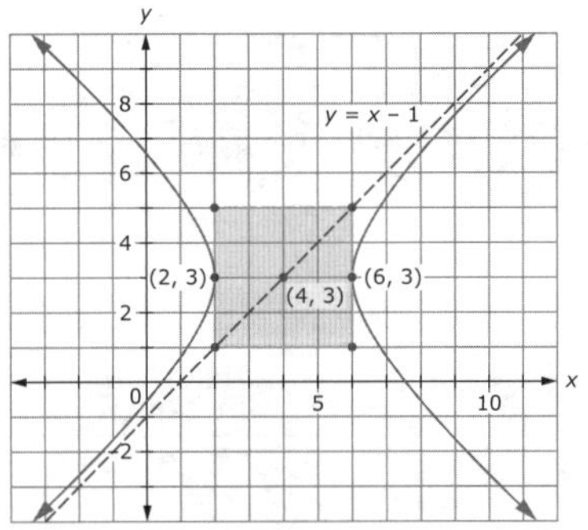

My Notes

EXAMPLE: Using the Graphing Form of an Equation

Graph $4x^2 - 25y^2 + 8x - 100y = -4$ and label the center, vertices, and foci.

SOLUTION:

First, write the equation in graphing form.

$$4x^2 - 25y^2 + 8x - 100y = -4$$

becomes:

$$\frac{(y+2)^2}{4} - \frac{(x+1)^2}{25} = 1$$

The hyperbola is centered at (−1, −2), and it opens up and down. The length of the transverse axis is $2b = 4$, and the length of the conjugate axis is $2a = 10$.

To locate the foci, use the equation $a^2 + b^2 = c^2$.

$$25 + 4 = c^2$$

$$29 = c^2$$

$$5.4 \approx c$$

So the foci are about 5.4 units above and below the center. They lie on a circle through the vertices of the auxiliary rectangle. The graph is shown.

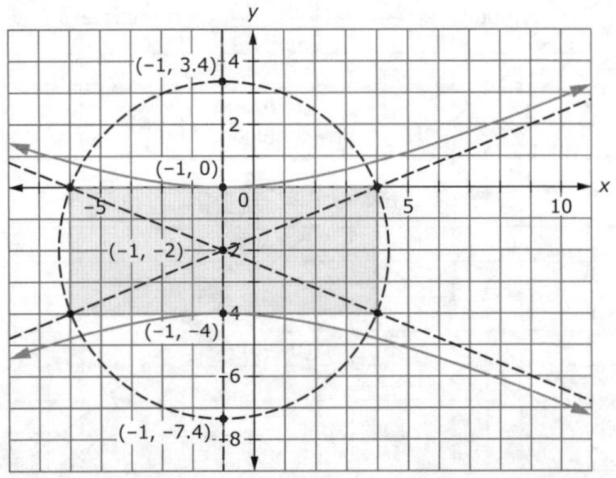

Practice Exercises

Review what you've learned using these practice problems. For practice problems with feedback, try the Coach and Play items in the Practice section online.

1. The equation $x^2 - y^2 + 4x - 8y = 0$ represents a hyperbola. Determine the graphing form of the equation of the hyperbola.

A. $\frac{(x+2)^2}{12} - \frac{(y-4)^2}{12} = 1$

B. $\frac{(y-4)^2}{12} - \frac{(x+2)^2}{12} = 1$

C. $\frac{(x+2)^2}{12} - \frac{(y+4)^2}{12} = 1$

D. $\frac{(y+4)^2}{12} - \frac{(x+2)^2}{12} = 1$

2. Which equation is the graphing form of a hyperbola with foci at the points $(h \pm c, k)$ and is characterized by the constants a and b, with a equal to one-half the difference in focal lengths from the foci to any point on the hyperbola, and $b^2 = c^2 - a^2$.

A. $\frac{(x-h)^2}{(a-h)^2} - \frac{(y-k)^2}{(b-k)^2} = 1$

B. $\frac{(x^2-h^2)}{a^2} - \frac{(y^2-k^2)}{b^2} = 1$

C. $\frac{(x-h)^2}{a^2} - \frac{(y-k)^2}{b^2} = 1$

D. $\frac{(x+h)^2}{a^2} - \frac{(y+k)^2}{b^2} = 1$

E. $\frac{(x^2+h^2)}{a^2} - \frac{(y^2+k^2)}{b^2} = 1$

3. The foci of a hyperbola are located at (–6, 0) and (6, 0), and the absolute difference in the lengths of the focal lengths joining the foci to any point on the hyperbola is 10. The equation of the hyperbola is $\frac{x^2}{a^2} - \frac{y^2}{b^2} = 1$.

Write the values of a and b.

$a =$ ____________

$b =$ ____________

4. The foci of a hyperbola are located at (0, –25) and (0, 25). The positive y-intercept of the hyperbola is located at $y = 10$.

Write the values of a and b.

$a =$ ____________

$b =$ ____________

5. The foci of a hyperbola are located at (–8, –4) and (2, –4).

The x value of the vertex for the right-hand branch of the hyperbola is $x = 1$.

The equation of the hyperbola is $\frac{(x-h)^2}{a^2} - \frac{(y-k)^2}{b^2} = 1$.

Write the values of a, b, h, and k.

$a =$ ____________

$b =$ ____________

$h =$ ____________

$k =$ ____________

6. The figure shows the graph of a hyperbola with foci at (10, –30) and (10, 20).

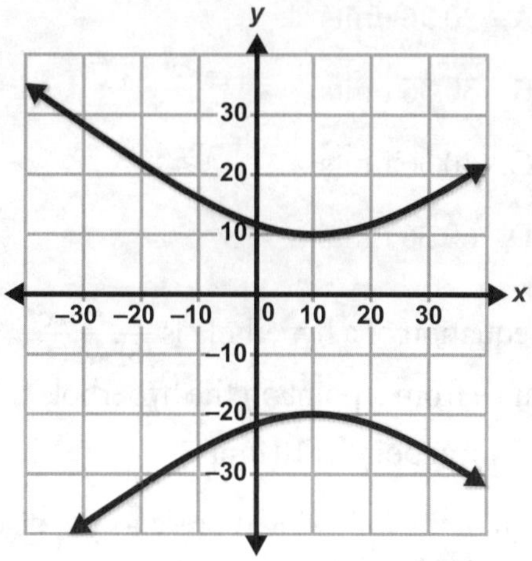

The hyperbola can be described by using the equation

$\frac{(y-k)^2}{b^2} - \frac{(x-h)^2}{a^2} = 1$.

Write the values of a, b, h, and k.

$a =$ ____________

$b =$ ____________

$h =$ ____________

$k =$ ____________

Practice Exercises *(continued)*

7. The figure shows a hyperbola with its two foci, F_1 and F_2, and a point on the hyperbola.

(Note: the horizontal scale is not the same as the scale shown for the vertical axis.)

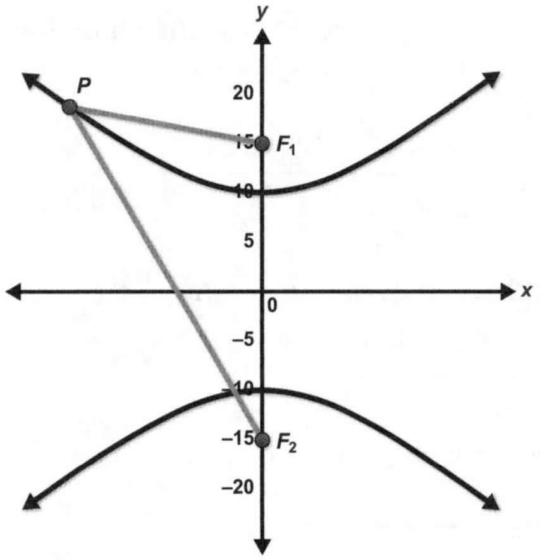

The length of line segment $\overline{PF_1}$ is 20.36 units.

Which of the following best represents the length of $\overline{PF_2}$?

A. 20.36 units

B. 30.36 units

C. 40.36 units

D. 60.36 units

8. The equation of a hyperbola is $\frac{y^2}{(8)^2} - \frac{x^2}{(4)^2} = 1$. The distance from a point on the hyperbola to the closer focal point is 10 units.

Which of the following represents the distance to the further focal point?

A. 14 units

B. 18 units

C. 26 units

D. 34 units

9. The figure shows a hyperbola with its two foci, F_1 and F_2, and point, P, on the hyperbola.

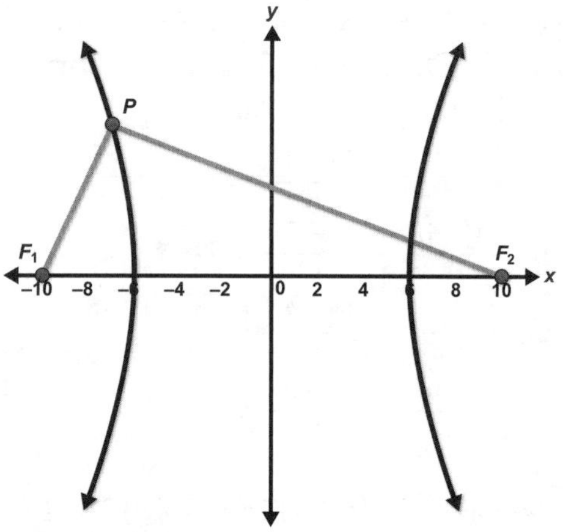

(Note: the vertical scale is not the same as the horizontal scale.)

The vertices of the hyperbola are located at (–6, 0) and (6, 0). The length of $\overline{PF_2}$ is 17 units.

Write the length of $\overline{PF_1}$.

$\overline{PF_1}$ = ______ units

10. A hyperbola is defined by the equation $\frac{(x+4)^2}{16} - \frac{(y-1)^2}{9} = 1$.

Which of the following are equations of its asymptotes? *Select all that apply.*

A. $4x - 3y = 16$

B. $y = -\frac{3}{4}x - 2$

C. $3x - 4y = -8$

D. $4(y - 1) = 3(x + 4)$

E. $3x + 4y = -8$

F. $x = \frac{4}{3}y - 8$

11. The equation of a hyperbola is $\frac{y^2}{(16)^2} - \frac{x^2}{(20)^2}$. The distance from a point on the hyperbola to the closer focal point is 31 units.

Write the distance from the point to the further focal point.

Distance to further focal point = ______ units

Apply

How Are Hyperbolas Used in Designing Structures?

A hyperboloid is a three-dimensional figure that has hyperbolas as its vertical cross sections. Hyperboloids are found in many different types of buildings all over the world.

How are hyperboloids used in architecture?

Hyperboloids

One use of a hyperboloid is the Warka water tower, which was designed to help provide water to arid communities in Africa. Find an image of the Warka water tower or a similar hyperbolic structure. Research its height to assign coordinates to the coordinate grid, plot any relevant points, and then find the foci and equation of the hyperbolic cross section. Discuss why the hyperboloid is a good choice for such a structure.

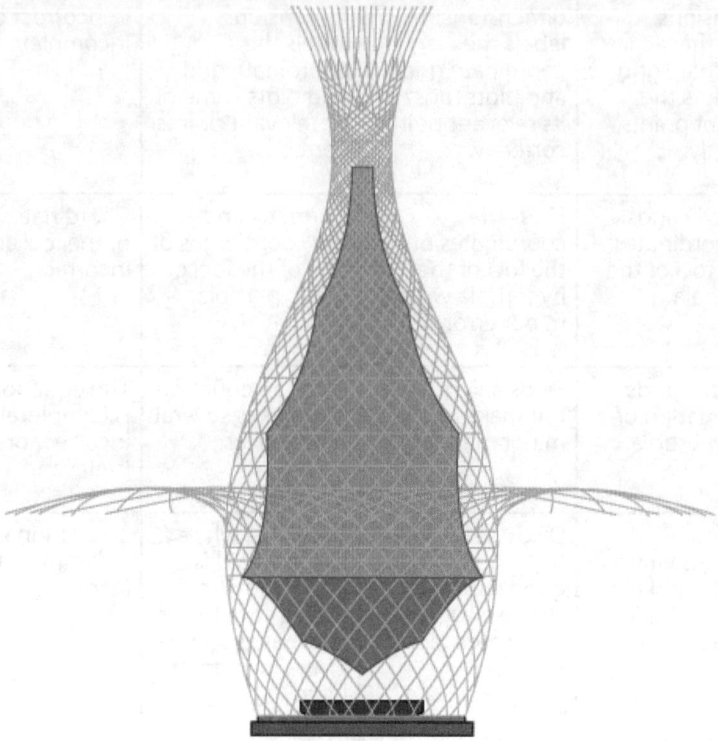

Show what you've learned by completing the other performance tasks in the online Apply section.

Apply *(continued)*

Your answer to Apply will be assessed on the following criteria:

1. Researching a structure, finding its dimensions, labeling the coordinate grid, and plotting relevant points, including the vertices
2. Finding the equation of the hyperbola
3. Finding the coordinates of the foci of the hyperbola
4. Discussing why a hyperboloid is a good choice for such a structure

Criteria \ Scale	4	3	2	1	0
	Exceeds Criteria	Meets Criteria	Progressing to Criteria	Below Expectations	No Expectation
Data and Diagram	Chooses a structure and finds its dimensions; labels the coordinate grid and plots the relevant points correctly.	Chooses a structure and finds its dimensions; labels the coordinate grid and plots most of its relevant points correctly.	Chooses a structure and finds its dimensions; labels the coordinate grid and plots some of its relevant points correctly.	Information about the chosen structure is incorrect or incomplete.	Does not attempt task.
Coordinates of Foci	Correctly finds the coordinates of the foci of the hyperbola.	Finds the coordinates of the foci of the hyperbola with a minor error.	Correctly finds the coordinates of one of the foci of the hyperbola.	Coordinates of the foci are incorrect.	Does not attempt task.
Equation	Correctly finds the equation of the hyperbola.	Finds an equation but makes a minor math error.	Finds an equation but makes several math errors.	The equation is completely incorrect or irrelevant.	Does not attempt task.
Discussion	Clearly discusses why a hyperboloid is a good choice for such a structure.	Discusses why a hyperboloid is a good choice for such a structure with some lack of clarity.	Discusses why a hyperboloid is a good choice for such a structure but makes some errors in math or logic.	Discussion is incomplete or incorrect.	Does not attempt task.

UNIT 11: Conic Sections

11.3 Analyze Graphs and Equations of Parabolas

photo: Getty Images

Lesson Objectives

- Derive the equation of a parabola.
- Translate between the equation of a parabola and its graph.
- Explain the relationship between a focus and directrix and their effect on the graph of a parabola.

Essential Question

- How does the relationship between the focus and directrix relate to the parabola, graphically and algebraically?

photo: Getty Images

Investigations

Exploring Parabolas

Shift the plane, in relation to a double cone, and your whole perspective can change.

Modeling a Parabola

Use paper folding to define a parabola. Control the shape with the focus and the directrix.

Derive the Equation of a Parabola

Since parabolas are about distance, use the distance formula to derive an equation.

The Equation of a Parabola

Mirror mathematical and real-life situations. Identify information using parabola equations.

Key Vocabulary

axis of symmetry, conic section, directrix, focal length, focus (foci), parabola, vertex form/graphing form of a quadratic

Discover

As you complete Engage and the investigations, record the most important ideas you've learned.

Engage	**Investigation 1**
Investigation 2	**Investigation 3**

Name ____________________ Date __________

Check for Understanding

Analyze Graphs and Equations of Parabolas: Investigation 1

1. Which distance is the same as the distance from the focus to the vertex of a parabola?

   A. the perpendicular distance from the focus to the directrix of the parabola

   B. the perpendicular distance from the vertex to the directrix of the parabola

   C. the distance from the focus of the parabola to any other point on the parabola

   D. the distance from the vertex of the parabola to any other point on the parabola

2. Nicholas and Genevieve each drew a different parabola. Nicholas' parabola has a focus at the point (5, 8) and a directrix of $y = 2$. Genevieve's parabola has a directrix of $y = -4$ and is narrower than Nicholas' parabola. Which of these points could be the focus of Genevieve's parabola? *Select all that apply.*

   A. (–10, –8) **B.** (–5, 4) **C.** (–1, –2)

   D. (0, 12) **E.** (3, –10) **F.** (6, 0)

3. Below is the graph of a parabola with its directrix shown. Point P lies on the parabola. *Fill in the blanks in the sentence that follows.*

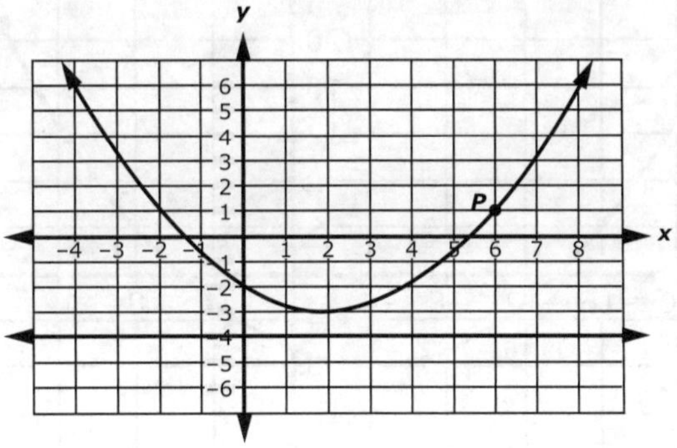

The distance from the parabola's focus to its directrix is ______ units, and the distance from the parabola's focus to point P is ______ units.

4. In which quadrants of a coordinate plane could a parabola lie if its directrix is the x-axis? In which quadrants could the parabola lie if its directrix is the y-axis? Explain.

Name ________________________________ **Date** ______________

Check for Understanding

Analyze Graphs and Equations of Parabolas: Investigation 2

1. Martina graphed a parabola with a focus at (8, 6) and a directrix of $y = 4$ on a coordinate plane. From the focus, if she moves to the right and then up, she arrives at the point (x, y) on the parabola. *Fill in the blanks to complete the sentence below.*

The number of units from (x, y) to the focus can be represented by the expression ______________________________, and the number of units from (x, y) to the directrix can be represented by the expression __________, so setting these expressions equal to each other and simplifying will give the parabola's equation.

2. What is the equation of a parabola that has a focus at (5, 10) and a directrix of $y = 2$?

A. $y = \frac{1}{16}(x - 5)^2 + 6$ **B.** $y = \frac{1}{16}(x + 5)^2 + 6$

C. $y = \frac{1}{4}(x - 5)^2 + 6$ **D.** $y = \frac{1}{4}(x + 5)^2 + 6$

3. Explain how you can find the focus and the directrix of the parabola graphed.

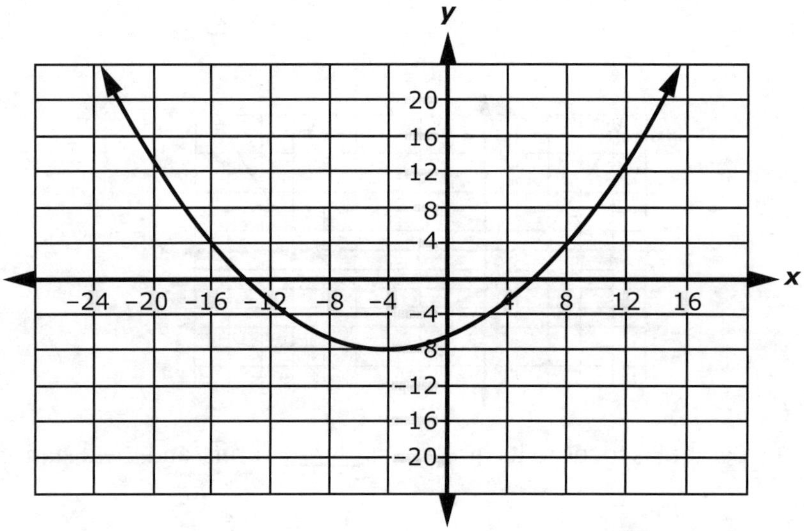

Name ______________________________ Date ____________

Check for Understanding

Analyze Graphs and Equations of Parabolas: Investigation 3

1. A radio telescope has a parabolic reflector that brings incoming radio waves to a common collection point, which is the focus of the reflector's cross-section. If the cross-section of the reflector is modeled by the equation $x^2 = 200y$, with the units in meters, how many meters from the vertex is the collection point for the radio waves?

   A. 25 **B.** 50 **C.** 100 **D.** 200

2. Consider the parabola with the equation $28x + 63 = y^2 - 14y$. *Fill in the blanks.*

   The vertex of the parabola is at the point (______, ______), the focus is at the point (______, ______), and the equation of the directrix is ______ = ______.

3. An electric heater has been designed so that its heat source is at the focus of the cross-section of a parabolic dish. Suppose the dish's cross-section opens to the left and that its vertex is located at the origin of a coordinate plane that has units in inches. The equation of the cross-section depends on the distance between the heat source and the cross-section's vertex. Which statements are correct? *Select all that apply.*

   A. If the distance is 8 inches, the equation of the cross section is $x = -\frac{1}{40}y^2$.

   B. If the distance is 9 inches, the equation of the cross section is $x = -\frac{1}{45}y^2$.

   C. If the distance is 5 inches, the equation of the cross section is $x = -\frac{1}{20}y^2$.

   D. If the distance is 4 inches, the equation of the cross section is $x = -\frac{1}{12}y^2$.

   E. If the distance is 6 inches, the equation of the cross section is $x = -\frac{1}{24}y^2$.

4. The standard form of the equation of a parabola is $y = ax^2 + bx + c$, where a, b, and c are real numbers, and $a \neq 0$. Explain how to find the values of a, b, and c in the equation of a parabola with a focus at the point (12, 1) and a directrix of $y = 1.5$.

Summary

Before you attempt the Practice Exercises, review what you've learned.

A **parabola** is the set of points such that the distance to the **focus** equals the distance to the **directrix**.

Standard Form versus Vertex Form

A parabola can be in a vertical orientation where x is the independent variable, resulting in a curve that opens up or down.

When a parabola is in a horizontal orientation, the y–variable is independent, which causes the curve to open left or right.

Standard Form of a Parabola	
Concave up or concave down	Concave left or concave right
$y = ax^2 + bx + c$	$x = ay^2 + by + c$

Vertex Form

	Vertical	Horizontal
Equation	$y=\frac{1}{4p}(x-h)^2+k$	$x=\frac{1}{4p}(y-k)^2+h$
Axis of Symmetry	$x = h$ (vertical line)	$y = k$ (horizontal line)
Vertex	(h, k)	(h, k)
Focus Point	$(h, k+p)$	$(h+p, k)$
Direction of Opening	Opens up if $\frac{1}{4p}>0$ Opens down if $\frac{1}{4p}<0$	Opens right if $\frac{1}{4p}>0$ Opens left if $\frac{1}{4p}<0$
Equation of Directrix	$y=k-p$	$x=h-p$
$\frac{1}{4p}>0$	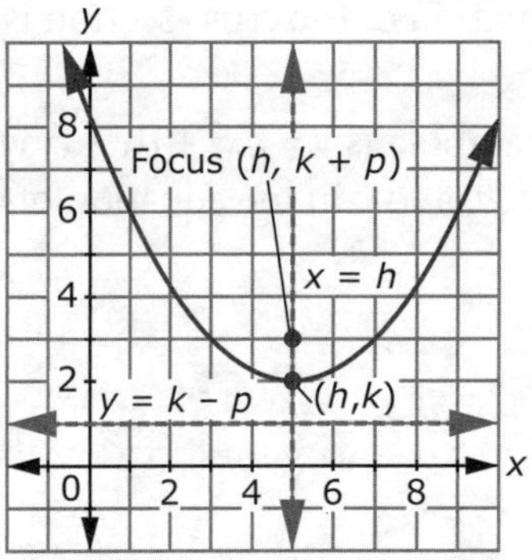	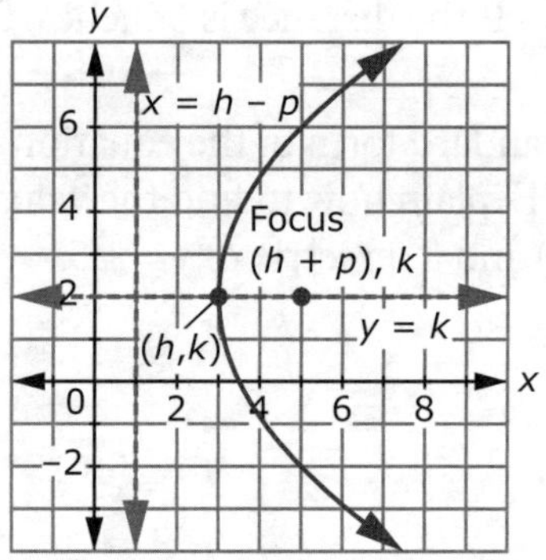
$\frac{1}{4p}<0$	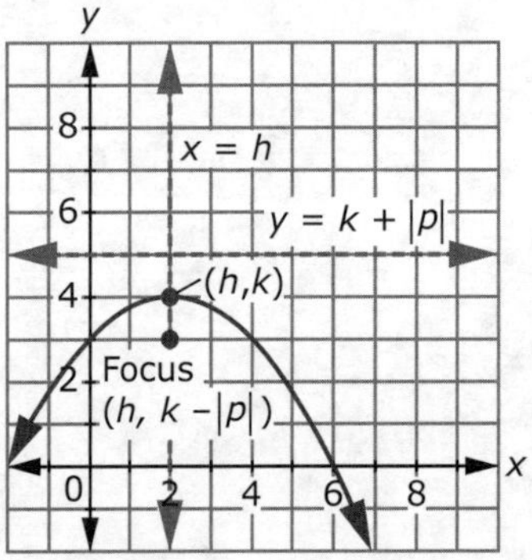	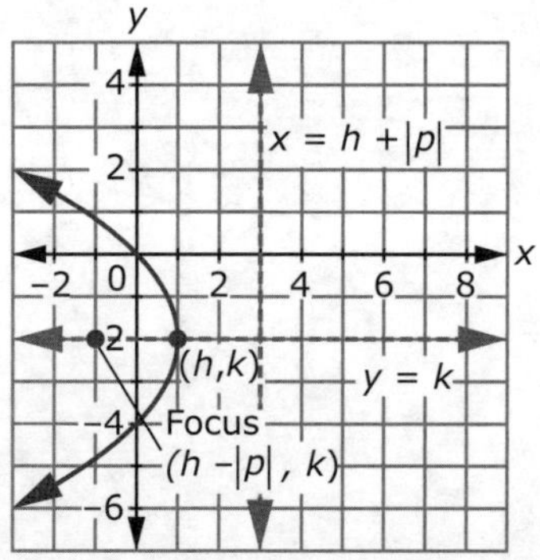

Summary *(continued)*

Every point (x, y) that is the same distance from the focus as the perpendicular distance from that point to the directrix is a point on the parabola. Suppose you are given a focus at (3, 4) and a directrix of $y = 1$.

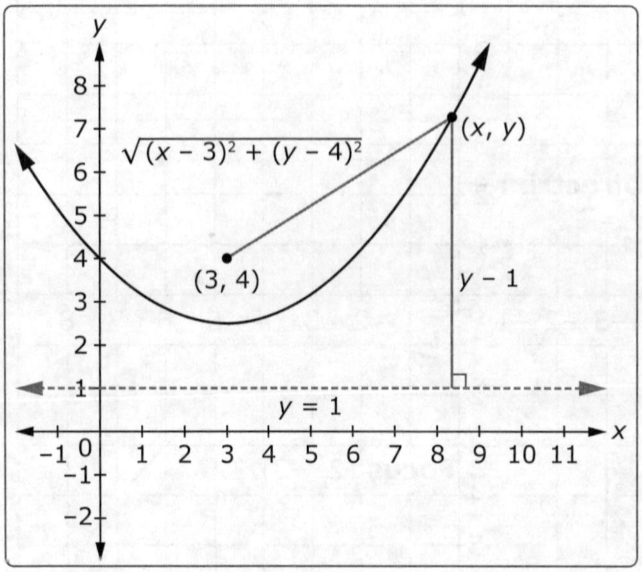

Use the distance formula or Pythagorean theorem to derive the equation of a parabola:

$$\sqrt{(x-3)^2+(y-4)^2} = y - 1$$

$$(x-3)^2+(y-4)^2 = (y-1)^2$$

$$x^2 - 6x + 9 + y^2 - 8y + 16 = y^2 - 2y + 1$$

$$x^2 - 6x + 9 + 16 - 1 = 6y$$

$$6y = x^2 - 6x + 24$$

$$6y = (x^2 - 6x + 9) + 24 - 9$$

$$6y = (x-3)^2 + 15$$

$$y = \frac{1}{6}(x-3)^2 + 2.5$$

EXAMPLE

Find the vertex, direction, focus, and directrix of the parabola $x = \frac{1}{12}(y-5)^2$.

- The parabola is horizontal.
- $a = \frac{1}{4p} = \frac{1}{12}$, which is greater than 0, so the parabola opens to the right.
- Since $\frac{1}{4p} = \frac{1}{12}$, $p = 3$.
- The value of h is 0, and the value of k is 5, so the vertex is at (0, 5).
- The focus is (0 + 3, 5), or (3, 5).
- The directrix is $x = 0 - 3$, or $x = -3$.

The graph of the parabola is shown.

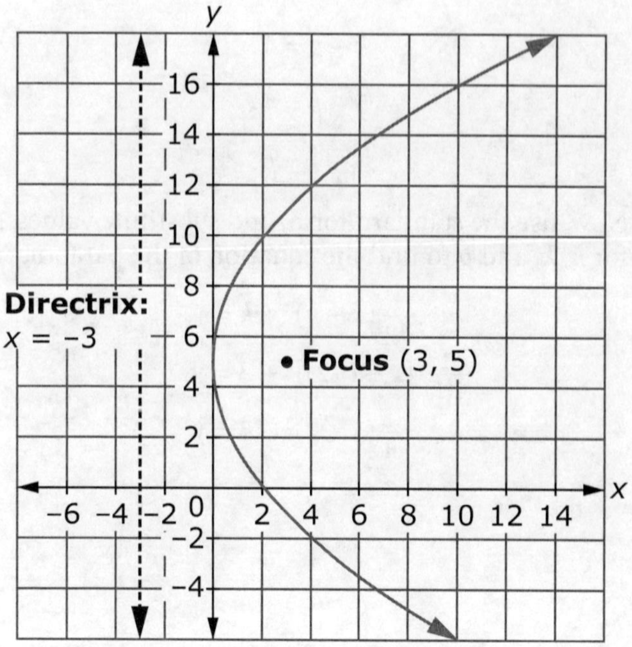

My Notes

Summary *(continued)*

EXAMPLE

What is the equation of a parabola with directrix $y = 4$ and vertex $(-2, 5)$?

If the directrix is a horizontal line, the parabola is vertical. So, $h = -2$ and $k = 5$. You know $y = k - p$, and since the directrix is $y = 4$, you know by substitution that $k - p = 4$. Substitute 5 for k and solve for p:

$$k - p = 4$$
$$5 - p = 4$$
$$1 = p$$

Now, use the standard form, and substitute values for h, k, and p to find the equation of the parabola.

$$\frac{1}{4p}(x - h)^2 + k = y$$
$$\frac{1}{4(1)}[x - (-2)]^2 + 5 = y$$
$$\frac{1}{4}(x + 2)^2 + 5 = y$$

EXAMPLE

Write the equation of the parabola shown in the graph.

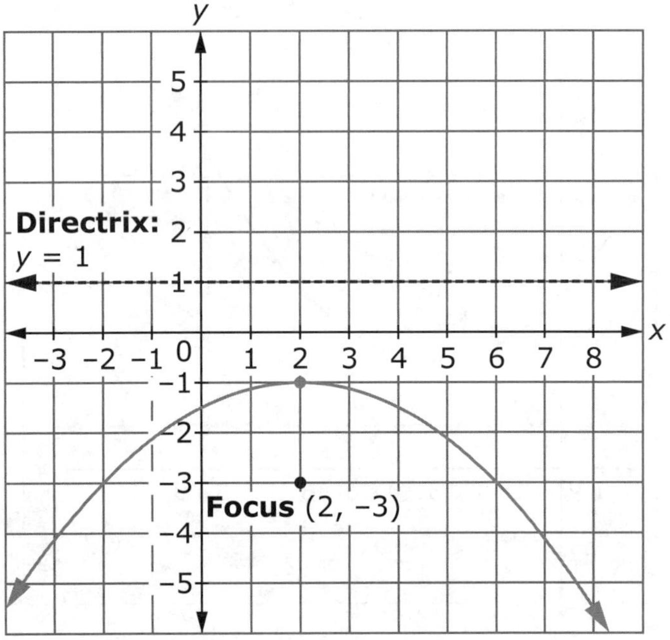

The parabola is vertical and opens down, so the equation is of the form: $\frac{1}{4p}(x - h)^2 + k = y$, where $p < 0$.

The vertex is located at $(2, -1)$, so substitute 2 for h and -1 for k.

$$\frac{1}{4p}(x - 2)^2 - 1 = y$$

The equation for the directrix is $y = k - p$. The directrix is $y = 1$, so $k - p = 1$. Substitute -1 for k and solve for p.

$$k - p = 1$$
$$-1 - p = 1$$
$$p = -2$$

Substitute -2 for p:

$$\frac{1}{4(-2)}(x - 2)^2 - 1 = y$$
$$-\frac{1}{8}(x - 2)^2 - 1 = y$$

My Notes

Practice Exercises

Review what you've learned using these practice problems. For practice problems with feedback, try the Coach and Play items in the Practice section online.

1. Given a parabola with a focus of (4, −6) and a directrix of $y = -8$, find the y-coordinate when $x = 0$.

   $y =$ ____________

2. In the set of points equidistant from (−2, 8) and the y-axis, there is a point of the form $(x, 0)$.

   What is the x value of that point?

   $x =$ ____________

3. Marty graphed a parabola given in vertex form, but then he forgot what equation he used. He knows that the focus is at (5, 7) and that the vertex is at (5, 3).

   What is the equation for Marty's parabola?

   A. $y = \frac{1}{4}(x - 5)^2 + 3$

   B. $y = 4(x + 5)^2 - 3$

   C. $y = \frac{1}{16}(x - 5)^2 + 3$

   D. $x = \frac{1}{16}(y - 3)^2 + 5$

   E. $y = \frac{1}{8}(x - 5)^2 + 7$

4. A certain parabola has a vertex of (−4, −6) and a directrix of $y = 4$. Determine the y value of the point where the parabola intersects the y-axis.

   The parabola intersects the y-axis at

   $y =$ ____________.

5. A student graphs a parabola with its focus at (2, 8) and its vertex on the x-axis. He graphs a second parabola with the same focus, but he moves the directrix down 6 units.

   What is the y-coordinate of the vertex of the second parabola?

   $y =$ ____________

6. The graph of a quadratic equation has a y-intercept at 8 and x-intercepts at 2 and 6.

   Name the line that is the directrix of the parabola.

   $y =$ ____________

7. The line of symmetry of a vertically oriented parabola and its directrix intersect at (4, 2).

   If the focus is on the line $y = 8$, where is the vertex located?

   The vertex is at the point (________, ________).

8. A horizontally oriented parabola has its vertex at the origin and passes through (−1, 4).

   Name the directrix of the parabola.

   $x =$ ____________

Practice Exercises *(continued)*

9. Charity is exploring transformations of parabolas. She is currently considering the family of quadratic functions with a vertex at (1, 2) that are vertically oriented.

Which of the following equations represent parabolas in that family of functions? *Select all that apply.*

A. $y - 2 = (x - 1)^2$

B. $x^2 + 3y = 4x - 1$

C. $y^2 = x + 4y - 5$

D. $x^2 - 3y = 2x - 7$

E. $y = -\frac{1}{2}x^2 + 2x + 1$

F. $2y = -x^2 + 2x + 3$

10. A parabolic satellite dish is concave to the left. Its receiver is at the focus, at point (–3, –1) on a coordinate graph. The directrix of the parabolic cross section on the graph is the line $x = 3$.

What equation represents the cross section of the satellite dish?

A. $x = -\frac{1}{12}y^2 + 1$

B. $x = -\frac{1}{12}y^2 - 1$

C. $x = -\frac{1}{12}(y + 1)^2$

D. $x = -\frac{1}{12}(y - 1)^2$

11. Classify the functions as represented by the graph of a parabola that opens up, a parabola that opens down, or not a parabola.

Parabola Opens Up	Parabola Opens Down	Not a Parabola

A. $g(x) = -9 + x^2$

B. $h(x) = 3x^2 - 2x - 5$

C. $f(x) = -x^2$

D. $m(x) = 3x - x^2 - 4$

E. $j(x) = x^3 - 4x^2$

F. $k(x) = 2x^2 + x^4$

Apply

Can a Parabola Help You Hear Better?

What properties of parabolas make them helpful for sound applications?

Research the use of parabolas to enhance audio input, past and present. Write a description of one object you found; include an image, a graph of the parabola, and an equation to model its cross section. Be sure to include the focus and directrix.

How does a parabolic microphone work?

Show what you've learned by completing the other performance tasks in the online Apply section.

Apply *(continued)*

Your answer to Apply will be assessed on the following criteria:

1. Researching and citing an appropriate application, including photo and dimensions
2. Explaining the object's use and how the parabolic shape helps it work
3. Writing and graphing the equation of the parabola that models the object
4. Determining the focus and directrix and adding them to the graph

Criteria \ Scale	4	3	2	1	0
	Exceeds Criteria	**Meets Criteria**	**Progressing to Criteria**	**Below Expectations**	**No Expectation**
Research	Finds a real-world application of parabolas for sound enhancement; includes a photo that clearly illustrates it and includes dimensions.	Finds a real-world application of parabolas for sound enhancement; includes a photo that clearly illustrates it or includes dimensions.	Gives an application that does not clearly illustrate the use of parabolas for sound enhancement, but includes a photo that suggests it.	Gives an application that does not clearly illustrate the use of parabolas for sound enhancement and is lacking a photo.	Does not attempt task.
Explanation	Explains clearly how the object is used and how the parabolic shape makes it work.	Explains how the object is used and how the parabolic shape makes it work, with minor errors.	Explains how the object is used but does not explain how the parabolic shape makes it work.	Explains the use of the object, with major errors.	Does not attempt task.
Graph	Describes the parabola and sketches an accurate graph of it, with key features identified.	Describes the parabola and sketches a graph of it with key features, but has minor errors.	Describes the parabola and sketches a graph of it, but with key features missing or identified improperly.	Describes the parabola inadequately and sketches a graph that does not correspond to it.	Does not attempt task.
Equation	Writes an equation that matches the description and graph, and identifies the focus and directrix.	Writes an equation that matches the description and graph, and identifies the focus and directrix, with minor errors.	Writes an equation for the graph with errors and identifies either the focus or the directrix.	Writes an equation that does not match the description and graph and does not correctly identify the focus and directrix.	Does not attempt task.

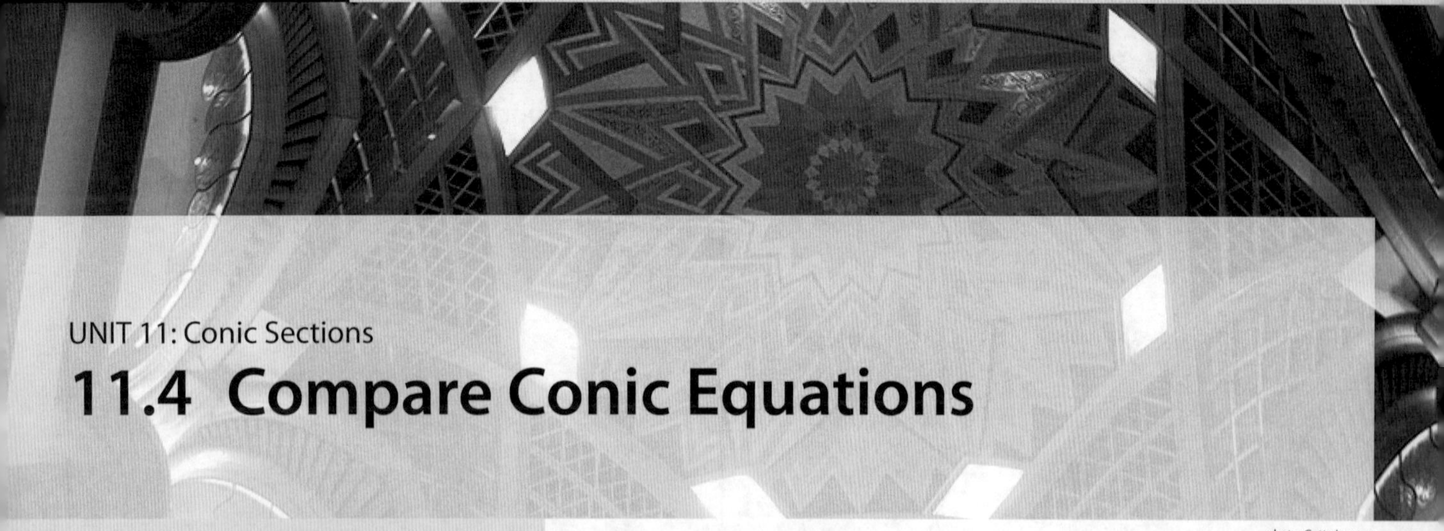

UNIT 11: Conic Sections

11.4 Compare Conic Equations

photo: Getty Images

Lesson Objective

- Compare the graphs and equations of parabolas, circles, ellipses, and hyperbolas.

Essential Question

- What are the similarities and differences between the graphs and equations of the various conic sections?

photo: Getty Images

Investigations

Comparing Conic Sections

Equations of conic sections can be very similar. Make one slight change and the shape may morph.

Conic Section Equations

Which conic section does an equation model? Look for similarities and tease out the differences.

Conic Section Graphs

Build a 2-D sphere. Can you use circles and ellipses?

Find Intersections

Systems of equations get conic! Can you use your toolbox of methods to find intersections?

Key Vocabulary

asymptote, axes, axis of symmetry, center, circle, cone, conic section, conjugate axis, directrix, ellipse, expanded form, focus (foci), hyperbola, major axis, minor axis, parabola, radius, transverse axis, vertex

Discover

As you complete Engage and the investigations, record the most important ideas you've learned.

Engage	Investigation 1
Investigation 2	**Investigation 3**

Name ______________________________ Date __________

Check for Understanding

Compare Conic Equations: Investigation 1

1. Which equations represent conic sections? *Select all that apply.*

   A. $2x^2 - 3y = 2x^2 + 5x$ B. $6y^2 - 4y = 9 - 3x^2$ C. $y^2 - 9y + 1 = y^2 + x$

   D. $y^2 + 7y = 8x^2 - x - 1$ E. $9x^2 + 2y = 8 + 9x^2$ F. $3x^2 - y = 2 + 10x$

2. Decide which conic each equation represents and put a checkmark in the correct column.

Equation	Circle	Ellipse	Hyperbola	Parabola
$3x^2 - 9x + 1 = 12y^2 + 2y + 15$				
$2y^2 + y = 10x^2 + 2y^2 - 9x + 26$				
$5x^2 + 6x - 7 = 18 - y - 5y^2$				
$y^2 - 22y - 3 = 42 + 4x - 8x^2$				
$4x^2 - 11x - 33 = 6y^2 + 4x^2 - 3x$				
$3y^2 - 20y + 2x = 17 - 3x^2 + 9y$				
$2x^2 + 7x - 50 = 5 - 13y^2 + 2y$				
$y^2 + 8y - 36 = 5x^2 + x + 4$				

3. In a conic equation of the form $Ax^2 + By^2 + Cx + Dy = E$, $A \neq 0$, $B \neq 0$, and $A + B = 0$. Which conic section could the equation represent?

   A. circle B. ellipse C. hyperbola D. parabola

4. Julius was given a conic equation. Both sides had an x^2 term and a y^2 term, and according to the flowchart that he made, the equation should represent a circle, ellipse, or hyperbola. However, when he graphed the equation, he got a parabola. What mistake could Julius have made in his flowchart? Give an example to support your answer.

Name ______________________ Date ____________

Check for Understanding

Compare Conic Equations: Investigation 2

1. LaToya represented the two-dimensional sketch of the three-dimensional image of a sphere by graphing a circle and ellipse with the same center on the same coordinate plane. The ellipse's equation is $\frac{x^2}{144} + \frac{y^2}{16} = 1$. Which could be the circle's equation?

   A. $x^2 + y^2 = 4$ **B.** $x^2 + y^2 = 12$ **C.** $x^2 + y^2 = 16$ **D.** $x^2 + y^2 = 144$

2. A sphere has been represented below by graphing a circle and an ellipse that both have the same center. Which statements are correct? *Select all that apply.*

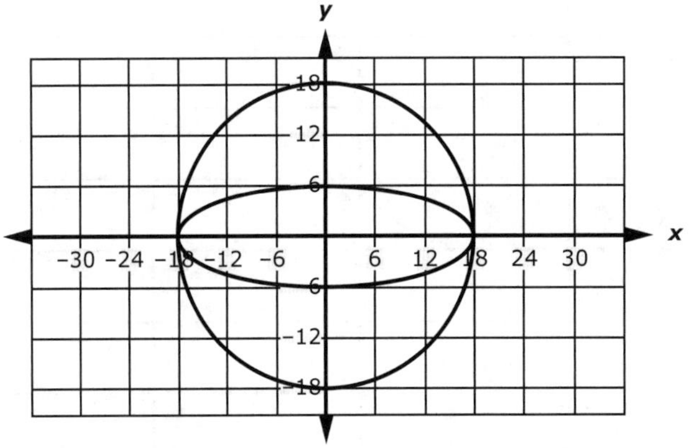

   A. The ellipse's foci are $12\sqrt{2}$ units apart.

   B. The circle's equation is $x^2 + y^2 = 324$.

   C. One focus of the ellipse is $(12\sqrt{2}, 0)$.

   D. The ellipse's equation is $\frac{x^2}{36} + \frac{y^2}{324} = 1$.

   E. The circle has a radius of 36 units.

   F. The ellipse's vertices are 12 units apart.

3. Julio represented the two-dimensional sketch of the three-dimensional image of a sphere by graphing a circle and ellipse with the same center on the same coordinate plane. The equation of the ellipse was $\frac{x^2}{625} + \frac{y^2}{81} = 1$. If Julio represented the sphere to scale on the coordinate plane, how many units was the sphere's radius? Explain.

Name ________________________________ Date ____________

Check for Understanding

Compare Conic Equations: Investigation 3

1. For each pair of conic sections given in the table, assume that the system of equations has a finite number of solutions. Fill in the maximum number of solutions for each.

System of Equations	Maximum Number of Solutions
ellipse and ellipse	
hyperbola and hyperbola	
parabola and circle	
parabola and hyperbola	

2. An ellipse and a hyperbola both have the same center. How many solutions could the system of equations for the two conic sections have? *Select all that apply.*

   A. 0 **B.** 1 **C.** 2

   D. 3 **E.** 4 **F.** 5

3. Which systems of equations of two circles have two solutions? *Select all that apply.*

   A. $x^2 + y^2 = 4$ and $x^2 + (y + 1)^2 = 144$ **B.** $x^2 + y^2 = 25$ and $(x - 3)^2 + y^2 = 9$

   C. $x^2 + y^2 = 100$ and $(x + 2)^2 + y^2 = 49$ **D.** $x^2 + y^2 = 36$ and $x^2 + (y - 4)^2 = 81$

   E. $x^2 + y^2 = 64$ and $x^2 + (y - 6)^2 = 1$ **F.** $x^2 + y^2 = 16$ and $x^2 + (y + 5)^2 = 121$

4. What are the solutions to the system of equations $2y + 6 = 3x^2$ and $3x^2 + y^2 = 21$?

   A. $(-2, -3)$ and $(-2, 3)$ **B.** $(-2, -3)$ and $(2, -3)$

   C. $(-2, 3)$ and $(2, 3)$ **D.** $(2, -3)$ and $(2, 3)$

5. Suppose a system of equations has the solutions $(-5, -12)$, $(-5, 12)$, $(5, -12)$, and $(5, 12)$. If one of the equations in the system is $x^2 + y^2 = 169$, give an example of what the other equation in the system could be. Explain how you got your answer.

Summary

Before you attempt the Practice Exercises, review what you've learned.

A **conic section** is formed by the intersection of a double cone and a plane.

Conic sections include circles, ellipses, hyperbolas, and parabolas.

Knowing the graphing forms of each type of conic section can help you identify a conic section when only given its equation.

Graphing Form of a Circle
$(x-h)^2+(y-k)^2=r^2$, where (h, k) is the center and r is the radius.

Graphing Form of an Ellipse	
Vertical	**Horizontal**
$\frac{(x-h)^2}{a^2}+\frac{(y-k)^2}{b^2}=1$ $b>a$ Center: (h, k) End points of major axis: $(h, k \pm b)$	$\frac{(x-h)^2}{a^2}+\frac{(y-k)^2}{b^2}=1$ $a>b$ Center: (h, k) End points of major axis: $(h \pm a, k)$

Graphing Form of a Hyperbola	
Vertical	**Horizontal**
(Opens Up and Down) $\frac{(y-k)^2}{b^2}-\frac{(x-h)^2}{a^2}=1$ Center: (h, k) Vertices: $(h, k \pm b)$	(Opens Left and Right) $\frac{(x-h)^2}{a^2}-\frac{(y-k)^2}{b^2}=1$ Center: (h, k) Vertices: $(h \pm a, k)$

Graphing Form of a Parabola	
Vertical	**Horizontal**
$y=a(x-h)^2+k$ if $a>0$: Concave Up if $a<0$: Concave Down Axis of Symmetry: $x=h$ Vertex: (h, k)	$x=a(y-k)^2+h$ if $a>0$: Opens Right if $a<0$: Opens Left Axis of Symmetry: $y=k$ Vertex: (h, k)

You can look for clues when analyzing a given equation, especially if the equation is not written in graphing form.

If the equation contains	Then the equation represents
x^2 or y^2 terms, but not both	a parabola
both x^2 and y^2 terms, and they have the same coefficients	a circle
both x^2 and y^2 terms, and they have different coefficients but with the same sign	an ellipse
both x^2 and y^2 terms, and they have coefficients with different signs	a hyperbola

EXAMPLE: Identify the Type of Conic Section

Identify the conic section defined by each equation.

$$x^2-y^2=9$$

$$x^2+y^2=81$$

$$x=y^2+7$$

My Notes

Summary *(continued)*

SOLUTION:

$x^2 - y^2 = 9$ is an equation that defines a hyperbola. The equation contains both x^2 and y^2 terms, and the coefficients of these terms have different signs.

$x^2 + y^2 = 81$ is an equation that defines a circle. The equation contains both x^2 and y^2 terms. The coefficients of these terms are the same, and both are positive.

$x = y^2 + 7$ is an equation that defines a parabola. The equation contains only a y^2 term.

You can also rewrite each equation in its standard form to confirm its type.

$$x^2 - y^2 = 9$$
$$\frac{x^2}{9} - \frac{y^2}{9} = 1$$
$$x^2 + y^2 = 81$$
$$x^2 + y^2 = 9^2$$
$$x = y^2 + 7$$

You can identify key features for the graph of a conic section by analyzing its equation.

EXAMPLE: Determine a Relationship

Determine a relationship, if any, between the two conic sections defined by the equations $x^2 - y^2 = 9$ and $\frac{x^2}{25} + \frac{y^2}{36} = 1$.

SOLUTION:

$x^2 - y^2 = 9$ is an equation that defines a hyperbola. It can be written in standard form as $\frac{(x-0)^2}{3^2} - \frac{(y-0)^2}{3^2} = 1$. It is a horizontal hyperbola, so it will open to the left and right. Its center is at (0, 0). It will have vertices at (−3, 0) and (3, 0).

$\frac{x^2}{25} + \frac{y^2}{36} = 1$ is an equation that defines an ellipse. It can be written in standard form as $\frac{(x-0)^2}{5^2} + \frac{(y-0)^2}{6^2} = 1$. Its center is at (0, 0). The length of the horizontal radius is 5 units, and the length of the vertical radius is 6 units.

If you graph them together, the vertices of the hyperbola will lie in the interior of the ellipse. Since the hyperbola opens to the left and right, the hyperbola and ellipse intersect each other.

You can use graphing to verify.

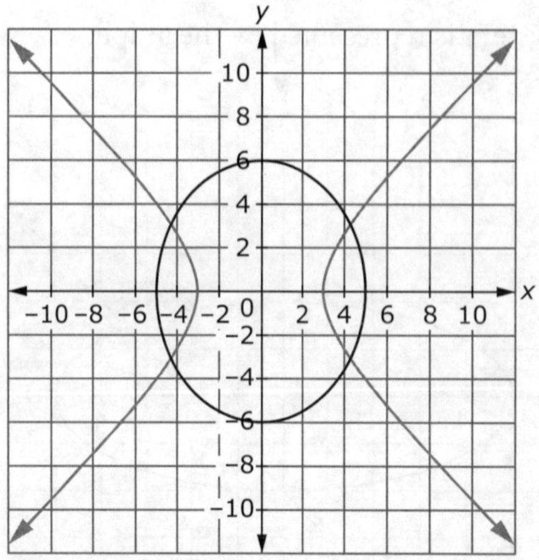

A quadratic system of equations consists of two or more equations, at least one of which must be an equation that represents a conic section. You can solve quadratic systems by graphing and use algebra to verify the solutions.

My Notes

Summary *(continued)*

EXAMPLE: Solve a Quadratic System

Solve the quadratic system defined by the equations $4x^2 + 25y^2 = 100$ and $y = x - 2$ by graphing.

SOLUTION:

This system is represented by the graph.

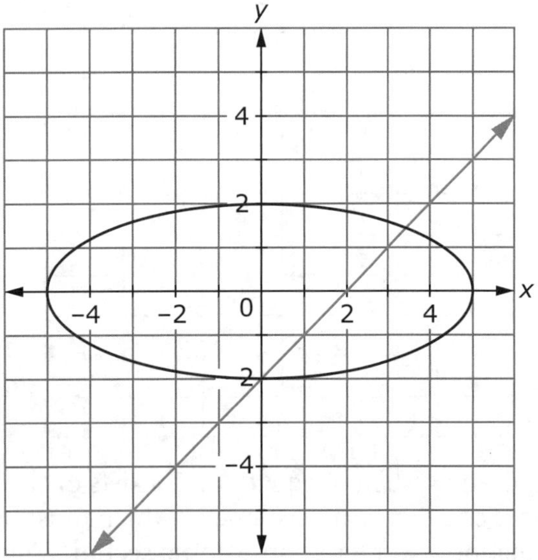

The graphs intersect at two points, with coordinates (0, −2) and (3.45, 1.48). The point with coordinates (0, −2) is easy to determine from the graph. The point with coordinates (3.45, 1.48) is not as easy to determine with confidence.

You can use substitution to verify the solutions you identified by graphing.

$4x^2 + 25y^2 = 100$	$y = x - 2$
$4(0) + 25(-2)^2 = 100$	$-2 = 0 - 2$
$4(0) + 25(4) = 100$	$-2 = -2$
$0 + 100 = 100$	
$100 = 100$	

My Notes

Practice Exercises

Review what you've learned using these practice problems. For practice problems with feedback, try the Coach and Play items in the Practice section online.

1. Identify the conics that are circles. *Select all that apply.*

   A. $4(x + 2)^2 + 2(y + 4)^2 = 20$

   B. $\frac{(y+1)^2}{10} - \frac{(x+2)^2}{10} = 1$

   C. $3x^2 + 7x - 5 = 3y^2 + 5y + 1$

   D. $\frac{5(x-1)^2}{40} + \frac{y^2}{8} = 1$

   E. $x^2 + y^2 + 1 = 0$

   F. $(2x - 3)^2 + 4y^2 = 20$

2. Which of the following equations are parabolas? *Select all that apply.*

   A. $4x^2 + (y - 5)^2 = 16$

   B. $y^2 - 5y - 2 = x$

   C. $\frac{8x^2}{2} + \frac{y}{2} = \frac{1}{2}$

   D. $\frac{2x^2}{4} - \frac{6y^2}{9} = 1$

   E. $\frac{y^2}{25} + \frac{6x^2}{36} = 1$

   F. $\frac{(x-2)}{12} - \frac{(y+1)^2}{9} = 1$

3. A hyperbola with the foci at the points $(h \pm c, k)$ is characterized by the constants a and b with a equal to one-half the difference in focal lengths from the foci to any point on the hyperbola, and $b^2 = c^2 - a^2$.

   Which of the following represents the equation of this hyperbola?

   A. $\frac{(x-h)^2}{(a-h)^2} - \frac{(y-k)^2}{(b-k)^2} = 1$

   B. $\frac{(x^2-h^2)}{a^2} - \frac{(y^2-k^2)}{b^2} = 1$

   C. $\frac{(x-h)^2}{a^2} - \frac{(y-k)^2}{b^2} = 1$

   D. $\frac{(x+h)^2}{a^2} - \frac{(y+k)^2}{b^2} = 1$

   E. $\frac{(x^2+h^2)}{a^2} - \frac{(y^2+k^2)}{b^2} = 1$

4. A circle centered at the origin has a radius of 6 units.

   An ellipse has the equation $\frac{x^2}{(4)^2} + \frac{y^2}{(9)^2} = 1$.

   Write the approximate coordinates of the two points of intersection between the circle and ellipse that have positive y values.

   Point 1: (__________, __________)

   Point 2: (__________, __________)

5. The equation for a particular ellipse is $\frac{x^2}{(5)^2} + \frac{y^2}{(50)^2} = 1$. A parabola defined by an equation of the form $y = Ax^2 - B$ intersects the ellipse at both its x-intercepts and at its negative y-intercept.

   Which of the following best represents the values of A and B?

   A. $A = \frac{1}{10}$; $B = 5$

   B. $A = \frac{1}{10}$; $B = 50$

   C. $A = \frac{1}{2}$; $B = 50$

   D. $A = 2$; $B = 50$

   E. $A = 10$; $B = 50$

6. A certain ellipse is defined by the equation $\frac{x^2}{(9)^2} + \frac{y^2}{(8)^2} = 1$, and a certain hyperbola is defined by the equation $\frac{y^2}{(3)^2} - \frac{x^2}{(5)^2} = 1$.

   Write the coordinates of the two points of intersection between the ellipse and the hyperbola that have positive y-coordinates.

   Point 1: (__________, __________)

   Point 2: (__________, __________)

Practice Exercises *(continued)*

7. The figure shows the upper half of an ellipse, the upper half of a circle, and a parabola, all centered at the origin.

The lengths of the major and minor axis are 16 and 8 units, respectively.

The diameter of the circle is 12 units. The parabola is defined by the equation $y = Ax^2$ and passes through the intersection points of the ellipse and circle as shown.

Write the coordinates of the intersection point in quadrant I and the value of A.

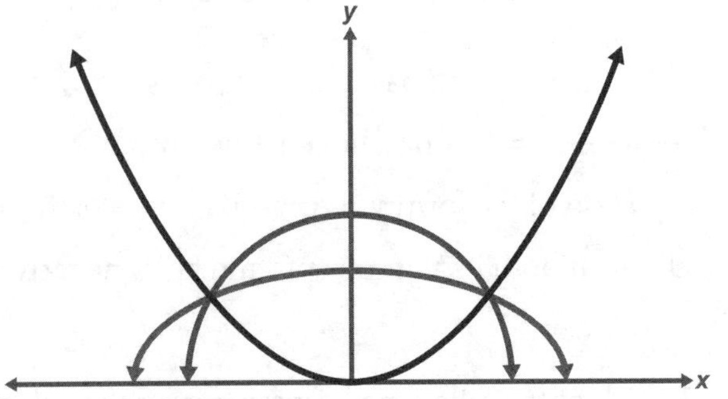

Intersection point: (______, (______)

Value of A: ______

8. The equation of a conic section can be given in the form $Ax^2 + By^2 + Cx + Dy = E$.

Identify the conic section whose equation is determined by each of the following conditions.

$A = 0$ or $B = 0$ ______

$A = 0$ and $B = 0$ ______

$A > 0$ and $B > 0$, $A \neq B$ ______

$A = B$, $A \neq 0$, $B \neq 0$ ______

If $A > 0$, then $B < 0$

If $A < 0$, then $B > 0$ ______

A. line

B. parabola

C. circle

D. ellipse

E. hyperbola

9. Bridgette has drawn a representation of a sphere with a radius of 3 units, using the graphs of a circle and an ellipse, as shown.

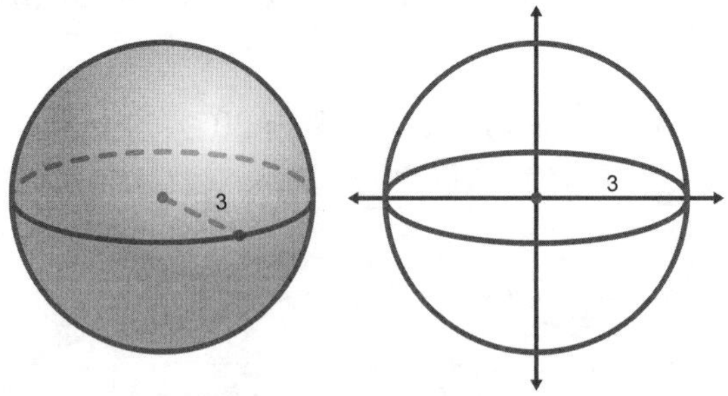

She needs to determine the equations of the graphs in the 2-dimensional x-y–coordinate system, shown above right, that will generate the representation of the 3-dimensional sphere.

What are possible equations of the circle and ellipse that Bridgette can use to create her 2-dimensional drawing?

A. Circle: $x^2 + y^2 = 9$ Ellipse: $\frac{x^2}{1} + \frac{y^2}{9} = 1$

B. Circle: $x^2 + y^2 = 9$ Ellipse: $\frac{x^2}{9} + \frac{y^2}{10} = 1$

C. Circle: $x^2 + y^2 = 9$ Ellipse: $\frac{x^2}{9} - \frac{y^2}{4} = 1$

D. Circle: $x^2 + y^2 = 9$ Ellipse: $\frac{x^2}{9} + \frac{y^2}{1} = 1$

E. Circle: $x^2 + y^2 = 9$ Ellipse: $\frac{x^2}{9} + \frac{y^2}{4} = 1$

10. Consider the quadratic system defined by the equations: $y^2 - 4x^2 = 16$ and $\frac{x^2}{100} + \frac{y^2}{N} = 1$ for some positive number N, such that $N < 50$.

Determine a value of N at which the system will have exactly 4 solutions, a value of N at which the system will have exactly 2 solutions, and a value of N at which the system has no solutions.

The system will have exactly 4 solutions if $N =$ ☐.

The system will have exactly 2 solutions if $N =$ ☐.

The system will not have any solutions if $N =$ ☐.

Apply

What Path Must the Ball Travel to Score?

Objects projected into the air follow a trajectory in the shape of a parabola. This "parabolic path" is seen in many sports: a baseball coming off the end of a bat, a soccer ball drop kicked by the goalie, or a jump shot in basketball from the top of the key. How can these paths be represented on the coordinate plane?

How can you predict where the basketball will go?

You have studied different forms of equations that model a parabola. In physics, the parabolic path of a projectile is given by the equation, $y = y_0 + x \tan\theta + \frac{gx^2}{2(v_0 \cos\theta)^2}$, where θ is the angle of inclination of the projectile, g is acceleration due to gravity, y_0 is the initial height, and v_0 is the initial velocity of the projectile.

- Research a parabolic path as it relates to a sport of your choosing; identify the parameters listed above.
- Use the Graphing Calculator and the equation above to graph the object's trajectory.
- Label the essential points on the graph and create another equation modeling the graph, expressed in one of the forms discussed in this unit.
- Explain and interpret your model.

Show what you've learned by completing the other performance tasks in the online Apply section.

Apply *(continued)*

Your answer to Apply will be assessed on the following criteria:

1. Reporting and analyzing the results of your research on parabolic paths related to your chosen sport
2. Graphing the physics equation and labeling the critical points
3. Finding the quadratic equation that models the parabolic path for an object in your chosen sport
4. Explaining and interpreting the graph, including the critical points of the parabola

Criteria / Scale	4 Exceeds Criteria	3 Meets Criteria	2 Progressing to Criteria	1 Below Expectations	0 No Expectation
Results	Chooses a sport that has an aspect that can be modeled by a quadratic equation and reports appropriate data for each of the four parameters.	Chooses a sport that has an aspect that can be modeled by a quadratic equation and reports appropriate data for three of the four parameters.	Chooses a sport that has an aspect that can be modeled by a quadratic equation and reports appropriate data for two of the four parameters.	Chooses a sport that has an aspect that can be modeled by a quadratic equation and reports appropriate data for one of the four parameters.	Does not attempt task.
Graph	Correctly uses the equation to create a graph and labels all of the critical points.	Correctly uses the equation to create a graph and labels all of the critical points with a minor error.	Correctly uses the equation to create a graph but does not label the critical points or graphs an incorrect equation with critical points labeled.	Incorrectly graphs the equation with no critical points labeled or discusses correct critical points with no equation graphed.	Does not attempt task.
Equation	Correctly uses the data found and the physics equation to determine the quadratic equation that models the path of the projectile.	Correctly uses the data found and equation to determine the quadratic equation that models the path of the projectile with minor errors.	Uses the data found and equation to determine a quadratic equation but does not have measurements in the same unit or substitutes the numbers into the incorrect parameters.	Uses the data found and equation to determine a quadratic equation but does not have measurements in the same unit and substitutes the numbers into the incorrect parameters.	Does not attempt task.
Explanation	Correctly uses the graph and critical points to explain the context in terms of the graph.	Correctly uses the graph and critical points to explain the context in terms of the graph with minor errors.	Uses the graph and critical points to explain the context but does not explain one or two critical points or incorrectly explains one or two critical points.	Uses the graph and critical points to explain the context but does not explain three critical points or incorrectly explains three critical points.	Does not attempt task.

Notes

Notes

Notes